Topological Embeddings

This is Volume 52 in
PURE AND APPLIED MATHEMATICS
A series of Monographs and Textbooks
Editors: PAUL A. SMITH AND SAMUEL EILENBERG
A complete list of titles in this series appears at the end of this volume

Topological Embeddings

T. BENNY RUSHING

Department of Mathematics
University of Utah
Salt Lake City

ACADEMIC PRESS 1973 New York and London

ACADEMIC PRESS, INC.
111 Fifth Avenue, New York, New York 10003

United Kingdom Edition published by
ACADEMIC PRESS, INC. (LONDON) LTD.
24/28 Oval Road, London NW1

LIBRARY OF CONGRESS CATALOG CARD NUMBER: 72-77344

AMS (MOS) 1970 Subject Classifications: 57A, 57C

PRINTED IN THE UNITED STATES OF AMERICA

To the Memory
of my
Mother
CHRISTINE WALTERS RUSHING (1913–1970)

Contents

Chapter 3 Flattening, Unknotting, and Taming Special Embeddings

Chapter 4 Engulfing and Applications

Chapter 5 Taming and PL Approximating Embeddings

Preface

Although the field of topological embeddings is an active, major area of topology, there has not been a comprehensive treatment of the theory available. Hence, it has been very difficult for students to become knowledgable in this area without close supervision. The related area of piecewise linear (PL) topology was in the same state of affairs until 1963 when Zeeman's notes entitled "Seminar on Combinatorial Topology" appeared. (Since that time a number of expositions on the fundamentals of PL topology have appeared, including those of Stallings, Hudson, and Glaser.) It is the author's hope that by structuring the theory of topological embeddings, this work will make the area more accessible to students and will serve as a useful source of reference for specialists in the field.

A knowledge of general topology as well as of the fundamentals of modern algebra is prerequisite for this book. Although an understanding of algebraic topology would be helpful, it should not be considered as a prerequisite. In fact, it might be desirable for a student to cover this book before studying algebraic topology since some of the material will serve as motivation.

The topics are organized so as to lead the student into the theory in a natural fashion. This development does not take the shortest possible route to the frontier, but presents a number of special cases and historical discussions. In general, the material covered is of a "grass roots" nature and involves techniques which have found numerous applications. The student who has mastered the topics presented should be ready to study the current literature in topological embeddings and undertake his own research. It will probably be the case that problems of current interest will occur to the alert student in the course of studying this book.

It is our purpose to provide a sound basis for development in the area of topological embeddings, and not to give an exhaustive treatment of the current status of every facet of the area. (Indeed such a treatment would soon be obsolete.) However, a concrete mathematical introduction to the main elements is given and the current status of each is discussed. Many references are included for the reader who wants to pursue a given subject further. In the Appendix, several topics are mentioned which are particularily appropriate as follow-up material to this book.

Topological embeddings is extremely geometrical in nature. In order to study or do research in the area successfully, it is almost a necessity that one develop a facility for visualizing the geometry and describing it via schematic figures. Many times a very complex idea can be conveyed rather quickly through use of a picture. Also, proofs are more easily remembered if one has descriptive pictures associated with them. This book contains many figures. We hope that they serve to help the reader understand the ideas presented more readily and guide him in describing his own ideas pictorically.

Acknowledgments

Although the list of people who have helped in the realization of this book is too long to enumerate completely here, I am grateful to all who have helped. I would like specifically to thank colleagues H. W. Berkowitz, C. E. Burgess, J. C. Cantrell, R. E. Chamberlin, R. J. Daverman, R. D. Edwards, L. C. Glaser, R. C. Lacher, and T. M. Price; and students F. Benson, O. Bierman, R. Dieffenbach, C. Mannes and G. Venema for their help. Portions of the manuscript of this book were covered in the classroom by the author in 1969–1970 and 1970–1971, by Cantrell in 1971–1972 and by Glaser in 1971–1972, and I appreciate all the suggestions made by members of those classes. Thanks are due C. H. Edwards who first introduced me to piecewise linear topology and who had an influence on the portions of this book concerning piecewise linear topology. I am particularily indebted to J. C. Cantrell who first introduced me to topological embeddings and whose guidance, while I was a student at the University of Georgia, and continued encouragement have been of immeasurable benefit. The splendid cooperation of all members of Academic Press involved with this project is appreciated. It is a pleasure to thank the National Science Foundation for its support of this work during the academic year 1970–1971 as well as other support during the three year period which this project spanned. Finally, I thank my wife, Gail, for her help in correcting the proof sheets and, most of all, for her patience and good humor.

CHAPTER **1**

Main Problem and Preliminary Notions

1.1. INTRODUCTION

Included in Sections **1.2** and **1.5** of this first chapter are brief discussions, in a general setting, of the key problem of topological embeddings. Also, in this chapter, we lay some foundations for studying this problem. A number of basic definitions are made in Section **1.3** and topological manifolds are considered. Several elementary properties of manifolds are included as exercises at the end of this section. Section **1.4** is devoted to a discussion of the fundamental theory of the category of polyhedra and piecewise linear maps. The proofs of a few standard theorems, although easy, are omitted because they appear frequently in the literature. Piecewise linear manifolds are defined in Section **1.6** and the basic notions associated with them are developed including general position, relative simplicial approximation, regular neighborhoods, relative regular neighborhoods, and handlebody decompositions. In Section **1.7**, we say what it means for a submanifold to be locally flat and establish a few basic facts, such as the transitivity of local flatness. Here we also prove that locally collared implies collared and discuss collars, bicollars, pinched collars, and pinched bicollars. Finally, in Section **1.8**, we consider cellular sets. First a decomposition theorem and the generalized Schoenflies theorem are proved. The section is concluded with several useful consequences of the generalized Schoenflies theorem.

1.2. MAIN PROBLEM

An **embedding** of a topological space X into a topological space Y is a homeomorphism $f: X \to Y$ of X onto a subset of Y. Two embeddings

$f: X \to Y$ and $g: X \to Y$ are said to be **equivalent** if there is a homeomorphism $h: Y \to Y$ of Y onto itself such that $hf = g$. This is an equivalence relation on the set of all embeddings of X into Y. We are now ready to state the main problem. (For further discussion of the main problem see Section **1.5**.)

Main Problem of Topological Embeddings. *What are the equivalence classes of embeddings of a space X into a space Y?*

In order to formulate an even stronger question than the one above, we make the following definition: An **isotopy** of a topological space Y is a collection $\{e_t\}$, $0 \leqslant t \leqslant 1$, of homeomorphisms of Y onto itself such that the mapping $e: Y \times [0, 1] \to Y$ defined by $e(x, t) = e_t(x)$ is continuous. A homeomorphism $h: Y \to Y$ is said to be **realized by an isotopy** $\{e_t\}$ of Y if $e_0 = 1$ and $e_1 = h$. Two embeddings f and g of a space X into a space Y are **isotopically equivalent** if they are equivalent by a homeomorphism h which can be realized by an isotopy of Y. Now we are ready to state the stronger question mentioned above: *What embeddings of a space X into a space Y are isotopically equivalent?*

1.3. TOPOLOGICAL MANIFOLDS

We view Euclidean n-space E^n, as a metric space in the usual way, that is, with metric

$$\mathrm{dist}((x_1, \ldots, x_n), (y_1, \ldots, y_n)) = \left[\sum_{i=1}^{n} (x_i - y_i)^2\right]^{1/2}$$

and as a vector space over the field of real numbers by defining addition and multiplication by scalars coordinatewise. **Upper n-Euclidean half-space** E^n_+ is the subspace of E^n consisting of points having non-negative last coordinates and **lower n-Euclidean half-space** E^n_- is the set of all points of E^n having last coordinates less than or equal to zero. We define the **standard n-sphere** S^n to be $\{x \mid x \in E^{n+1} \text{ and } \| x \| = 1\}$; $S^n_+ = S^n \cap E^{n+1}_+$ and $S^n_- = S^n \cap E^{n+1}_-$.

Let I denote the interval $[-1, 1]$ and let the Cartesian product of I with itself n times, denoted by I^n, be called the **standard n-cube**. We define the **interior of** I^n, denoted by Int I^n, to be the Cartesian product of $(-1, 1)$ with itself n times and the **boundary of** I^n, denoted by either $Bd\, I^n$, ∂I^n, or $\dot{I}^n$ to be $I^n - \mathrm{Int}\, I^n$. A space homeomorphic to I^n is called a **(closed) n-cell**. An **n-dimensional topological manifold** M is a separable metric space each point of which has a closed neighborhood which is an n-cell. A space homeomorphic with Int I^n is called an **open n-cell.** The **interior of M,** denoted by Int M, is the set of all points which have neighborhoods which are open n-cells and the **boundary of M,** denoted by either $Bd\ M$, ∂M, or $\dot{M}$, is M — Int M. We say that M is **unbounded** if $\dot{M} = \emptyset$, **closed** if M is compact and unbounded, and **open** if M is noncompact and unbounded.

EXERCISE 1.3.1. A topological space is said to be **locally Euclidean** if each point has a neighborhood which is homeomorphic to Euclidean n-space. (a) Convince yourself that the long line is a locally 1-Euclidean space which is neither separable nor metric. (You may refer to [Hocking and Young, 1, p. 55] to do this.) (b) Construct a locally n-Euclidean, non-Hausdorff space for $n \geqslant 1$. (Hint: Let A be a subspace of a topological space X. By a space obtained through **splitting the subspace** A of X we mean, as given in the remark in [Rushing, 7], any space homeomorphic to the space $(X - A) \cup (A \times \{0, 1\})$ having open sets of the forms $(G - A) \cup ((G \cap A) \times 0)$ and $(G - A) \cup ((G \cap A) \times 1)$, where G is open in X. Now show that the space you seek can be obtained by splitting a nonempty, proper, closed (as a subset) subspace of a connected, locally n-Euclidean space.)

EXERCISE 1.3.2. Show that if a point x of a manifold M has a neighborhood N such that there is a homeomorphism $h: N \twoheadrightarrow I^n$, where $h(x) \in \text{Bd}\, I^n$, then x has no neighborhood homeomorphic to Int I^n. To do this you may use the following classical theorem. For a proof of this theorem refer to p. 303 of [Eilenberg and Steenrod, 1] or p. 278 of [Hocking and Young, 1]. Proofs of this theorem exist which do not use algebraic topology, for example, Chapter III of [Cantrell, 5].

Invariance of Domain Theorem. *Let U_1 and U_2 be homeomorphic subsets of the locally n-Euclidean spaces M_1 and M_2. If U_1 is open in M_1, then U_2 is open in M_2.*

EXERCISE 1.3.3. If $h: M \twoheadrightarrow Q$ is a homeomorphism of the manifold M onto the manifold Q, then show that $h(\partial M) = \partial Q$. (Hint: Use invariance of domain.)

EXERCISE 1.3.4. Show that the boundary of an n-dimensional manifold is an $(n - 1)$-dimensional manifold, without boundary. (Hint: Use Exercise **1.3.2.**)

EXERCISE 1.3.5. If M and Q are manifolds then show that $\partial(M \times Q) = (M \times \partial Q) \cup (\partial M \times Q)$. (The derivative of a product is the first times the derivative of the second plus the derivative of the first times the second!)

EXERCISE 1.3.6. Show that the dimension of a manifold is well-defined, that is, show that two manifolds M^m and Q^q (superscripts denote dimensions) such that $m > q$, cannot be homeomorphic. (Hint: Use invariance of domain.)

EXERCISE 1.3.7. A space X is **homogeneous** if for any pair of points x and y in X there is a homeomorphism h of X onto itself such that $h(x) = y$. Show that every connected n-manifold M without boundary is homogeneous.

EXERCISE 1.3.8. Associated with any isotopy $\{h_t\}$ of a space Y is the mapping $H: Y \times [0, 1] \rightarrow Y \times [0, 1]$ defined by $H(x, t) = (h_t(x), t)$. This mapping is one-to-one, onto, and continuous. We say that an isotopy $\{h_t\}$. $0 \leqslant t \leqslant 1$, is **invertible** if the collection $\{h_t^{-1}\}$, $0 \leqslant t \leqslant 1$, of inverse homeomorphisms is an isotopy. Obviously an isotopy $\{h_t\}$ is invertible if and only if the associated mapping H is a homeomorphism. By using invariance of domain, prove that isotopies of unbounded manifolds are invertible. (See [Crowell, 1] for a proof that every isotopy of a locally compact Hausdorff space is invertible.)

1.4. THE CATEGORY OF POLYHEDRA AND PIECEWISE LINEAR MAPS

This section develops the basic definitions and properties related to the category of polyhedra and piecewise linear maps. It is the intent that the organization will give a "feeling" for this category as well as present a concrete development. Some of the proofs of standard theorems are omitted because they appear elsewhere in the literature and because we must conserve space in order to include all of the topics desired. References will be given for these elementary omitted proofs although the reader could probably derive his own proofs without too much effort. Some good supplementary references for the material covered in this section are: Chapter I of [Glaser, 3], Chapter I of [Zeeman, 1], Chapter I of [Hudson, 1], Chapters I and II of [Stallings, 4], Chapter V of [Hocking and Young, 1], Chapter IV of [Singer and Thorpe, 1], Chapter I of [C.H. Edwards, 1] and Chapter IV of [Alexandroff, 1].

Points $x_0, x_1, \ldots, x_k$ in E^n are said to be **pointwise independent** if the vectors $x_1 - x_0, x_2 - x_0, \ldots, x_k - x_0$ are linearly independent.

EXERCISE 1.4.1. Show that the above definition is independent of which point is chosen as x_0.

The ***k*-dimensional hyperplane** H^k determined by the $k + 1$ pointwise independent points $x_0, x_1, \ldots, x_k$ is

$$\left\{x \mid x = \sum_{i=0}^{k} t_i x_i \quad \text{and} \quad \sum_{i=0}^{k} t_i = 1\right\}.$$

EXERCISE 1.4.2. Show that the hyperplane H^k determined by the pointwise independent points $x_0, x_1, ..., x_k$ is the same set as a translation by x_0 of the vector space spanned by $x_1 - x_0, x_2 - x_0, ..., x_k - x_0$, that is, show that

$$H^k = \left\{ x \mid x = x_0 + \sum_{i=1}^{k} \gamma_i(x_i - x_0), \quad \gamma_i \in E^1 \right\}.$$

The unique numbers $t_0, t_1, ..., t_k$ are called the **barycentric coordinates with respect to** $x_0, x_1, ..., x_k$ of the point $x = \sum_{i=0}^{k} t_i x_i$ in H^k. The set of points in H^k which have nonnegative barycentric coordinates with respect to the pointwise independent set $x_0, x_1, ..., x_k$ is called the k-**dimensional simplex** spanned by $x_0, x_1, ..., x_k$ and is denoted by $\langle x_0, x_1, ..., x_k \rangle$. The points $x_0, x_1, ..., x_k$ are called the **vertices** of $\langle x_0, x_1, ..., x_k \rangle$. A simplex τ spanned by a subset of the vertices of a simplex σ, is called a **face** of σ, written $\tau < \sigma$. The one-dimensional faces, that is, faces spanned by two vertices, are called edges. The empty set is a (-1)-dimensional simplex and is a face of each simplex. Those faces of a simplex other than the simplex itself are called **proper faces**. The collection of all proper faces of a simplex is the **boundary** of the simplex.

A subset X of E^n is **convex** if for each pair of points x and y in X the simplex $\langle x, y \rangle$ is in X. The **convex hull** of a set X in E^n is the intersection of all convex subsets of E^n that contain X.

EXERCISE 1.4.3. If the points $x_0, x_1, ..., x_k$ are pointwise independent in E^n then $\langle x_0, x_1, ..., x_k \rangle$ is the convex hull of $\{x_0, x_1, ..., x_k\}$.

EXERCISE 1.4.4. Show that the diameter of the convex hull of any set is equal to the diameter of the set itself. (In particular the diameter of a simplex is the length of its longest edge.)

A map f of a simplex $\langle x_0, x_1, ..., x_k \rangle$ into E^n is **linear** if $f(\sum t_i x_i) = \sum t_i f(x_i)$ for each point $\sum t_j x_j \in \langle x_0, x_1, ..., x_k \rangle$. Since a linear map of $\tau = \langle x_0, x_1, ..., x_k \rangle$ is determined by its values on the vertices, it follows that if $\sigma = \langle y_0, y_1, ..., y_m \rangle$ is an m-simplex $m \leqslant k$, then there is a unique linear map f of τ onto σ such that each $f(x_i)$ is some fixed y_{j_i} and $\bigcup_{i=1}^{k} y_{j_i} = \bigcup_{i=1}^{m} y_i$. If $k = m$, then f would be a linear homeomorphism. Notice, however, that the image under a linear map of a simplex into E^n need not be a simplex. (This can happen if the images of the vertices are not pointwise independent.)

A (rectilinear) **simplicial complex** is a finite collection K of simplexes (all in E^n) such that

(1) $\sigma \in K$ and $\tau < \sigma \Rightarrow \tau \in K$,
(2) $\sigma \in K$ and $\tau \in K \Rightarrow \sigma \cap \tau < \sigma$ and $\sigma \cap \tau < \tau$.

A complex K is said to be **k-dimensional** (or a k-complex) if k is the maximum dimension of the simplexes of K. The set $|K| = \bigcup_{\sigma \in K} \sigma$ is called a **polyhedron** and K is called a (rectilinear) **triangulation** of $|K|$. The **dimension** of a polyhedron P is defined to be the dimension of a complex which triangulates it.

EXERCISE 1.4.5. Show that the dimension of a polyhedron is well-defined, that is, show that any two triangulations will have the same dimension. (Hint: Use invariance of domain.)

A **subcomplex** of a simplicial complex K is a subcollection of K which is itself a complex. A **subpolyhedron** of a polyhedron P is a set which is the polyhedron of some subcomplex of some triangulation of P.

Theorem 1.4.1. *The intersection of two polyhedra is a subpolyhedron of each. The product of two polyhedra is a polyhedron.* (See Lemma 1.4 of [Hudson, 1] or Corollary 2, p. 5, Chapter I of [Zeeman, 1].)

The simplicial complex L is said to be a **subdivision** of the complex K if $|K| = |L|$ and every simplex of L is contained in a simplex of K. The **barycenter** $\hat{\sigma}$ of a p-simplex σ is the point whose barycentric coordinates are all equal to $1/(p+1)$. The **first barycentric subdivision** of a complex K with simplexes $\sigma_1, \sigma_2, \ldots, \sigma_k$ is the complex

$$K' = \{\langle \hat{\sigma}_{i_0}, \hat{\sigma}_{i_1}, \ldots, \hat{\sigma}_{i_s} \rangle \mid \sigma_{i_0} < \sigma_{i_1} < \cdots < \sigma_{i_s}\}.$$

EXERCISE 1.4.6. Show that K' is indeed a subdivision of K. (You may refer to p. 78 of [Singer and Thorpe, 1].)

The nth **barycentric subdivision** $K^{(n)}$ is defined inductively by $K^{(n)} = (K^{(n-1)})'$. **Derived subdivisions** are defined in the same way, except that the barycenter of the simplex σ_i is replaced by an arbitrary interior point. The **mesh** of a complex is the maximum of the diameters of its simplexes. The following theorem is one reason for the importance of barycentric subdivisions.

Theorem 1.4.2. *If K is a complex of dimension k, then* mesh $K' \leqslant (k/(k+1)) \cdot$ mesh K. *In particular,* $\lim_{n \to \infty}$ mesh $K^{(n)} = 0$.

PROOF. Let $\langle \hat{\sigma}_0, \hat{\sigma}_1, \ldots, \hat{\sigma}_l \rangle$ be a simplex of maximal diameter. By Exercise **1.4.4** there exist m and n such that mesh K' = diameter

$\langle \hat{\sigma}_m , \hat{\sigma}_n \rangle$ where σ_m is a face of σ_n. Then, we can let $\sigma_m = \langle v_0 , ..., v_p \rangle$ and $\sigma_n = \langle v_0 , ..., v_p , v_{p+1} , ..., v_q \rangle$. Thus,

$$\text{mesh } K' = \| \hat{\sigma}_m - \hat{\sigma}_n \| = \left\| \frac{1}{p+1} \sum_{i=0}^{p} v_i - \frac{1}{q+1} \sum_{j=0}^{q} v_j \right\|$$

$$= \frac{1}{q+1} \left\| \frac{q+1}{p+1} \sum_{i=0}^{p} v_i - \sum_{j=0}^{q} v_j \right\| = \frac{1}{q+1} \left\| \sum_{j=0}^{q} \left(\frac{1}{p+1} \left(\sum_{i=0}^{p} v_i \right) - v_j \right) \right\|$$

$$= \frac{1}{p+1} \frac{1}{q+1} \left\| \sum_{j=0}^{q} \sum_{i=0}^{p} (v_i - v_j) \right\| \leqslant \frac{1}{p+1} \frac{1}{q+1} \sum_{j=0}^{q} \sum_{i=0}^{p} \| v_i - v_j \|$$

$$= \frac{1}{p+1} \frac{1}{q+1} \sum_{\substack{j=0 \\ j \neq i}}^{q} \sum_{i=0}^{p} \| v_i - v_j \| \leqslant \frac{1}{p+1} \frac{1}{q+1} \sum_{\substack{j=0 \\ j \neq i}}^{q} \sum_{i=0}^{p} \text{mesh } K$$

$$= \frac{1}{p+1} \frac{1}{q+1} ((p+1)(q+1) - (p+1)) \text{ mesh } K$$

$$= \frac{1}{p+1} \frac{1}{q+1} (p+1)q \text{ mesh } K = \frac{q}{q+1} \text{ mesh } K \leqslant \frac{k}{k+1} \text{ mesh } K.$$

Theorem 1.4.3. *If L is a subcomplex of K, then every subdivision of L can be extended to a subdivision of K, that is, given a subdivision L_* of L, there is a subdivision K_* of K such that L_* is a subcomplex of K_*.*

PROOF. Inductively, in order of increasing dimension, subdivide each simplex σ in $| K | - | L |$ that meets $| L |$ into the simplexes spanned by vertices of each simplex in $\dot{\sigma}_*$ along with a fixed point x in Int σ.

Theorem 1.4.4. *If K and L are simplicial complexes with $| L | \subset | K |$, then there is a derived subdivision of K which contains as a subcomplex some subdivision of L.* (See Theorem I.2 of [Glaser, 3] or Lemma 4 of [Zeeman, 1, p. 8].)

Corollary 1.4.1. *If K and L are two complexes such that $| K | = | L |$, then K and L have a common subdivision.*

Corollary 1.4.2. *The union of two polyhedra is a polyhedron.*

PROOF. Use the theorem to subdivide a large simplex containing them both, so that each appears as a subcomplex. Then, the union is also a subcomplex.

A **map** $f: K \to L$ of the simplicial complex K into the simplicial complex L is a triple $(|f|, K, L)$ where $|f| : |K| \to |L|$ is a continuous map of topological spaces. From now on we will identify $|f|$ and f. The map $f: K \to L$ is said to be **linear** if it maps each simplex of K linearly into some simplex of L. In the case that the image of each simplex of K is a simplex of L, the map f is said to be **simplicial**. A simplicial homeomorphism is called an **isomorphism**. A **piecewise linear** (PL) **map of the complex** K **into the complex** L is a map $f: K \to L$ for which there is a subdivision K' of K such that $f: K' \to L$ is linear.

REMARK 1.4.1. It can be shown that if f is a PL map of simplicial complexes, then the complexes can be subdivided so that the map is simplicial. Thus, if $f: K \to L$ is a PL homeomorphism, then so is $f^{-1}: L \to K$.

A map $f: X \to Y$ of the polyhedron X into the polyhedron Y is said to be a **piecewise linear** (PL) **map of polyhedra** if $f: K \to L$ is a PL map of complexes for every pair of triangulations K, L of X, Y.

REMARK 1.4.2. It can be shown that if the f of the above definition is PL for *some* pair of triangulations, then it is PL for *every* pair.

REMARK 1.4.3. It can be shown that the composition of two PL maps is a PL map.

1.5. METHOD OF ATTACKING THE MAIN PROBLEM

In this book the spaces X and Y of the main problem will usually be polyhedra. In this case, we see that we can break the problem of studying embeddings of X into Y down into two problems.

Problem 1 (The taming problem). *Which embeddings of X into Y are equivalent to* PL *embeddings?*

Problem 2 (The PL unknotting problem). *Which* PL *embeddings of X into Y are equivalent?*

We will usually be happy if we can show that the two embeddings under consideration are equivalent to PL embeddings, for we will have then reduced our topological problem to the PL category and life is relatively simple in that category. Of course, we will also be happy if we can show that one of the embeddings is equivalent to a PL embedding and the other is not, because then they would have to be in different equivalence classes. We will be concerned with Problem **1** more in the latter part of this book. At first most of the techniques will determine whether or not the embeddings under consideration are equivalent without passing to the PL category.

1.6. PIECEWISE LINEAR MANIFOLDS AND PIECEWISE LINEAR TOOLS

This section presents the fundamental theory of PL manifolds. Some supplementary references for this section are: [Glaser, 3], [Hudson, 1], [Stallings, 4], [Zeeman, 1], and [C.H. Edwards, 1].

A. Piecewise Linear Manifolds

In the preceding section we said that in this book the space X (called the **embedded space**) and the space Y (called the **ambient space**) of the main problem will usually be polyhedra. We are about to define infinite polyhedra; and the spaces X and Y will always be either finite or infinite polyhedra. In fact, the ambient space will always be a special type of (possibly infinite) polyhedron called a PL manifold, which we are also about to define. The embedded space will also be a PL manifold many times, although not always.

A (rectilinear) **locally finite simplicial complex** (or possibly infinite complex) is a (possibly infinite) collection K of simplexes in E^n such that

(1) $\sigma \in K$ and $\tau < \sigma \Rightarrow \tau \in K$,
(2) $\sigma \in K$ and $\tau \in K \Rightarrow \sigma \cap \tau < \sigma$ and $\sigma \cap \tau < \tau$, and
(3) each point in $|K| = \bigcup_{\sigma \in K} \sigma$ has a neighborhood which intersects only a finite number of simplexes of K nonvacuously.

An **infinite polyhedron** is a subset of E^n which can be triangulated as a locally finite simplicial complex having an infinite number of simplexes. **Triangulation** by locally finite complexes is defined analogously to the finite case. **Subcomplexes** and **linear maps** of infinite complexes and **subpolyhedra** of infinite polyhedra are also defined analogously to the finite case. A **piecewise linear (PL) map** of the possibly infinite complex K into the possibly infinite complex L is a map $f\colon K \to L$ such that there is a subdivision K' of K for which $f\colon K' \to L$ is linear and such that $f(|K|)$ is a possibly infinite polyhedron. (Notice that $f(|K|)$ need not be a subpolyhedron of L.) Now a **PL map** for infinite polyhedra is defined analogously to the finite case. (For an exposition of "weaker" definitions as well as a well written justification of those definitions see [Zeeman, 4].) A **PL n-ball** is a polyhedron which is PL homeomorphic to an n-simplex and a **PL n-sphere** is a polyhedron which is PL homeomorphic to the boundary of an $(n+1)$-simplex. If σ is a simplex of the (possibly infinite) complex K, then the **star of σ in K**, denoted by $\mathrm{St}(\sigma, K)$, and the **link of σ in K**, denoted by $\mathrm{Lk}(\sigma, K)$, are defined as follows:

$$\mathrm{St}(\sigma, K) = \{\tau \in K \mid \text{for some } \gamma \in K,\ \sigma < \gamma, \text{ and } \tau < \gamma\},$$
$$\mathrm{Lk}(\sigma, K) = \{\tau \in \mathrm{St}(\sigma, K) \mid \tau \cap \sigma = \emptyset\}.$$

A (possibly infinite) complex K is called a **combinatorial n-manifold** if the link of each of its vertices is either a PL $(n-1)$-sphere or a PL $(n-1)$-ball. We will now give two definitions of piecewise linear manifold and we will show that they are equivalent (Theorem **1.6.2**) a little later.

Definition PL-1. A (possibly infinite) polyhedron $M \subset E^m$ is said to be a **PL n-manifold** if it can be triangulated as a combinatorial n-manifold. (We will show in Corollary **1.6.2** that every triangulation of a PL manifold is combinatorial.)

Definition PL-2. A (possibly infinite) polyhedron $M \subset E^m$ is said to be a **PL n-manifold** if each point of M has a closed neighborhood which is PL homeomorphic with an n-simplex.

The **boundary** and **interior** of a PL manifold under Definition **PL-2** are defined analogous to the topological case. Equivalent definitions in terms of Definition **PL-1** will be given later. **Unbounded, closed,** and **open** PL manifolds are defined analogous to the corresponding topological definitions.

Two disjoint simplexes σ and τ in E^n are said to be **joinable** if there is a simplex γ which is spanned by the vertices of σ and τ. If this is the case σ and τ are called **opposite faces** of γ, and γ is called the **join** of σ and τ, denoted by $\sigma * \tau$.

EXERCISE 1.6.1. Show that $Lk(\sigma, K) = \{\tau \in K \mid \sigma * \tau \in K\}$.

EXERCISE 1.6.2. If $\sigma = \tau * \gamma \in K$, then $Lk(\sigma, K) = Lk(\tau, Lk(\gamma, K))$.

Two finite complexes K and L in E^n are **joinable** provided that (i) if $\sigma \in K$ and $\tau \in L$, then σ and τ are joinable, and (ii) if $\sigma, \sigma' \in K$ and $\tau, \tau' \in L$ then $\sigma * \tau \cap \sigma' * \tau'$ is a common face of $\sigma * \tau$ and $\sigma' * \tau'$. If K and L are joinable, their **join** is the complex $K * L = \{\sigma * \tau \mid \sigma \in K$

and $\tau \in L\}$. (Notice that K and L are contained in $K * L$, since the empty simplex is a face of every simplex, and so is in K and L and since $\sigma * \emptyset = \sigma$ for any simplex σ. Also notice that $K * L = K$ if L is the complex containing only the empty simplex.) Two disjoint polyhedra X and Y in E^n are said to be **joinable** if any two triangulations K and L of X and Y are joinable.

EXERCISE 1.6.3. Show that the polyhedra X and Y in E^n are joinable if and only if (i) if L_{pq} is the line segment from a point $p \in X$ to $q \in Y$, then $L_{pq} \cap X = p$ and $L_{pq} \cap Y = q$, and (ii) two different such line segments are either disjoint or intersect in an end-point.

The **join** of two joinable polyhedra X and Y is $|L * K|$, denoted by $X * Y$, where L and K triangulate X and Y, respectively. It is easy to see that $X * Y$ is simply the union of all of the line segments of Exercise **1.6.3.**

EXERCISE 1.6.4. If $X_1 \overset{\text{PL}}{\approx} X_2$ and $Y_1 \overset{\text{PL}}{\approx} Y_2$ ($\overset{\text{PL}}{\approx}$ denotes PL homeomorphic), then $X_1 * Y_1 \overset{\text{PL}}{\approx} X_2 * Y_2$.

The **cone** over the polyhedron X with **vertex** (or **cone point**) v, denoted by $\mathscr{C}(X)$, is the join of X and the point v. (Notice that there are points v such that $\mathscr{C}(X)$ is defined, because any point in $E^k - E^n$, $k > n$, will work if $X \subset E^n$.) The **suspension** of the polyhedron X, denoted by $\mathscr{S}(X)$, is the join of X with a pair of points (a 0-sphere) called the **suspension points.** (Notice that there are always suspension points so that the suspension is defined.)

EXERCISE 1.6.5. Let B^n be a PL n-ball and S^n be a PL n-sphere. Show that $\mathscr{C}(B^n) \overset{\text{PL}}{\approx} \mathscr{S}(B^n) \overset{\text{PL}}{\approx} \mathscr{C}(S^n) \overset{\text{PL}}{\approx} B^{n+1}$ and $\mathscr{S}(S^n) \overset{\text{PL}}{\approx} S^{n+1}$, where $\overset{\text{PL}}{\approx}$ denotes PL homeomorphic. (Notice that, by using Exercise **1.6.4,** it may be assumed that B^n [S^n] is an n-simplex [the boundary of an $(n + 1)$-simplex].)

EXERCISE 1.6.6. Show that $B^p * B^q \overset{\text{PL}}{\approx} B^{p+q+1}$, $B^p * S^q \overset{\text{PL}}{\approx} B^{p+q+1}$, and $S^p * S^q \overset{\text{PL}}{\approx} S^{p+q+1}$.

Theorem 1.6.1. *Let K be a simplicial complex, K' a subdivision of K, and v a vertex of K. Then,* Lk(v, K) *and* Lk(v, K') *are* PL *homeomorphic, as are* St(v, K) *and* St(v, K').

In view of the next exercise, it is appropriate to mention here **Zeeman's standard mistake** of confusing projective maps with piecewise linear maps. (Zeeman does not make this mistake; he just observed that other people often make it.) For example, the projection f into the base of a triangle (from the vertex opposite the base) of a segment which is not parallel or perpendicular to the base is not piecewise linear (see Fig. **1.6.1**).

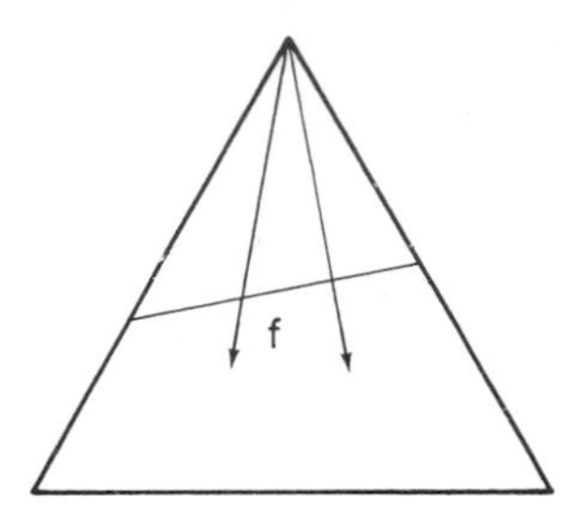

Figure 1.6.1

This difficulty can usually be circumvented by defining pseudo-radial projection as follows. Let K be a complex and let $|L|$ be a polyhedron with triangulation L contained in the cone $\mathscr{C}(|K|)$ over $|K|$ with vertex v such that for each point $p \in |K|$ the line segment L_p from v to p intersects $|L|$ in a single point. (The **radial projection** $f: |L| \to |K|$ is the homeomorphism defined by $f(x) = p$, where $p \in |K|$ is the point such that $L_p \cap |L| = x$. We have just seen that the radial projection is not always PL.) $\mathscr{C}(K)$ inherits a natural triangulation from K, and by Theorem **1.4.4**, there is a subdivision of $\mathscr{C}(K)$ that contains as a subcomplex some subdivision L_* of L. The **pseudo-radial projection** of $|L|$ into $|K|$ is defined to be the linear extension of the radial projection of the vertices of L_* into $|K|$.

EXERCISE 1.6.7. Prove Theorem **1.6.1**.

Theorem 1.6.2. *Definitions* **PL-1** *and* **PL-2** *of* PL *n-manifold are equivalent. Furthermore, if M is a* PL *n-manifold under Definition* **PL-2** *then every triangulation K of M is combinatorial.*

PROOF. It is trivial to see that if M is a PL n-manifold under Definition **PL-1**, then it is a PL n-manifold under Definition **PL-2**. Let K be a triangulation of M as a combinatorial manifold. Then, the stars of the vertices of K give a covering of M by PL n-balls. Since each point of M has one of these PL n-balls as a closed neighborhood, it follows that M is a PL n-manifold under Definition **PL-2**.

Now suppose that M is a PL n-manifold under Definition **PL-2**. Let J be a triangulation of M and let v be a vertex of J in Int M. By definition **PL-2**, there is a PL embedding $f: \Delta \to M$ where Δ is an n-simplex, such that $f^{-1}(v) \in \text{Int}\, \Delta$. By using Theorem **1.4.4**, Remark **1.4.1**, and Theorem **1.4.3** we can obtain subdivisions Δ' of Δ and J' of J such that $f: \Delta' \to J'$

is simplicial. Thus, Bd $\Delta \overset{\text{PL}}{\approx}$ Lk$(f^{-1}(v), \Delta')$ and Lk$(v, J') \overset{\text{PL}}{\approx}$ Lk(v, J) by Theorem **1.6.1**, and f gives a simplicial isomorphism between Lk$(f^{-1}(v), \Delta')$ and Lk(v, J'). Hence, Lk(v, J) is PL homeomorphic to Bd Δ, and is thus a PL $(n - 1)$-sphere.

If v is a vertex in Bd M the situation is similar except that $f^{-1}(v) \in$ Bd Δ, and so it follows that Lk(v, J) is a PL $(n - 1)$-ball.

Corollary 1.6.1. *Any subdivision of a combinatorial manifold is a combinatorial manifold.*

Corollary 1.6.2. *Every triangulation of a* PL *manifold is combinatorial.*

Let us now digress to mention a few classical questions.

Question 1.6.1. *Is every topological manifold homeomorphic to a* PL *manifold?*

The answer is yes for 1-manifolds and 2-manifolds [Radó, 1]. Moise showed in [1] that the answer is yes for 3-manifolds without boundary. This result was extended in [Bing, 4] and [Moise, 2] to show that 3-manifolds with boundaries can be triangulated. An alternative proof was given in [Bing, 5] based on the approximation theorem for 2-complexes. Kirby, Siebenmann, and Wall [1] have recently shown that the answer is yes for "most" manifolds of dimension greater than four; however, there are some in each dimension greater than four for which this is not the case. The answer is unknown for 4-manifolds.

Question 1.6.2. *Is every topological manifold homeomorphic to a polyhedron?*

The answer is yes for manifolds of dimensions less than four by the above references and is unknown for manifolds of dimensions greater than or equal to four.

Question 1.6.3. *Is every polyhedron which is a topological manifold, also a* PL *manifold?*

The answer is yes for manifolds of dimensions less than four, see Theorem 1 of [Moise, 1], and is unknown for manifolds of dimensions greater than or equal to four.

Question 1.6.4 (Hauptvermutung for polyhedra). *If two polyhedra are homeomorphic then are they* PL *homeomorphic?*

In 1961 Milnor [1] showed that the answer is no in high dimensions. (Another example which is obtained by somewhat easier methods appears in [Stallings, 1].)

Question 1.6.5 (Hauptvermutung for PL manifolds). *If two* PL *manifolds are homeomorphic then are they* PL *homeomorphic?*

This is trivially the case for PL 1-manifolds, is classically the case but nontrivally for PL 2-manifolds, was proved for triangulated 3-manifolds by Moise [1] in the 1950s, is unknown for 4-manifolds, and has recently been shown to be true for most manifolds of dimension greater than four and false for a few in every such dimension by Kirby, Siebenmann, and Wall.

Lemma 1.6.1. *Let K be a triangulation of the* PL *n-manifold M. If σ is a k-simplex f K, then either*

(1) $\mathrm{Lk}(\sigma, K)$ *is a* PL $(n - k - 1)$*-sphere and* $\mathrm{Int}\,\sigma \subset \mathrm{Int}\,M$, *or*
(2) $\mathrm{Lk}(\sigma, K)$ *is a* PL $(n - k - 1)$*-ball and* $\sigma \subset \mathrm{Bd}\,M$.

PROOF. The proof will be by induction on k. The case $k = 0$ follows from Theorem **1.6.2.** If $k > 0$, write $\sigma = v * \tau$, where v is a vertex of σ and τ is the opposite $(k - 1)$-face. Then, $\mathrm{Lk}(\sigma, K) = \mathrm{Lk}(v, \mathrm{Lk}(\tau, K))$ by Exercise **1.6.2.** But, by induction, $\mathrm{Lk}(v, \mathrm{Lk}(\tau, K))$ is the link of a vertex in an $(n - k)$-sphere or ball, and is therefore an $(n - k - 1)$-sphere or ball.

Any point of $\mathrm{Int}\,\sigma$ has $\sigma * \mathrm{Lk}(\sigma, K)$ as a closed neighborhood, and so lies in $\mathrm{Int}\,M$ or $\mathrm{Bd}\,M$ according to whether it lies in the interior or boundary of this neighborhood. But $\mathrm{Int}\,\sigma \subset \mathrm{Int}(\sigma * \mathrm{Lk}(\sigma, K))$ if $\mathrm{Lk}(\sigma, K)$ is a sphere and so $\mathrm{Int}\,\sigma \subset \mathrm{Int}\,M$ if $\mathrm{Lk}(\sigma, K)$ is a sphere. If $\mathrm{Lk}(\sigma, K)$ is a ball, then $\mathrm{Int}\,\sigma \subset \mathrm{Bd}(\sigma * \mathrm{Lk}(\sigma, K))$ and so $\mathrm{Int}\,\sigma \subset \mathrm{Bd}\,M$. Thus, $\sigma \subset \mathrm{Bd}\,M$ since $\mathrm{Bd}\,M$ is closed in M.

Corollary 1.6.3. *Each $(n - 1)$-simplex of a combinatorial n-manifold K is the face of exactly one or two n-simplexes of K.*

We define the **boundary** of a combinatorial manifold K to be the subcomplex consisting of all simplexes whose links are balls. We now define the interior and boundary of a PL manifold M in terms of Definition **PL-1** as follows: Let K be a triangulation of M (K is combinatorial by Corollary **1.6.2**). Then, the **boundary** of M is $|\,\mathrm{Bd}\,K\,|$ and the **interior** of M is $M - \mathrm{Bd}\,M$. It follows immediately from Lemma **1.6.1.** that these definitions are equivalent to the corresponding definitions in terms of Definition **PL-2**.

B. Regular Neighborhoods

If L is a subcomplex of the simplicial complex K, then we say that there is an **elementary simplicial collapse** from K to L if $K - L$ consists of two simplexes A and B, where $A = a * B$ with a a vertex of A. Then, $|K| = |L| \cup A$ and $|L| \cap A = a * (\text{Bd } B)$. This elementary simplicial collapse from K to L may be described by saying that "we collapse A onto $a * (\text{Bd } B)$" (see Fig. **1.6.2**).

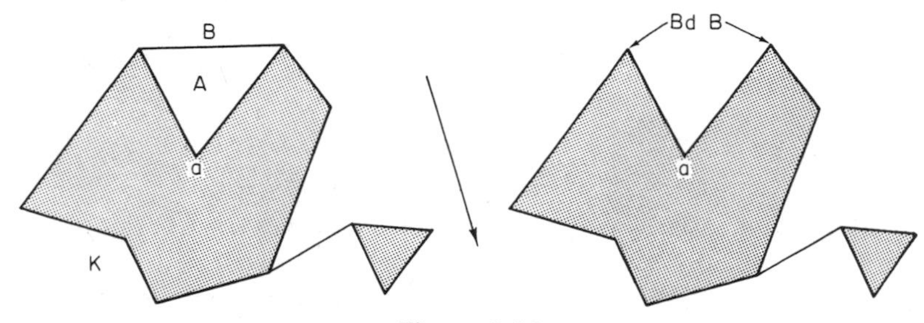

Figure 1.6.2

The complex K simplicially collapses to the subcomplex L, written $K \overset{s}{\searrow} L$, if there is a finite sequence of elementary simplicial collapses going from K to L. If K simplicially collapses to a vertex, then K is said to be **simplicially collapsible**, written $K \overset{s}{\searrow} 0$.

EXERCISE 1.6.8. (a) Show that $\mathscr{C}(K) \overset{s}{\searrow} 0$ for any complex K. (b) Show that if L is a subcomplex of K, then $\mathscr{C}(K) \overset{s}{\searrow} \mathscr{C}(L)$. (c) If $|K|$ is a connected 1-polyhedron containing no simple closed curve, show that $K \overset{s}{\searrow} 0$.

Example 1.6.1. By Exercise **1.6.8**, every contractible 1-complex is collapsible. We now give an example of a contractible 2-complex B (that is, the identity map of B onto itself is homotopic to a constant map) that is not collapsible. The complex B is known as **Bing's house with two rooms.**

Pictured in Fig. **1.6.3** we have a block of wood, I^3 with one termite situated on the front and another situated on the back. Now each of these termites is told that he can eat into this block of wood, however is warned that if he ever eats a hole through the wood he will fall and be killed. The front termite is told to start at point x and eat his way directly to the back half of the block of wood and then to eat all of the back half that he can. Likewise, the rear termite is told to eat his way directly to the front half and then to eat all of it that he can. At the right is pictured B which is what is left after the termites have finished their meal. In that picture, A_1 and A_3 are squares with an open square

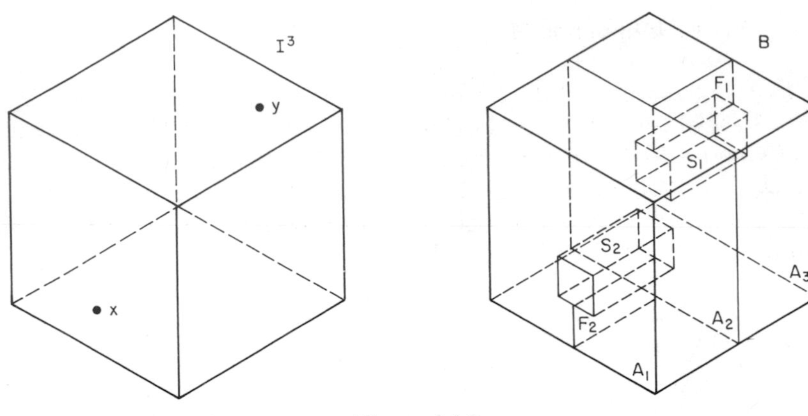

Figure 1.6.3

removed, A_2 is a square with two open squares removed, S_1 and S_2 are square cylinders, and F_1 and F_2 are rectangular disks.

As the termites ate their way into the wood they described a retraction (in fact, a strong deformation retraction) of I^3 onto B. It is trivial to prove that any retract of a contractible space is contractible (for instance, see p. 156, Theorem 4.11 of [Hocking and Young, 1]). Hence, since I^3 is contractible, B is contractible. However, no matter how you triangulate B it will not be collapsible because there can be no free 1-simplex to get started. (For other uses of Bing's house with two rooms see [Glaser, 2] and [Bing, 3].)

EXERCISE 1.6.9. Give an example of a subcomplex L of a complex K such that $K \overset{s}{\searrow} 0$ but such that L is not simplicially collapsible.

We now imitate for polyhedra the above definition of collapsing for complexes. Let X and Y be polyhedra with $Y \subset X$. We say that there is an **elementary collapse** from X to Y if there is a PL n-ball B^n and a PL $(n-1)$-ball $B^{n-1} \subset \text{Bd } B^n$ such that $X = Y \cup B^n$ and $Y \cap B^n = B^{n-1}$ (see Fig. **1.6.4**). We may describe this elementary collapse by saying, "We collapse B^n onto B^{n-1}".

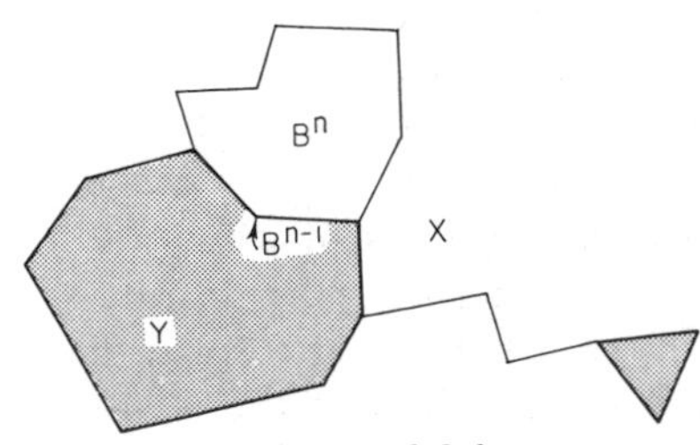

Figure 1.6.4

The polyhedron X **collapses** to the polyhedron Y, $X \searrow Y$, if there is a finite sequence of elementary collapses going from X to Y. We say that X is collapsible, $X \searrow 0$, if it collapses to a point. (For example, every PL ball is collapsible. Indeed, we shall see from the regular neighborhood theorem that this fact characterizes balls among PL manifolds.) Finally, we say that the complex K collapses to the complex L, written $K \searrow L$, if $|K| \searrow |L|$. It is obvious that $K \overset{s}{\searrow} L$ implies $K \searrow L$, however we will point out in the following example that the converse is false.

Example 1.6.2. We say that a triangulation T of an n-cell can be **shelled** if the n-simplexes of T can be ordered $\sigma_1, \sigma_2, \ldots, \sigma_k$ so that for each integer $m \leqslant k$, $\sigma_m \cup \sigma_{m+1} \cup \cdots \cup \sigma_k$ is an n-cell. Thus, if a triangulation of an n-cell can be shelled then it is collapsible in a nice way.

We now give an example due to Bing of a triangulation of a 3-cell that cannot be shelled. This example appeares as Example 2 of [Bing, 3]. Consider the cube with the plugged knotted hole shown in Fig. **1.6.5**.

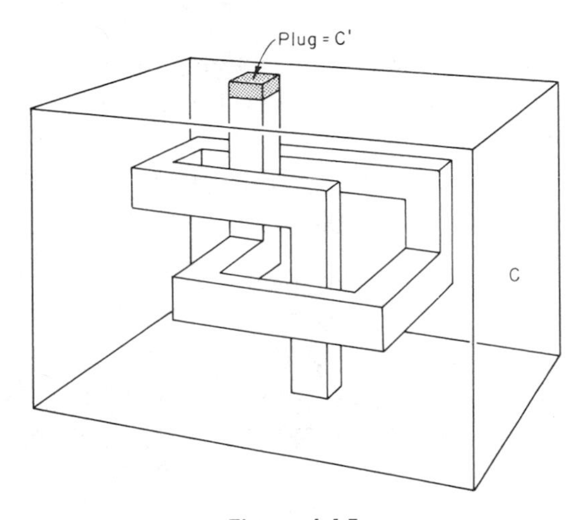

Figure 1.6.5

The object is topologically a cube since the hole is plugged at the upper end with a small cube C' and so may be viewed as a 3-ball minus a 3-ball which hits its boundary in a 2-ball. The resulting 3-cell C may easily be triangulated so that the edges of C' are 1-simplexes of the triangulation,

Consider a simple closed curve J made up of a spanning segment of C, which is an edge of C', and a polygonal arc of Bd C. Then, J is knotted as we shall prove in Chapter **2**. If we were to start shelling the triangulation of C, it is easy to see that at each stage there would be a knotted simple closed curve which lies, except possibly for one spanning simplex, on the boundary of the resulting 3-cell. It follows that the triangulation of C cannot be shelled for at the last stage there could be no such simple closed curve.

Although, it is not immediately obvious that the above triangulation is collapsible, in Example 3 of [Bing, 3], it is stated that a similar example with two knots instead of one will yield a noncollapsible example. In fact, Bing states that he can show that for each integer n there is a triangulation of a cube whose nth barycentric subdivision is not collapsible. The proof of this was written out in [Goodrick, 1]. Goodrick's proof that one can also obtain such triangulations for I^n, $n > 3$, by taking suspensions is incorrect. In [Lickorish and Martin, 1] it is shown that Bing's result is in some sense the best possible.

Even though $K \searrow L$ does not imply $K \overset{s}{\searrow} L$, the following result usually circumvents any difficulty. (This is Theorem 4 of [Zeeman, 1] and Theorem III.6(W) of [Glaser, 3]. A particularly nice proof is given in [Cohen, 4].)

Theorem 1.6.3. *If K and L are complexes such that $K \searrow L$, then there is a subdivision K_* of K such that $K_* \overset{s}{\searrow} L_*$ (where L_* is the induced subdivision of L).*

For example, every combinatorial ball (that is, triangulated PL ball) has a subdivision which is simplicially collapsible.

Lemma 1.6.2. *If X and Y are polyhedra such that $X \searrow Y$, then Y is a* PL *deformation retract of X.*

Proof. By Theorem **1.6.3**, there is a triangulation (K, L) of the PL pair (X, Y) such that $K \overset{s}{\searrow} L$. The lemma now follows easily by induction on the number of elementary simplicial collapses in $K \overset{s}{\searrow} L$, it being obvious if $K \overset{s}{\searrow} L$ consists of a single elementary collapse.

Lemma 1.6.3. *If P and Σ are subpolyhedra of a polyhedron Q where $Q \searrow P$, then there is a subpolyhedron Λ of Q such that $\Sigma \subset P \cup \Lambda$, $Q \searrow P \cup \Lambda \searrow P$ and* $\dim \Lambda \leqslant \dim \Sigma + 1$.

Proof. By Theorem **1.6.3**, there is a triangulation (K, L, N) of (Q, P, Σ) for which we have a sequence of elementary collapses as follows

$$K = K_r \overset{s}{\searrow} K_{r-1} \overset{s}{\searrow} \cdots \overset{s}{\searrow} K_0 = L.$$

Suppose that $K_i = K_{i-1} \cup v_i * \sigma_i \cup \sigma_i$ for $i = 1, \ldots, r$. We may assume that $\dim v_i * \sigma_i \geqslant \dim v_{i-1} * \sigma_{i-1}$ for $i = 1, ..., r$, for suppose that

$$K_i = K_{i-1} \cup v_i * \sigma_i \cup \sigma_i \searrow K_{i-1} = K_{i-2} \cup v_{i-1} * \sigma_{i-1} \cup \sigma_{i-1} \searrow K_{i-2}$$

and $\dim v_i * \sigma_i < \dim v_{i-1} * \sigma_{i-1}$. Then, $v_i * \mathrm{Bd}\, \sigma_i \subset K_{i-2}$ and so $K_{i-2} \cup v_i * \sigma_i \cup \sigma_i$ is a subcomplex of K_i. Since

$$K_i = (K_{i-2} \cup v_i * \sigma_i \cup \sigma_i) \cup v_{i-1} * \sigma_{i-1} \cup \sigma_{i-1},$$

it follows that we can reorder the above collapse to be

$$K_i \searrow K_{i-2} \cup v_i * \sigma_i \cup \sigma_i \searrow K_{i-2}.$$

Let $j \leqslant r$ be the largest integer such that

$$K_j = K_{j-1} \cup v_j * \sigma_j \cup \sigma_j \searrow K_{j-1}$$

and $\dim \sigma_j = \dim \Sigma$. Then, $\Lambda = \mathrm{Cl}(|K_j| - |K_0|)$ obviously has the desired properties.

Lemma 1.6.4. *Let C, P, and X^x be polyhedra such that $C \searrow P$. Then, there is a subpolyhedron X_0 of $C \cup X$ such that $P \subset X_0$, $C \cup X \searrow X_0$ and* $\dim \mathrm{Cl}(X_0 - P) \leqslant \dim X = x$.

PROOF. By Theorem **1.6.3**, there is a triangulation (K, L, N) of $(C \cup X, C, P)$ such that $L \overset{s}{\searrow} N$. By the proof of the above lemma, we can assume the sequence of elementary simplicial collapses is in order of nonincreasing dimension. We claim that it is possible to perform all of those of dimension greater than x on the complex $C \cup X$, collapsing it to a subcomplex which we call X_0 such that $\dim X_0 = x$. There is no trouble during collapses of dimensions greater than $x + 1$, because X cannot get in the way. Also, there is no trouble in performing $(x + 1)$-dimensional collapses, because the free face in such a collapse, if it belongs to X, will be a principal simplex in X.

EXERCISE 1.6.10. Show that a connected PL n-manifold with nonempty boundary collapses onto some $(n - 1)$-subpolyhedron. (Hint: Use Corollary **1.6.3**.)

If $K \subset J$ are complexes, we say that K is **full** in J if no simplex of $J - K$ has all of its vertices in K.

EXERCISE 1.6.11. (a) If $K \subset J$ and $J^{(1)}$ is a first derived subdivision of J, then $K^{(1)}$ is full in $J^{(1)}$. (b) If K is full in J and J_* is a subdivision of J, then K_* is full in J_*.

If J is a complex and $X \subset |J|$, then the **simplicial neighborhood** $N(X, J)$ of X in J is the minimal subcomplex of J containing all simplexes of J that hit X.

Suppose that P is a polyhedron in the PL n-manifold M. Let M_0 be a polyhedron in M which contains a topological neighborhood in M of P. If M is compact, pick $M_0 = M$. Let (J, K) be a triangulation of the pair (M_0, P) such that K is full in J. If $J^{(1)}$ is a first derived subdivision of J, then the polyhedron $|N(P, J^{(1)})|$ is called a **derived neighborhood** of P in M. It follows from Exercise **1.6.11a** that if (J, K) is any triangulation of (M_0, P) and $J^{(2)}$ is a second derived subdivision of J, then $|N(P, J^{(2)})|$ is a derived neighborhood of P in M. (We will see in the regular neighborhood theorem below that a derived neighborhood of P in M is a PL n-manifold which collapses to P.)

If P is a polyhedron, by a **PL isotopy** of P is meant a PL homeomorphism H of $P \times I$ onto $P \times I$ ($I = [0, 1]$ here) which is level preserving; that is, $H(P \times t) = P \times t$ for each $t \in I$. Let the PL homeomorphism $H_t: P \twoheadrightarrow P$ be defined by $H(x, t) = (H_t(x), t)$ for each $x \in P$, $t \in I$. Then, H is also called a **PL isotopy** between H_0 and H_1. We call H an **ambient isotopy** if $H_0 = 1$. Finally, if $X, Y, Z \subset P$, then X and Y are said to be **ambient isotopic leaving Z fixed** if there is an ambient isotopy H of P such that $H_1(X) = Y$ and $H_t | Z = 1$ for all $t \in I$.

Let P be a polyhedron in the PL n-manifold M. Then a **regular neighborhood of P in M** is a polyhedron N such that

(a) N is a closed neighborhood of P in M,
(b) N is a PL n-manifold, and
(c) $N \searrow P$.

The existence and uniqueness properties of regular neighborhoods are given in the following theorem due to Whitehead [Whitehead, 1] which appears as Theorem 8 of [Zeeman, 1], Theorems II.15n and II.16n of [Glaser, 3], and Theorem 2.11 of [Hudson, 1].

Regular Neighborhood Theorem 1.6.4. *If P is a polyhedron in the* PL *manifold M, then*

1. (Existence). *Any derived neighborhood of P in M is a regular neighborhood of P in M.*

2. (Uniqueness). *If N_1 and N_2 are any two regular neighborhoods of P in M, then there is a* PL *homeomorphism h of N_1 onto N_2 such that $h | P = 1$.*

3. (Uniqueness). *If $P \subset$* Int *M, then any two regular neighborhoods of P in* Int *M are ambient isotopic leaving $P \cup$* Bd *M fixed.*

Corollary 1.6.4. *Every collapsible* PL *n-manifold M is an n-ball.*

PROOF. *If p* is a point to which M collapses, then M is a regular neighborhood of p. Let K be a triangulation of M which has p as a vertex. Then, $\mathrm{St}(p, K) = p * \mathrm{Lk}(p, K)$ collapses to p by Exercise **1.6.8** and so is also a regular neighborhood of p in M. But, then M and $\mathrm{St}(p, K)$ are PL homeomorphic by Part 2 of Theorem **1.6.4.**

EXERCISE 1.6.12. Give a proof using the regular neighborhood theorem and also a direct proof of the following fact: Let M^n be a connected PL n-manifold without boundary and let $Q \subset P$ be polyhedra in M such that $P \searrow Q$. Then, given an open subset U of M such that $Q \subset U$ there is an ambient isotopy e_t of M such that $P \subset e_1(U)$ and $e_t = 1$ outside some compact subset of M and on Q. [Hint for the direct proof: Let (K, L) triangulate (P, Q) so that K simplicially collapses to L. By induction on the number of elementary collapses, you need only consider the case of one elementary collapse from K to L, pushing $A = a * B$ to $a * \partial B$ as in the definition of elementary simplicial collapse. By hypothesis U contains L, and also contains a neighborhood of $a * \partial B$ in A. Therefore, in the closed star of B, you can perform an elementary isotopy (keeping the boundary of the star, and everything outside, fixed), that stretches this neighborhood over A.]

C. Relative Regular Neighborhoods

Our discussion of relative regular neighborhoods will be based on [Hudson and Zeeman, 1], [Tindell, 2], [Husch, 1] and [Cohen, 1]. Relative regular neighborhoods are like air conditioning, power steering, and color TV in the sense that once one gets accustomed to using them, it is difficult to manage without them. Hudson and Zeeman [1] were the first to relativize the concept of regular neighborhoods. Unfortunately, their work contained an error as was evidenced by an example of Tindell [2]. However, Husch [1] formulated a correct version of Hudson and Zeeman's work. Finally, Cohen [1] conceived a more general notion of relative regular neighborhoods than that of Hudson and Zeeman and developed a very useful theory of his relative regular neighborhoods.

If X and Y are subpolyhedra of some larger polyhedron (see Fig. **1.6.6**), then

$$X_{\mathrm{R}} = \mathrm{Cl}(X - Y) \qquad \text{and} \qquad Y_{\mathrm{R}} = \mathrm{Cl}(X - Y) \cap Y.$$

Given two subcomplexes K, L of some larger complex, let K_R and L_R be defined in the obvious way. We say that K is **link-collapsible** on L if $\mathrm{Lk}(A, K_R)$ is collapsible for each simplex A in L_R. Given two subpolyhedra X, Y of some larger polyhedron, we say that X is **link-collapsible** on Y if for some triangulation (K, L) of (X, Y) we have K link-collapsible on L.

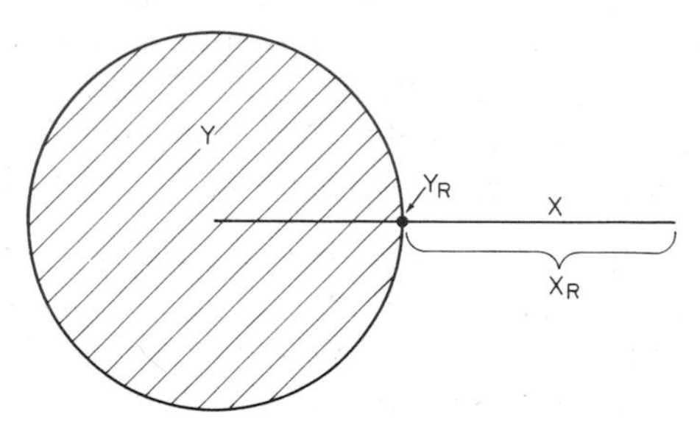

Figure 1.6.6

EXERCISE 1.6.13. Show that the definition of link-collapsibility for polyhedra is independent of the triangulation.

Example 1.6.3

(a) Any polyhedron is link-collapsible on itself and on the empty set.
(b) A simplex is link-collapsible on any subcomplex.
(c) A PL manifold is link-collapsible on its boundary and on any subpolyhedron of the boundary.
(d) A *PL* manifold is *not* link-collapsible on an interior point.
(e) A cone is link-collapsible on its base, and on any subpolyhedron of the base.
(*f*) X is link-collapsible on Y if and only if X_R is link-collapsible on Y_R.

Let X, Y, N be compact polyhedra in the PL manifold M^m (see Fig. 1.6.7). We say that N is a **HZ-regular neighborhood of X mod Y in M** if

(1) N is an m-manifold,
(2) N is a topological neighborhood of $X - Y$ in M, and

$$N \cap Y = \partial N \cap Y = Y_R,$$

(3) $N \searrow X_R$.

We say that N **meets the boundary regularly** if, further

(4) $\mathrm{Cl}((N \cap \partial M) - Y)$ is a HZ-regular neighborhood of $X \cap \partial M$ mod $Y \cap \partial M$ in ∂M.

If N_1 is another HZ-regular neighborhood of X mod Y in M, we say that N_1 is smaller than N if N contains a topological neighborhood of $N_1 - Y$ in M.

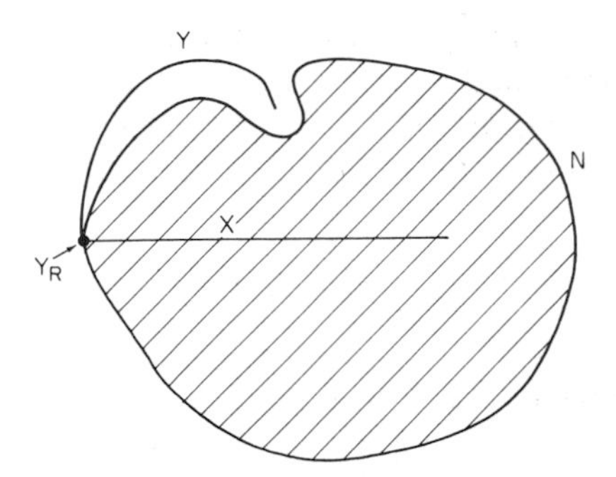

Figure 1.6.7

REMARK 1.6.1. If we put $Y = \emptyset$ in the relative definition, then we recover the absolute definition, and so the relative definition is a generalization.

REMARK 1.6.2. Any HZ-regular neighborhood of X mod Y is also an HZ-regular neighborhood of X_R mod Y_R, but not conversely in general, because of Condition 2.

REMARK 1.6.3. If M is unbounded Condition 4 is vacuous, and so trivially true. If M is bounded and $X \subset \text{Int } M$, then Condition 4 is the same as saying $N \subseteq \text{Int } M$.

REMARK 1.6.4. In Condition 4 the term $\text{Cl}((N \cap \partial M) - Y)$ is necessary instead of $N \cap \partial M$ in order to allow possibilities such as the one indicated in Fig. **1.6.8**.

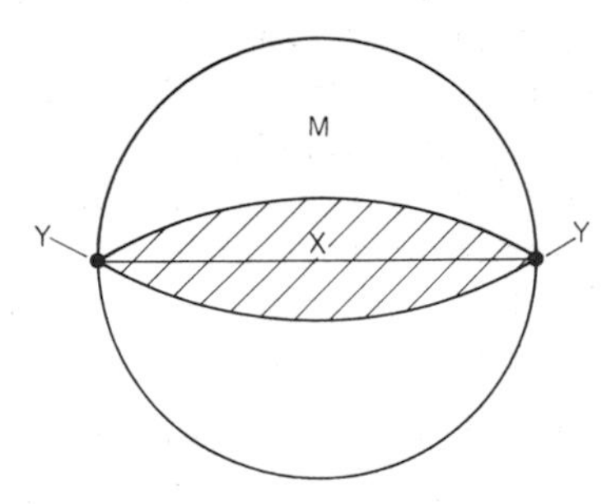

Figure 1.6.8

Suppose that X and Y are polyhedra in the PL manifold M. A **second derived neighborhood** N **of** X **mod** Y in M is constructed as follows: choose a triangulation J of M that contains subcomplexes triangulating X and Y; then choose a second derived complex $J^{(2)}$ of J and define $N = |N(X - Y, J^{(2)})|$.

Let X, Y be compact polyhedra in a PL manifold M, and let N be a regular neighborhood of X mod Y in M.

(β) N **satisfies condition** β if there exists a triangulation J of M containing subcomplexes G, K, and L which triangulate N, X, and Y, respectively such that

$$\mathrm{Lk}(\sigma, G) \searrow \mathrm{Lk}(\sigma, K_R)$$

for each σ in L_R.

(γ) If N **satisfies condition** γ if X is a manifold, Y is the boundary of X, and there exists a triangulation J of M containing subcomplexes G, K, and L which triangulate N, X, and Y, respectively such that the ball pairs $(\mathrm{Lk}(\sigma, G), \mathrm{Lk}(\sigma, K))$ are unknotted for each vertex σ of L_R.

HZ-Relative Regular Neighborhood Theorem 1.6.5. *Let X and Y be polyhedra in the* PL *manifold M where X is link-collapsible on Y, then*

1. (Existence). *Any second derived neighborhood N of X* mod *Y in M is a* HZ-*regular neighborhood of X* mod *Y in M. If, further, $X \cap \partial M$ is link-collapsible on $Y \cap \partial M$, then N meets the boundary regularly.*
2. (Uniqueness). *Let N_1 and N_2 be regular neighborhoods of X* mod *Y in M such that there is a triangulation J of M for which N_1 and N_2 satisfy condition β; then there is a smaller regular neighborhood N_3 and a* PL *homeomorphism of N_1 onto N_2 keeping N_3 fixed. Further, the homeomorphism can be realized by an isotopy in M moving N_1 onto N_2 through a continuous family of regular neighborhoods and keeping N_3 fixed.*
3. (Uniqueness). *Let $Q \subseteq M$ be a compact* PL *manifold. (If $Q \cap \mathrm{Bd}\, M \neq \emptyset$, we require that* $\mathrm{Bd}\, Q \subset \mathrm{Bd}\, M$.) *Let N_1 and N_2 be two regular neighborhoods of Q* mod $\mathrm{Bd}\, Q$ *in M which meet the boundary regularly, and suppose that N_1 and N_2 satisfy condition γ. Let P be the closure of the complement of a second derived neighborhood of $N_1 \cup N_2$* mod $\mathrm{Bd}\, Q$ *in M. Then, there exists an ambient isotopy of M moving N_1 onto N_2 and keeping P fixed.*

Hudson and Zeeman's original statement of 2 above did not include the requirement that there be a triangulation J of M for which N_1 and N_2 satisfy condition β. The following example shows that that requirement is necessary. (Since this example uses a couple of elementary concepts not discussed until later, it may be skipped for the time being and considered later.)

Example 1.6.4 (Tindell). Let (B^3, B^1) be a knotted (3, 1)-ball pair in E^3 and let $B^4 = a * B^3$ and $B^2 = a * B^1$ where $a = (0, 0, 0, 1) \in E^4$.

The (4, 2) ball pair (B^4, B^2) is locally knotted at the vertex a [that is, the first local homotopy group, introduced in Section **2.3**, is bad] and hence (B^4, B^2) is knotted [that is, is not homeomorphic to the pair (I^4, I^2)]. However, it is easy to see that B^2 is unknotted in E^4 [that is, there is an onto PL homeomorphism h: $E^4 \to E^4$ such that $h(B^2) = \Delta$ is a 2-simplex]. B^4 collapses conewise to B^2 so that $h(B^4)$ collapses to $h(B^2) = \Delta$. Also, $\partial\Delta \subset h(\partial B^4)$ and Int $\Delta \subset h(\text{Int } B^4)$, so that $h(B^4)$ is an HZ-regular neighborhood of Δ mod $\partial\Delta$ in E^4. Let Σ be the two-fold suspension of Δ in E^4; then Σ is an HZ-regular neighborhood of Δ mod $\partial\Delta$ in E^4. If the first uniqueness result of Theorem **1.6.5** were true without condition β, there would be a homeomorphism carrying $(h(B^4), \Delta)$ onto (Σ, Δ) which implies that $(h(B^4), \Delta) = (h(B^4), h(B^2))$ is an unknotted ball pair. This is clearly false since it is a homeomorphism of the knotted ball pair (B^4, B^2). This type of argument can be carried out in E^n for every $n \geqslant 4$ so the β condition is necessary for dimensions greater than three.

We will now discuss Cohen's conception of relative regular neighborhoods. HZ-relative regular neighborhoods were of polyhedra in PL manifolds. Now we consider regular neighborhoods of X mod Y in Z where X and Y are (possibly infinite) polyhedra in the (possibly infinite) polyhedron Z. V is defined to be a **C-regular neighborhood of X mod Y in Z** if J is a complex containing K and L as full subcomplexes and h: $(|J|, |K|, |L|) \to (Z, X, Y)$ is a PL homeomorphism such that $V = h(|N(|K - L|, J')|)$.

C-Relative Regular Neighborhood Theorem 1.6.6. *Let X and Y be polyhedra in the polyhedron Z and let J be a complex containing K and L as subcomplexes and h: $(|J|, |K|, |L|) \to (Z, X, Y)$ a* PL *homeomorphism.*

1. (Existence). *By definition, $h(|N(|K - L|, J^{(2)})|)$ is a C-regular neighborhood of X mod Y in Z.*
2. (Uniqueness). *If V and W are* C*-regular neighborhoods of X mod Y in Z, then there is an ambient isotopy G: $Z \times I \to Z \times I$ such that*

(a) $G_0 = 1$,
(b) $G|(X \cup Y) \times I = 1$,
(c) $G_1(V) = W$.

Remark 1.6.5. The following fact is an interesting unpublished result of Cohen concerning C-relative regular neighborhoods: *If $Y \subset X$ are polyhedra, then a* C*-regular neighborhood of $X \times 0$* mod *$Y \times 0$ in $X \times I$ is topologically homeomorphic to*

$$X_R \times I/[(y, t) = (y, 0) \quad \text{if} \quad y \in Y_R, \quad 0 \leqslant t \leqslant 1].$$

REMARK 1.6.6. For more about C-relative regular neighborhoods, see Lemmas **5.3.2** and **5.3.3**.

D. General Position

In order to develop a "feeling" for general position arguments, we first prove a simple embedding theorem (Theorem **1.6.8**). This theorem will follow as a corollary of our main general position theorem (Theorem **1.6.11**). It is also a special case of an embedding theorem for n-dimensional separable metric spaces (see Theorem V3 [Hurewicz and Wallman, 1]). It is probably impossible to state a general position theorem that will apply to all situations involving general position, thus it is important to understand the techniques of general position. However, the statement of our general position theorem will suffice for many situations. It is often thought (for example, Chapter IV, Part C of [Glaser, 3]) that Corollary **1.6.5** suffices to do Stallings' engulfing, however, in Chapter 4 we shall find it necessary to use the full power of our main general position theorem (Theorem **1.6.10**) to do Stallings' engulfing.

For other discussions of general position see Chapter 6 of [Zeeman, 1], Chapter VI of [Hudson, 1], Chapter IV part A of [Glaser, 3] and [Henderson, 1].

A set of points X in E^n is said to be in **general position** in E^n if no $r + 2$ points of X lie on an r-dimensional hyperplane, $r = 1, 2, ..., n - 1$. That is, every subset of X with less than $n + 2$ points is pointwise independent.

Theorem 1.6.7. *E^n contains a dense set of points in general position.*

PROOF. Let $\{x_i\}_{i=1}^{\infty}$ be a countable dense subset of E^n, and define $y_1 = x_1$. Inductively having chosen points $y_1, ..., y_{k-1}$ in general position such that $\text{dist}(x_i, y_i) < 1/i$ for $i < k$, choose a point y_k within $1/k$ of x_k such that y_k does not lie on any of the finitely many hyperplanes of E^n which are determined by subsets containing at most n of the points $y_1, ..., y_{k-1}$. The sequence $\{y_i\}_{i=1}^{\infty}$ defined inductively in this manner is then a dense set of points in general position in E^n.

Theorem 1.6.8. *If K is a k-complex, then there is a homeomorphism $f: |K| \to E^{2k+1}$ which is linear on each simplex of K.*

PROOF. Let $x_1, ..., x_p$ be the vertices of K, and use Theorem **1.6.7** to find points $y_1, ..., y_p$ in general position in E^{2k+1}. Define $f(x_i) = y_i$ and extend f to $|K|$ linearly over each simplex of K. Since the image under f of the vertices of any simplex of K span a simplex of the same dimension in E^{2k+1}, it follows that f is one-to-one on each simplex of K.

Now suppose that x and y are two points of $|K|$ not lying in a single simplex of K. If σ and τ are simplexes of K containing x and y, respectively, then the union X of their vertices contains at most $2k + 2$ vertices. Since the points $y_1, ..., y_p$ are in general position in E^{2k+1}, it follows, that the set $f(X)$ spans a simplex η in E^{2k+1}. Then, $f(x)$ and $f(y)$ lie on the faces $f(\sigma)$ and $f(\tau)$ of η and neither lies on the face $f(\sigma) \cap f(\tau)$. It follows that $f(x) \neq f(y)$. Thus, f is one-to-one and is therefore a homeomorphism onto $f(|K|)$.

The following example indicates that the dimension $2k + 1$ in Theorem **1.6.8** is the best possible. (Flores [1] establishes a generalization of Theorem **1.6.7** by showing that the complex consisting of all faces of dimension $\leqslant n$ of a $(2n + 2)$-simplex cannot be embedded in E^{2n}.)

Example 1.6.5. *The* 1-*skeleton of a* 4-*simplex cannot be topologically embedded in* E^2.

PROOF. It suffices to show that five points in the plane cannot each be connected to every other point by an arc such that the arcs intersect only in end-points (see Fig. **1.6.9**). We will start with the only two

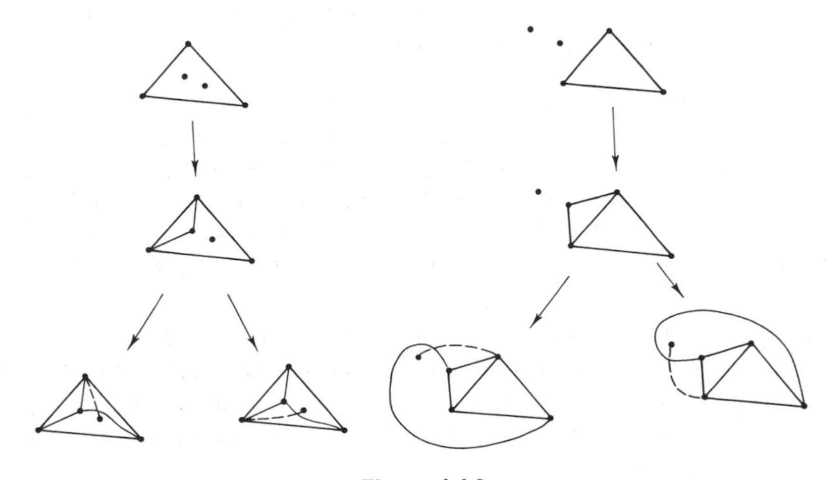

Figure 1.6.9

possible cases (by the Jordan curve theorem) of three points and their arcs union the other two points. We will then proceed to add arcs in every possible manner (restrictions imposed by the Jordan curve theorem) until we reach a stage where it becomes apparent that the necessary structure cannot be formed (again by the Jordan curve theorem). (Actually some of the restrictions mentioned above are imposed by the so-called θ-curve theorem which is a consequence of the Jordan curve theorem.)

Given polyhedra P and Q and a map $f: P \to Q$, define

$$S'(f) = \{x \in P \mid f^{-1}f(x) \neq x\}.$$

Then, the closure $S(f) = \text{Cl}\, S'(f)$ is called the **singular set** of f. The basic idea of general position approximation is to replace a given PL map with one whose singular set is of minimal dimension.

Theorem 1.6.9. *If $f: P \to Q$ is a* PL *map, then $S(f)$ is a subpolyhedron of P.*

PROOF. Let K and L be triangulations of P and Q such that $f: K \to L$ is simplicial, and consider a simplex A of K such that $\text{Int}\, A \cap S'(f) \neq \emptyset$. Then, $f^{-1}f(\text{Int}\, A)$ is a disjoint union of open simplexes of K. Now either $f \mid A$ is **degenerate** (that is, $\dim f(A) < \dim A$) or there is another simplex $B \in K$ with $f(A) = f(B)$. In either case it follows that $\text{Int}\, A \subset S'(f)$. Thus, $S'(f)$ is a union of interiors of simplexes of K and so it follows that $S(f) = \text{Cl}\, S'(f)$ is the polyhedron of a subcomplex of K.

A linear map $f: K \to Q$ of a complex K into a complex Q is **nondegenerate** if it embeds each simplex of K.

General Position Theorem 1.6.10. Part 1. *Let $f: K \to M$ be a linear map of the k-complex K into the unbounded combinatorial n-manifold M where $k \leqslant n$. Let L be a subcomplex of K such that $f \mid L$ is nondegenerate. Then, given $\epsilon > 0$, there are subdivisions (K', L') of (K, L) and M' of M and a nondegenerate linear map $g: K' \to M'$ such that g is ϵ-homotopic to f through a homotopy which agrees with f on $\mid L \mid$ throughout.*

Part 2. *Let $f: K \to M$ be a* PL *map of the k-complex K into the unbounded combinatorial n-manifold M which embeds each simplex of K. Let L be a subcomplex of K such that $f \mid \mid L \mid$ is an embedding. Then, given $\epsilon > 0$, there is a* PL *map* (not necessarily linear) *$g: K \to M$ such that*

(a) *$g \mid \mid L \mid = f \mid \mid L \mid$,*
(b) *g is* PL *ϵ-homotopic to f leaving $f(\mid L \mid)$ fixed,*
(c) *g embeds each simplex of K and*

$$\dim S(g \mid \sigma \cup \tau) \leqslant \dim \sigma + \dim \tau - n$$

for each two simplexes $\sigma, \tau \in K$.

Corollary 1.6.5 (Common general position theorem). *Let L be a p-dimensional subcomplex of the k-complex K such that* $\dim(|K| - |L|) = q$. *Let $f: K \to M$ be a* PL *map of K into the unbounded combinatorial n-manifold M such that $f \mid |L|$ is a* PL *homeomorphism. Then, given* $\epsilon > 0$, *there is a* PL *map $g: K \to M$ such that*

(a) $g \mid |L| = f \mid |L|$,
(b) $\text{dist}(g, f) < \epsilon$,
(c) $\dim S(g) \leqslant k + q - n$.

Proof of Part 1. Let $\{v_j\}_{j=1}^{\infty}$ be the vertices of M and let $B_j = \text{St}(v_j, M)$. Then, $M = \bigcup_{j=1}^{\infty} \text{Int } B_j$. By Theorem **1.4.2**, there is a subdivision (K_0, L_0) of (K, L) such that if $\sigma \in K_0$ then $f(\sigma) \subset \text{Int } B_j$ for some j. Let $\{\sigma_i\}_{i=1}^{r}$ enumerate the simplexes of $K_0 - L_0$ in order of nondecreasing dimension and let $L_i = L_0 \cup \{\sigma_1, ..., \sigma_i\}$, $i > 0$. Then, L_i is a subcomplex of K_0. We are going to inductively define linear maps $f_i = K^i \to M$, for subdivisions K^i of K_0, $i = 0, 1, ..., r$, such that

(1) $f \mid L^i$ is nondegenerate (L^i is the subcomplex of K^i triangulating L_i),
(2) f_i is (ϵ/r)-homotopic to f_{i-1} keeping $f(|L|)$ fixed,
(3) if $\sigma \in K_0$, then $f_i(\sigma) \subset \text{Int } B_j$ for some j.

Start with $f_0 = f$ and $K^0 = K$ and suppose that f_{i-1} is defined. Then, $f_{i-1}(\sigma_i) \subset \text{Int } B_j$ for some j. Let $K_1{}^i$ be a subdivision of K^{i-1} such that $N(\sigma_i, K_1{}^i) \subset f_{i-1}(\text{Int } B_j)$. Let $h_j: B_j \to I^n$ be a PL homeomorphism. Let $T = N(\sigma_i, K_1{}^i)$, and $L_* = \{\sigma \in T \mid \sigma \subset |L_{i-1}|\}$,

$$\text{Fr}(T) = \{\sigma \in T \mid \sigma \subset \text{frontier}_{|K|} \mid T \mid\}$$

and $\sigma_{i*} = \{\sigma \in T \mid \sigma \subset \sigma_i\}$.

Suppose that for $\delta > 0$, we can find a linear map g_i: $T' \to I^n$, where T' subdivides T, such that

(1) $g_i \mid \sigma'_{i*}$ is nondegenerate, and
(2) g_i is δ-homotopic to $h_j f_{i-1} \mid T$, keeping $L_* \cup \text{Fr}(T)$ fixed.

This will complete the inductive step, for by picking δ small enough, f_i can be taken to be the extension of $h_j^{-1} g_i$ to $|K|$ by the identity. Clearly, f_i will be linear for an appropriate subdivision K^i of $K_1{}^i$ and $f \mid L^i$ will be nondegenerate.

We now obtain g_i. Let T' be a subdivision of T on which $h_j f_{i-1}$ is linear and such that $L_* \cup \text{Fr}(T)$ is triangulated as a full subcomplex. Let $w_1, ..., w_r$ denote the vertices of T' in $L_* \cup \text{Fr}(T)$ and let

$w_{r+1}, \ldots, w_s$ denote the remaining vertices of T'. Define x_k by $f_{i-1}(w_k)$ for $k \leqslant r$. Having defined $w_1, \ldots, w_r, \ldots, w_k$ $(k \geqslant r)$, choose a point x_{k+1} near $f_{i-1}(w_{k+1})$ and not lying on any of the proper hyperplanes of E^n determined by the points $x_1, \ldots, x_k$. Then, let $g(w_k) = x_k$ for $k \leqslant s$ and define g_i: $T' \to I^n$ by linear extension over the simplexes of T'. It is easy to see that this is the desired map.

EXERCISE 1.6.14. Prove Part 2 of Theorem **1.6.10**. [Hint: The proof is similar to that of Part 1 and uses the fact that if V_1 and V_2 are finite-dimensional subspaces of a vector space V, then $V_1 + V_2$ is finite dimensional and

$$\dim V_1 + \dim V_2 = \dim(V_1 \cap V_2) + \dim(V_1 + V_2).$$

Try the case $M = E^n$ first.]

EXERCISE 1.6.15. Let S^p and S^q be disjoint PL spheres in E^n. If $n \geqslant p + q + 2$, show that S^p and S^q are **unlinked** in the sense that there is a n-ball $B^n \subset E^n$ with $S^p \subset B^n$ and $S^q \subset E^n - B^n$. [Hint: Embed a cone on one of the spheres. All cases except the case $p = q = \frac{1}{2}n - 1$, n even, follows from Corollary **1.6.5**. The exceptional case will follow from an easy argument.]

EXERCISE 1.6.16. Let P_0 and P_1 be PL homeomorphic k-dimensional polyhedra in E^n, with $n \geqslant 2k + 3$. If N_0 and N_1 are regular neighborhoods of P_0 and P_1, respectively, show that N_0 and N_1 are PL homeomorphic. [Hint: For the case $N_0 \cap N_1 = \emptyset$ define an embedding f: $P_0 \times I \to E^n$ such that $f(P_0 \times i) = P_i$, $i = 0, 1$.]

E. Relative Simplicial Approximation

The absolute simplicial approximation theorem, which dates back to Alexander [1], states that there is a simplicial approximation g to any given continuous map f between two finite simplicial complexes. (For a proof see [Hocking and Young, 1, p. 210] or [Singer and Thorpe, 1, p. 81].) The relative theorem permits one to leave f unchanged on any subcomplex on which f happens to already be simplicial. The usual technique of relative simplicial approximation was developed by Zeeman [2] and is similar to our proof of Part 1 of Theorem **1.6.11**. (Also, Zeeman's proof as well as Zeeman's example, which shows that the usual techniques for proving the absolute simplicial approximation theorem do not suffice to prove the relative simplicial approximation theorem, appear in Chapter IV of [Glaser, 3].) Here we will content outselves by deducing the relative simplicial approximation theorem as a consequence of a weak form of the existence part of the regular neighborhood theorem.

Relative Simplicial Approximation Theorem 1.6.11. *Let P, Q, and R be polyhedra with $Q \subset P$ and let $f\colon P \to R$ be a continuous map such that $f \mid Q$ is* PL. *Then, given $\epsilon > 0$, there is a* PL *map $g\colon P \to R$ such that*

(1) $f \mid Q = g \mid Q$,
(2) $\operatorname{dist}(f, g) < \epsilon$, *and*
(3) *f and g are homotopic keeping Q fixed.*

Proof. By Theorem **1.6.8**, we may assume that $R \subset E^n$ for some n. Let N be a polyhedral neighborhood of R in E^n for which there is a PL retraction $r\colon N \to R$. (Such a neighborhood N may be obtained from the existence part of Theorem **1.6.4** and Lemma **1.6.2**.) Let $\epsilon' \leqslant \epsilon$ be small enough that the ϵ'-neighborhood of R lies in N. By the uniform continuity of r, there is a $\delta > 0$ such that for each $x \in R$ the δ-neighborhood $N_\delta(x) \subset N$ and $\operatorname{diam} r(N_\delta(x)) < \epsilon'$. By uniform continuity of f, choose $\eta > 0$ such that each subset of P of diameter less than η is mapped by f onto a set of diameter less than $\delta/2$. Let (K, L) be a triangulation of (P, Q) such that K has mesh less than η. If $f^1\colon P \to E^n$ is the linear map of K which agrees with f on the vertices of K, then

$$\operatorname{dist}(f, f^1) < \delta, \qquad f^1 \mid Q = f \mid Q \qquad \text{and} \qquad f^1(P) \subset N.$$

Let $g = rf^1\colon P \to R$. Certainly, $\operatorname{dist}(f, g) < \epsilon' \leqslant \epsilon$ and $g \mid Q = f \mid Q$. Let $h\colon P \times I \to N$ be the obvious homotopy between f and g such that $h(x \times I)$ is the line segment (possibly degenerate) from $f(x)$ to $g(x)$ for each $x \in P$. Then,

$$H = rh\colon P \times I \to R$$

is the required homotopy between f and g.

Corollary 1.6.6. *Let P be a k-polyhedron, Q a subpolyhedron and M a* PL *n-manifold with $n \geqslant 2k + 1$. Then, given $\epsilon > 0$ and a map $f\colon P \to \operatorname{Int} M$ such that $f \mid Q$ is a* PL *embedding, there is a* PL *embedding $g\colon P \to \operatorname{Int} M$ with $\operatorname{dist}(f, g) < \epsilon$ and $f \mid Q = g \mid Q$.*

The above corollary follows from Theorem **1.6.11** and Theorem **1.6.10**.

F. Handlebodies

Handlebody Decomposition Theorem 1.6.12. *Let M^n be an n-dimensional* PL *manifold and let $\epsilon > 0$ be given. Then, $M = \bigcup_{i=0}^{r} H_i$ where*

$$\left(H_j,\, H_j \cap \left(\bigcup_{i=0}^{j-1} H_i\right)\right) \overset{\text{PL}}{\approx} (I^n, \operatorname{Bd} I^k \times I^{n-k})$$

for some $k \leqslant n$, *for each* $j = 0, 1, ..., r$. *Furthermore*, diam $H_j < \epsilon$ *for* $j = 0, 1, ..., r$. (H_j *is called a* **handle of index** k.)

PROOF. Let K be a triangulation of M and let K'' be the second barycentric subdivision of K. Then, K and K'' are combinatorial manifolds by Theorem **1.6.2**. Let $\sigma_0, \sigma_1, ..., \sigma_r$. be the simplexes of K in order of nondecreasing dimension. Let $H_i = \text{St}(\hat{\sigma}_i, K'')$, when $\hat{\sigma}_i$ is the barycenter of σ_i. Then each H_i is an n-ball. It follows from our construction and the way we ordered the simplexes σ_i that

$$H_j \cap \bigcup_{i=0}^{j-1} H_i = \text{St}(\hat{\sigma}_j, K'') \cap N(\text{Bd}\,\sigma_j, K'') = N(\text{Lk}(\hat{\sigma}_j, \sigma_j''), \text{Lk}(\hat{\sigma}_j, K'')).$$

By pseudo-radial projection from $\hat{\sigma}_j$, there is a PL homeomorphism h_1 of $(\text{Lk}(\hat{\sigma}_j, K''), \text{Lk}(\hat{\sigma}_j, \sigma_j''))$ onto $(\text{Bd}\,\sigma_j * \text{Lk}(\sigma_j, K), \text{Bd}\,\sigma_j)$.

First, we assume that σ_j is a k-simplex contained in Int M. Let $\mathscr{S}^{n-k}(\text{Bd}\,I^k)$ be an $(n-k)$-suspension of Bd I^k which is contained in I^n and let Σ^{n-k-1} be the $(n-k-1)$-suspension sphere [that is, Σ^{n-k-1} is obtained by taking the $(n-k-1)$-suspension of the first pair of suspension points where all suspension points are chosen as before]. (See Fig. **1.6.10**.) Then, $\mathscr{S}^{n-k}(\text{Bd}\,I^k) = \text{Bd}\,I^k * \Sigma^{n-k-1}$. Since

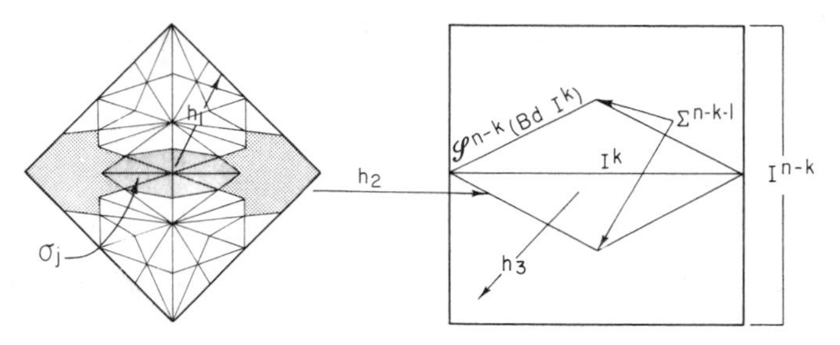

Figure 1.6.10

$\sigma_j \subset \text{Int}\,M$, it follows that $\text{Lk}(\sigma_j, K)$ is an $(n-k-1)$-sphere by Lemma **1.6.1**. Thus, by Exercise **1.6.4** we can construct a PL homeomorphism h_2 which takes $(\text{Bd}\,\sigma_j * \text{Lk}(\sigma_j, K), \text{Bd}\,\sigma_j, \text{Lk}(\sigma_j, K))$ onto $(\text{Bd}\,I^k * \Sigma^{n-k-1}, \text{Bd}\,I^k, \Sigma^{n-k-1})$. By pseudo-radial projection from the origin, we can get a PL homeomorphism h_3 which takes $(\mathscr{S}^{n-k}(\text{Bd}\,I^k), \text{Bd}\,I^k)$ onto $(\text{Bd}\,I^n, \text{Bd}\,I^k)$. Now consider the PL homeomorphism $h_3h_2h_1: (\text{Lk}(\hat{\sigma}_j, K''), \text{Lk}(\hat{\sigma}_j, \sigma_j'')) \twoheadrightarrow (\text{Bd}\,I^n, \text{Bd}\,I^k)$. Since $N(\text{Lk}(\hat{\sigma}_j, \sigma_j''), \text{Lk}(\hat{\sigma}_j, K''))$ is a regular neighborhood of $\text{Lk}(\hat{\sigma}_j, \sigma_j'')$ in $\text{Lk}(\hat{\sigma}_j, K'')$, by Theorem **1.6.4** (existence part), it follows that $h_3h_2h_1(N(\text{Lk}(\hat{\sigma}_j, \sigma_j''), \text{Lk}(\hat{\sigma}_j, K'')))$ is a regular neighborhood of Bd I^k in Bd I^n. Also, $\text{Bd}\,I^k \times I^{n-k}$ is a regular neighborhood of Bd I^k in Bd I^n. Hence, by

Theorem **1.6.4** (uniqueness part) there is a PL homeomorphism h_4 of $(\text{Bd } I^n, \text{Bd } I^k)$ onto itself such that

$$h_4 h_3 h_2 h_1(N(\text{Lk}(\hat{\sigma}_j, \sigma_j''), \text{Lk}(\hat{\sigma}_j, K''))) = \text{Bd } I^k \times I^{n-k}.$$

Then, the conewise extension h: $\text{St}(\hat{\sigma}_j, K'') \twoheadrightarrow I^n$ is the desired PL homeomorphism.

Notice that H_i could be made arbitrarily small by using Theorem **1.4.2** to make the mesh of K small.

Exercise 1.6.17. Complete the proof of Theorem **1.6.12**, that is, consider the case $\sigma_j \subset \text{Bd } M$.

1.7. LOCAL FLATNESS, (PINCHED) COLLARS, AND (PINCHED) BICOLLARS

It will turn out that many times we can show that two embeddings f and g of a space X into a space Y are equivalent if they both satisfy some "niceness" condition. In the case X and Y are topological manifolds the niceness condition most often imposed is that of local flatness which we now define. An m-manifold M contained in the interior of an n-manifold N is **locally flat** at $x \in \text{Int } M$, $[x \in \text{Bd } M]$, if there is a neighborhood U of x in N such that $(U, U \cap M)$ is homeomorphic to (E^n, E^m), $[(E^n, E_+^m)]$. An embedding f: $M \to N$ is **proper** if $f(\text{Bd } M) \subset \text{Bd } N$ and $f(\text{Int } M) \subset \text{Int } N$. We say that (M, N) is a **proper manifold** pair if the inclusion of M into N is proper. If (M, N) is a proper manifold pair, then M is **locally flat** $x \in \text{Int } M$, $[x \in \text{Bd } M]$, if there is a neighborhood U of x in N such that $(U, U \cap M)$ is homeomorphic to (E^m, E^m), $[(E_+^n, E_+^m)]$. An embedding f: $M \to N$ such that $f(M) \subset \text{Int } N$ or such that f is proper is said to be **locally flat** at a point $x \in M$ if $f(M)$ is locally flat at $f(x)$. Embeddings and submanifolds are **locally flat** if they are locally flat at every point.

Theorem 1.7.1 (Transitivity of local flatness). *Let L, M, and N be manifolds of dimensions l, m, and n, respectively. Suppose that either* (1) $L \subset \text{Int } M \subset \text{Int } N$, (2) (M, L) *is a proper manifold pair such that* $M \subset \text{Int } N$, *or* (3) (M, L) *and* (N, M) *are proper manifold pairs. Then, if L is locally flat in M and M is locally flat in N, it follows that L is locally flat in N.*

Proof. (We will show that L is locally flat in N at an arbitrary point $p \in \text{Int } L$, and this proof works for all three situations. The proofs for $p \in \text{Bd } L$ are modifications of this proof and are left as an exercise.)

Let U be a neighborhood of p in N such that there is a homeomorphism h_1: $(U, U \cap M) \twoheadrightarrow (E^n, E^m)$. Then, $h_1(L \cap U)$ is locally flat in E^m. Let V be a neighborhood of $h_1(p)$ in E^m such that there is a homeomorphism h_2: $(V, V \cap h_1(L \cap U)) \twoheadrightarrow (E^m, E^l)$. Define

$$\bar{h}_2: (V \times E^{n-m}, V \cap h_1(L \cap U)) \to (E^n, E^l)$$

to be the product extension of h_2. Let $h = \bar{h}_2 h_1$. Then, $h^{-1}(E^n)$ is the desired neighborhood of p such that $(h^{-1}(E^n), h^{-1}(E^n) \cap L)$ is homeomorphic to (E^n, E^l).

EXERCISE 1.7.1. Prove Theorem **1.7.1** for each of the three cases when $p \in \mathrm{Bd}\, L$.

Theorem 1.7.2. *If f: $M^k \to \mathrm{Int}\, N^n$ is a* PL *embedding of the* PL *k-manifold M into the* PL *n-manifold N, $n - k \neq 2$, then f is locally flat.*

EXERCISE 1.7.2. Prove Theorem **1.7.2**. To do so you may use the following two facts both of which will be proved later.

Fact 1. If f: $I^k \to E^n$ (or S^n) is a locally flat embedding, then there exists a homeomorphism h: $E^n \twoheadrightarrow E^n$ such that hf: $I^k \to E^n$ (or S^n) is the inclusion map.

Fact 2. If f: $S^k \to E^n$ (or S^n), $n - k \neq 2$, is a locally flat embedding, then there exists a homeomorphism h: $E^n \twoheadrightarrow E^n$ such that hf: $S^k \to E^n$ (or S^n) is the inclusion map.

REMARK 1.7.1. We will show later (Example **2.3.2**) that the PL disk in E^4 which is the cone from the origin over a trefoil knot in $\mathrm{Bd}\, I^4$ is not locally flat in E^4. Thus, Theorem **1.7.2** would be false without the hypothesis $n - k \neq 2$.

Let X be a subspace of the topological space Y. Then, X is said to be **collared** in Y if there is a homeomorphism h carrying $X \times [0, 1)$ onto an open neighborhood of X such that $h(x, 0) = x$ for all $x \in X$. If X can be covered by a collection of open subsets (relative to X) each of which is collared in Y, then X is **locally collared** in Y. If there is a homeomorphism h carrying $X \times (-1, 1)$ onto an open neighborhood of X such that $h(x, 0) = x$ for all $x \in X$, then X is **bicollared** in Y. Finally, if X can be covered by a collection of open subsets (relative to X) each of which is bicollared in Y, then X is **locally bicollared** in Y.

EXERCISE 1.7.3. Show that the boundary of every manifold is locally collared.

EXERCISE 1.7.4. Let (M, N) be a proper manifold pair where N is n-dimensional and M is $(n - 1)$-dimensional. Show that M is locally flat in N if and only if it is locally bicollared in N.

EXERCISE 1.7.5. Give an example of a proper manifold pair (M, N) such that M is locally bicollared in N, but not bicollared. (Hint: Consider the center 1-sphere of a Möbius band. Construct higher-dimensional examples by crossing the Möbius band with $(0, 1)$. Also consider the natural inclusion of each odd-dimensional projective space into the next higher even-dimensional projective space.)

Although locally bicollared does not imply bicollared by the above exercise, the next theorem asserts that locally collared does imply collared. Our first proof of this theorem will be by techniques of [Brown, 3, 4] which were developed in the original proof. At the end of this section (Theorem **1.7.7**) we will give a shorter and more recent proof developed by Connelly. Most readers might be well-advised to skip the proof of Theorem **1.7.3** for the time being, and go directly to the proof of Theorem **1.7.7**. (Brown's basic idea is to put the local collars together to get a collar, whereas Connelly's idea is to add a collar to the manifold and then use the local collars to push that collar into the manifold.)

Collaring Theorem 1.7.3 (Brown). *If the manifold M is contained in the manifold N and M is locally collared in N, then M is collared in N.*

We will first assume the following two lemmas and prove Theorem **1.7.3** and then we will prove the lemmas.

Lemma 1.7.1. *Suppose that the manifold M is contained in the manifold N. If, $\{U_\alpha\}_{\alpha\in A}$ is a pairwise disjoint collection of open subsets of M each of which is collared in N, then $\bigcup_{\alpha\in A} U_\alpha$ is collared in N.*

Lemma 1.7.2. *Suppose that the manifold M is contained in the manifold N. If M is the union of two open subsets U_1 and U_2 each of which is collared in N, then M is also collared in N.*

Proof of Theorem 1.7.3. (Notice that Theorem **1.7.3** follows from Lemma **1.7.2** in the case that M is compact. However, we have to do a little more work in the case that M is not compact. This proof is an adaptation and simplification of results of [Michael, 1] when reduced to the current situation.)

For each point $p \in M$, let U_p be an open set which is collared in N. Since M is a separable metric space, there is a countable subcover $\{O_i\}$ of the cover $\{U_p\}_{p\in M}$ of M. Let $K_n = \bigcup_{i=1}^{n} O_i$. By Lemma **1.7.2**, K_n is collared in N. Let $V_n = \{x \in M \mid \mathrm{dist}(x, M - K_n) > 1/n\}$. Then, $V_n \subset \mathrm{Cl}\, V_n \subset K_n$ and certainly V_n is collared in N since it is an open subset of K^n which is collared. Let $W_1 = V_1$, $W_2 = V_2$, $W_3 = V_3$, and $W_n = V_n - \mathrm{Cl}\, V_{n-3}$ for $n \geqslant 4$ (see Fig. **1.7.1**). Each W_n is collared in N since $W_n \subset V_n \subset K_n$. Now the two collections of open sets $\{W_{4n-1}\}_{n=1}^{\infty}$ and $\{W_{4n+1}\}_{n=1}^{\infty}$ are pairwise disjoint (see Exercise **1.7.6**).

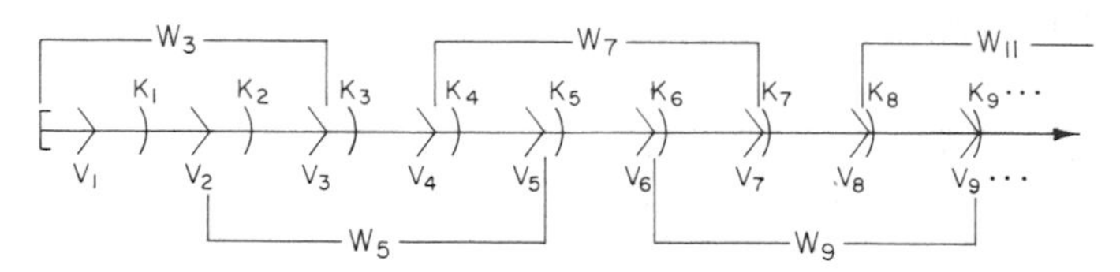

Figure 1.7.1

Hence, by Lemma **1.7.1**, $\bigcup_{n=1}^{\infty} W_{4n-1}$ and $\bigcup_{n=1}^{\infty} W_{4n+1}$ are both collared in N. But, $M = (\bigcup_{n=1}^{\infty} W_{4n-1}) \cup (\bigcup_{n=1}^{\infty} W_{4n+1})$ (see Exercise **1.7.6**) and so by Lemma **1.7.2** M is collared in N as desired.

Exercise 1.7.6. Show that the two collections $\{W_{4n-1}\}_{n=1}^{\infty}$ and $\{W_{4n+1}\}_{n=1}^{\infty}$ of the above proof are pairwise disjoint and that

$$M = \left(\bigcup_{n=1}^{\infty} W_{4n-1}\right) \cup \left(\bigcup_{n=1}^{\infty} W_{4n+1}\right).$$

Proof of Lemma 1.7.1. Suppose that h_α is the homeomorphism of $U_\alpha \times [0, 1)$ onto a neighborhood of U_α in N such that $h_\alpha(x, 0) = x$ for all $x \in U_\alpha$. Let

$$W_\alpha = h_\alpha(U_\alpha \times [0, 1)) \cap \left\{x \in N \mid \mathrm{dist}(x, U_\alpha) < \mathrm{dist}\left(x, \bigcup_{\beta \neq \alpha} U_\beta\right)\right\}.$$

Then, $\{W_\alpha\}_{\alpha \in A}$ is a pairwise disjoint collection of open subsets of N such that $U_\alpha \subset W_\alpha \subset h_\alpha(U_\alpha \times [0, 1))$ for all $\alpha \in A$. Let $\mathcal{O} = \bigcup_{\alpha \in A} h_\alpha^{-1}(W_\alpha)$. Then, $\mathcal{O}$ is an open subset of $M \times [0, 1)$ such that $\bigcup_{\alpha \in A} U_\alpha \times 0 \subset \mathcal{O}$. Now define the continuous, positive, real-valued function g on $\bigcup_{\alpha \in A} U_\alpha$ as follows:

$$g(x) = \mathrm{Min}\left[1, \mathrm{dist}\left(x, \left(\bigcup_{\alpha \in A} U_\alpha \times [0, 1)\right) - \mathcal{O}\right)\right].$$

Let $\Gamma\colon \bigcup_{\alpha \in A} U_\alpha \times [0, 1) \to \mathcal{O}$ be the homeomorphism defined by $\Gamma(x, t) = (x, tg(x))$. Let $f_\alpha\colon U_\alpha \times [0, 1) \to N$ be defined by $f_\alpha(x, t) = h_\alpha \Gamma(x, t)$. Then, $h\colon \bigcup_{\alpha \in A} U_\alpha \times [0, 1) \to N$ defined by $h \mid U_\alpha \times [0, 1) = f_\alpha$ is the desired collaring of $\bigcup_{\alpha \in A} U_\alpha$.

Before proving Lemma **1.7.2**, we will state a couple of preliminary definitions.

Let M be a manifold, let U be an open subset of $M \times 0$ and let λ: Cl $U \to [0, 1]$ be a continuous map such that $\lambda(x) = 0$ if and only if $x =$ Cl $U - U$. The **spindle neighborhood** $S(U, \lambda)$ of U in $M \times [0, 1)$ is $\{(x, t) \in M \times [0, 1) | (x, 0) \in U, t < \lambda(x, 0)\}$. It is easy to see that $S(U, \lambda)$ is a neighborhood of U in $M \times [0, 1)$ and that the spindle neighborhoods form a neighborhood basis for the neighborhoods of U in $M \times [0, 1)$. For, suppose that V is an open subset of $M \times [0, 1)$ containing U. Let $\lambda: \bar{U} \to [0, 1]$ be defined by

$$\lambda(x, 0) = \text{Min}[\text{dist}((x, 0), \text{Cl } U - U), \text{dist}((x, 0), (M \times [0, 1)) - V), 1].$$

Then, $S(U, \lambda) \subset V$.

We define the **map** $\pi_{S(U,\lambda)} : M \times [0, 1) \twoheadrightarrow M \times [0, 1)$ by

$$\pi(x, t) = \begin{cases} (x, t), & (x, t) \notin S(U, \lambda), \\ (x, 0), & (x, t) \in S(U, \lambda/2), \\ (x, 2t - \lambda(x)), & (x, t) \in S(U, \lambda) - S(U, \lambda/2), \end{cases}$$

where $\lambda/2$ is defined by $(\lambda/2)(x) = \frac{1}{2}\lambda(x)$.

Proof of Lemma 1.7.2. Consider the following two statements.

Statement A. *Let M be a manifold, and U an open subset of $M \times 0$. Let N_* be a neighborhood of U in $M \times [0, 1)$, and f a homeomorphism of* Cl N_* *onto the closure of a neighborhood of U such that $f |$ Cl $U = 1$. Then, there is a homeomorphism f':* Cl $N_* \to M \times [0, 1)$ *and a neighborhood V of U in N such that*

(a) $f' |(\text{Cl } N_* - N_*) = f |(\text{Cl } N_* - N_*)$,
(b) $f'(\text{Cl } N_*) = f(\text{Cl } N_*)$, and
(c) $f' | V = 1$.

(See Fig. 1.7.2.)

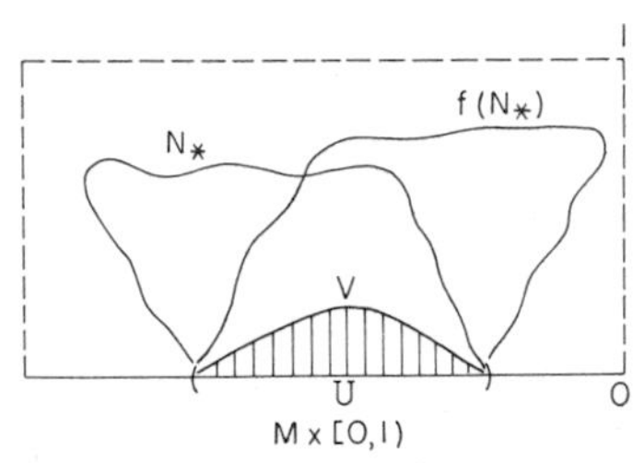

Figure 1.7.2

Statement B. *Let M be a manifold and let h: $M \to N$ be an embedding into the manifold N. Suppose that U_1 and U_2 are open subsets of M such that $M = U_1 \cup U_2$ and suppose that K is a closed subset relative to M*

of $U_1 \cap U_2$. *Suppose also that for* $i = 1, 2$, *there is a homeomorphism* h_i of $U_i \times [0, 1)$ *onto a neighborhood of* $h(U_i)$ *in* N *such that* $h_i(x, 0) = h(x)$ *for all* $x \in U_i$. *Then, there is a homeomorphism*

$$h_2': U_2 \times [0, 1) \twoheadrightarrow h_2(U_2 \times [0, 1))$$

such that $h_2'(x, 0) = h(x)$ *for all* $x \in U_2$ *and* $h_2' \mid V = h_1 \mid V$ *for some neighborhood* V *of* $K \times 0$ *in* $(U_1 \cap U_2) \times [0, 1)$ (see Fig. **1.7.3**).

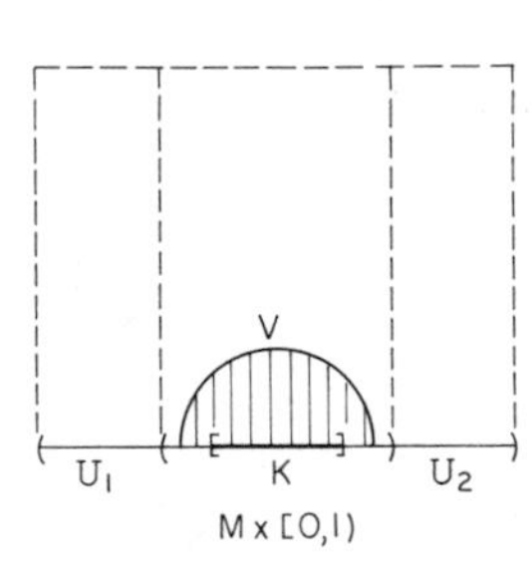

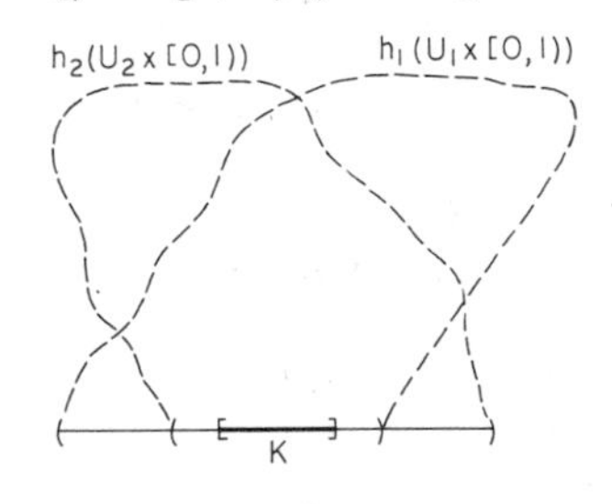

Figure 1.7.3

The proof of Lemma **1.7.2** will proceed as follows: We will show that Statement **A** is true, then we will show that Statement **A** implies Statement **B**, and finally we will show that Statement **B** implies Lemma **1.7.2**.

Proof of A. Let $S(U, \lambda)$ be a spindle neighborhood of U such that $S(U, \lambda) \subset N_* \cap f(N_*)$. Let π be the associated mapping $\pi_{S(U,\lambda)}$ and let f': Cl $N_* \to M \times [0, 1)$ be defined by

$$f'(x) = \begin{cases} x, & x \in \mathrm{Cl}(S(U, \lambda/2)), \\ \pi^{-1} f \pi(x), & x \in \mathrm{Cl}\, N_* - S(U, \lambda/2). \end{cases}$$

EXERCISE 1.7.7. Show that f' defined as above is well-defined and that f' and $V = (S(U, \lambda/2)$ satisfy the conclusion of Statement **A**.

Proof that A ⇒ B. Let U be an open subset of $U_1 \cap U_2$ such that $K \subset U \subset \mathrm{Cl}\, U \subset U_1 \cap U_2$. Then there is a neighborhood N_* of $U \times 0$ in $M \times [0, 1)$ such that

$$\mathrm{Cl}\, N_* \subset h_2^{-1} h_1(U_1 \times [0, 1)) \cap ((U_1 \cap U_2) \times [0, 1)).$$

Hence, the map $f = h^{-1} h_2 \mid \mathrm{Cl}\, N_*$ is a well-defined homeomorphism such that $f \mid \mathrm{Cl}\, U \times 0 = 1$ and $f(N_*)$ is open in $M \times [0, 1)$. By applying Statement **A**, we obtain a homeomorphism

$$f': \mathrm{Cl}\, N_* \to M \times [0, 1)$$

and a neighborhood V of $U \times 0$ such that

(a) $f' | (\text{Cl } N_* - N_*) = f | (\text{Cl } N_* - N_*)$,
(b) $f'(\text{Cl } N_*) = f(\text{Cl } N_*)$, and
(c) $f' | V = 1$.

Then, define h_2': $U_2 \times [0, 1) \to N$ by

$$h_2' = \begin{cases} h_1 f'(x), & x \in \text{Cl } N_* \cap (U_2 \times [0, 1)), \\ h_2(x), & x \in (U_2 \times [0, 1)) - N_* . \end{cases}$$

Exercise 1.7.8. Show that h_2' is a well-defined homeomorphism and that h_2' and V satisfy the conclusion of Statement **B**.

Proof that B ⇒ Lemma 1.7.2. Observe that if $U_1 \cap U_2 = \emptyset$, then Lemma **1.7.2** follows from Lemma **1.7.1**. Hence we assume that $U_1 \cap U_2 \neq \emptyset$. Since M is a normal space there are open subsets O_1 and O_2 of M such that $\text{Cl } O_1 \subset U_1$, $\text{Cl } O_2 \subset U_2$ and $M = O_1 \cup O_2$. Let $K = \text{Cl } O_1 \cap \text{Cl } O_2$. Then, K is a closed subset rel M of $U_1 \cap U_2$ (see Fig. **1.7.4**). By hypothesis there are homeomorphisms h_i, $i = 1, 2$,

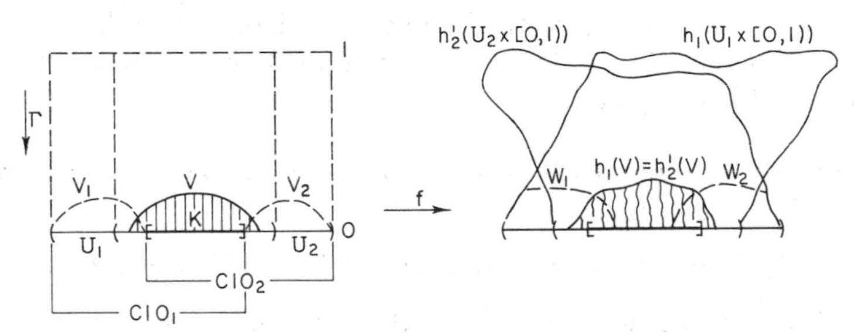

Figure 1.7.4

of $U_i \times [0, 1)$ onto a neighborhood of U_i in N such that $h_i(x, 0) = x$ for all $x \in U_i$. By applying Statement **B** with h the identity map we get a homeomorphism h_2': $U_2 \times [0, 1) \twoheadrightarrow h_2(U_2 \times [0, 1))$ and a neighborhood V of $K \times 0$ in $(U_1 \cap U_2) \times [0, 1)$ such that $h_2' | V = h_1 | V$ and $h_2' | U_2 \times 0 = h_2 | U_2 \times 0$.

Obviously

$$(O_1 - O_2) \cap \text{Cl}(O_2 - O_1) = \text{Cl}(O_1 - O_2) \cap (O_2 - O_1) = \emptyset.$$

that is $O_1 - O_2$ and $O_2 - O_1$ are completely separated in N. Since N is a metric space there exist disjoint open subsets W_1 and W_2 of N such that

$$O_1 - O_2 \subset W_2 \subset h_1(U_1 \times [0, 1)) \quad \text{and} \quad O_2 - O_1 \subset W_2 \subset h_2'(U_2 \times [0, 1)).$$

Let V_1 and V_2 be neighborhoods of

$$(O_1 - \mathrm{Cl}\, O_2) \times 0 \qquad \text{and} \qquad (O_2 - \mathrm{Cl}\, O_1) \times 0,$$

respectively, such that $h_1(V_1) \subset W_1$ and $h_2'(V_2) \subset W_2$. Let

$$f\colon V_1 \cup V_2 \cup V \to N$$

be defined by

$$f(x) = \begin{cases} h_1(x), & x \in V_1, \\ h_2'(x), & x \in V_2, \\ h_1(x) = h_2'(x), & x \in V. \end{cases}$$

Clearly f is a well-defined homeomorphism and $f(x, 0) = x$ for all $x \in M$. Since $V_1 \supset (O_1 - \mathrm{Cl}\, O_2) \times 0$, $V_2 \supset (O_2 - \mathrm{Cl}\, O_1) \times 0$, and $V \supset (\mathrm{Cl}\, O_1 \cap \mathrm{Cl}\, O_2) \times 0$, it follows that $V_* = V_1 \cup V_2 \cup V$ is a neighborhood of $M \times 0$ in $M \times [0, 1)$. Now define the continuous, positive, real-valued function g on M as follows:

$$g(x) = \mathrm{Min}[1, \mathrm{dist}(x, M \times [0, 1) - V_*)].$$

Let $\Gamma\colon M \times [0, 1) \to V_*$ be the homeomorphism defined by $\Gamma(x, t) = (x, tg(x))$. Then, $f\Gamma$ is the desired collaring of M in N and the proof of Lemma **1.7.2** is complete.

Collaring Theorem 1.7.4 (Brown). *The boundary of an n-manifold with boundary is collared.*

Proof. This follows from Exercise **1.7.3** and Theorem **1.7.3**.

Let M be a proper submanifold of the manifold N, that is, boundary contained in boundary and interior in interior. The pair (Bd N, Bd M) is said to be **collared** in (N, M) if there is a homeomorphism λ from (Bd $N \times [0, 1)$, Bd $M \times [0, 1)$) into (N, M) such that $\lambda(\mathrm{Bd}\, N \times [0, 1))$ [respectively, $\lambda(\mathrm{Bd}\, N \times [0, 1))$] is a neighborhood of Bd M [respectively, Bd M] in N [respectively, M], and $\lambda(x, 0) = x$ for each $x \in \mathrm{Bd}\, N$.

Exercise 1.7.9. Specify the modifications of the proof of Theorem **1.7.4** necessary to establish the following fact: If M is a proper submanifold of N which is locally flat at each point of Bd M, then the pair (Bd N, Bd M) is collared in (M, N). (For an alternate proof of this exercise see Theorem **1.7.7**.)

A connected m-dimensional manifold M^m without boundary in the interior of an n-dimensional manifold N^n, $n - m = 1$, is **two-sided** if there is a connected open neighborhood U of M in N such that $U - M$ has exactly two components each of which is open in N and each of which has M as its frontier relative to U. A connected m-manifold M^m with boundary contained in the interior of an n-manifold, $n - m = 1$, is **two-sided** Int M is two-sided.

For a discussion of two-sidedness and the proofs of some results concerning two-sidedness see [Rushing, 6].

Bicollar Theorem 1.7.5 (Brown). *Let M be a locally flat, connected, two-sided $(n - 1)$-submanifold without boundary of an n-manifold N. Then, M is bicollared in N.*

PROOF. Let U be a connected neighborhood of M in N such that $U - M$ has two components C_1 and C_2. It is easy to show that all points in M have arbitrarily small neighborhoods which intersect both C_1 and C_2. Since M is also locally flat in U, $C_1 \cup M$ and $C_2 \cup M$ are manifolds with boundary M. It follows from Theorem **1.7.4** that M is collared in each. Hence, M is bicollared in N.

Suppose that X is a subset of the submanifold M of the manifold N. Then, if a homeomorphism

$$h\colon M \times [0, 1]/[(x, t) = (x, 0) \quad \text{if} \quad x \in X, 0 \leqslant t \leqslant 1] \to N$$

is such that $h([(x, 0)]) = x$, we call

$$h(M \times [0, 1]/[(x, t) = (x, 0) \quad \text{if} \quad x \in X, 0 \leqslant t \leqslant 1])$$

a **collar of M pinched at X** if it is a neighborhood of $M - X$ in N. Similarly, if a homeomorphism

$$h\colon M \times [-1, 1]/[(x, t) = (x, 0) \quad \text{if} \quad x \in X, -1 \leqslant t \leqslant 1] \to N$$

is such that $h([(x, 0)]) = x$, we call

$$h(M \times [-1, 1]/[(x, t) = (x, 0) \quad \text{if} \quad x \in X, -1 \leqslant t \leqslant 1])$$

a **bicollar of M pinched at X** if it is a neighborhood of $M - X$ in N. (For a discussion of pinched collars and the following theorem, see [Rushing, 8].)

Pinched (Bi)collar Theorem 1.7.6. (i) *Let X be a closed subset of a manifold M which is contained in a manifold N. If $M - X$ is locally collared in N, then there is a collar of M in N pinched at X.*

(ii) *Let X be a closed subset of the boundary of a manifold M. Then, there is a collar of* Bd M *in M pinched at X.*

(iii) *Let M be an $(n - 1)$-submanifold of an n-manifold N and let X be a closed subset of M. If $M - X$ is a two-sided, connected, locally flat submanifold of the interior of N, then there is a bicollar of M in N pinched at X.*

Exercise 1.7.10. Prove Theorem **1.7.6**. (Be very careful and include all details.)

As mentioned earlier, we shall give a short proof, due to Connelly [2], of a relative version of Theorem **1.7.3**. Before stating and proving that relative version, let us generalize the definition given before Exercise **1.7.9**. A topological embedding $f\colon M^k \to N^n$ of the k-manifold M into the n-manifold N is said to be **allowable** if $f^{-1}(\partial N)$ is a $(k - 1)$-submanifold (possible empty) of ∂M. A manifold pair (N, M) is said to be **allowable** if the inclusion of M into N is allowable. *A* k-manifold M^k which is allowably contained in an n-manifold N^n is said to be **locally flat** at $x \in M - \partial N$, $x \in \operatorname{Int}(\partial N \cap M)$, $x \in \partial(\partial N \cap M)$, respectively, if there is a neighborhood U of x in N such that $(U, U \cap M)$ is homeomorphic to (E^n, E^k), (E^n_+, E^k_+), $(E^n_+, E^{k-1}_+ \times E^1_+)$, respectively. Let (N, M) be an allowable manifold pair. Then we say that the pair $(\partial N, \partial N \cap M)$ is **locally collared** in (N, M) if M is locally flat at each point of $\partial N \cap M$. We will say that the pair $(\partial N, \partial N \cap M)$ is **collared** in (N, M) if there is a homeomorphism C from $(\partial N, \partial N \cap M) \times [0, 1)$ onto a neighborhood pair (U, V) of $(\partial N, \partial N \cap M)$ in (N, M) such that $C(x, 0) = x$ for each $x \in \partial N$.

Relative Collaring Theorem 1.7.7. *Let (N, M) be an allowable manifold pair. If the pair $(\partial N, \partial N \cap M)$ is locally collared in (N, M), then it is collared in (N, M).*

Proof. First assume that M and N are compact. Since $(\partial N, \partial N \cap M)$ is locally collared in (N, M), there is an open cover $U_1, U_2, \ldots, U_s$ of ∂N such that for each i, $1 \leqslant i \leqslant s$, there is a closed embedding $h_i\colon \bar{U}_i \times [0, 1] \to N$ where $h_i^{-1}(\partial N) = \bar{U}_i \times 0$, $h_i(x, 0) = x$ for $x \in \bar{U}_i$ and

$$h_i(\bar{U}_i \times [0, 1]) \cap M = h_i((\partial N \cap M \cap \bar{U}_i) \times [0, 1]).$$

Let $V_1, V_2, \ldots, V_s$ be another cover of ∂N such that $\bar{V}_i \subset U_i$, $i = 1, \ldots, s$. Let $N^+ = N \cup (\partial N \times [-1, 0])$ where $(x, 0)$ is identified with x. Finally, let $H_i\colon \bar{U}_i \times [-1, 1] \to N^+$, $i = 1, \ldots, s$ be the embeddings defined by

$$H_i(x) = \begin{cases} h_i(x) & \text{for } x \in \bar{U}_i \times [0, 1], \\ x & \text{for } x \in \bar{U}_i \times [-1, 0]. \end{cases}$$

Inductively, we shall define maps f_i: $\partial N \to [-1, 0]$ and embeddings g_i: $N \to N^+$, $i = 0, 1, \ldots, s$ such that

(a) $f_i(x) = -1 \quad$ if $\; x \in \bigcup_{j \leqslant 1} \bar{V}_j$,
(b) $g_i(x) = (x, f_i(x)) \quad$ if $\; x \in \partial N$,
(c) $g_i(N) = N \cup \{(x, t) | \; t \geqslant f_i(x)\}$, and
(d) $g_s(M) = (\partial N \cap M) \times [-1, 0]$.

Note that since the V_i's cover ∂N, $g_s(N) = N^+$ and thus g_s^{-1} will give the required collar.

Define $g_0 = 1$, and inductively suppose g_{i-1} has been defined. Let ϕ_i: $H_i^{-1} g_{i-1}(N) \to \bar{U}_i \times [-1, 1]$ be an embedding that pushes down along fibers such that $\phi_i H_i^{-1} g_{i-1}(\bar{V}_i) = \bar{V}_i \times \{-1\}$ and

$$\phi_i \mid (\bar{U}_i - U_i) \times [-1, 1] \cup \bar{U}_i \times \{1\} = 1.$$

Such a ϕ_i can be defined as follows: Let λ_i: $\bar{U}_i \to [0, 1]$ be a Urysohn function which is 0 on $\bar{U}_i - U_i$ and 1 on $\bar{V}_i$. Let

$$s_x: [f_{i-1}(x), 1] \to [(1 - \lambda_i(x)) f_{i-1}(x) + \lambda_i(x)(-1), 1]$$

be the homeomorphism given by $s_x(t) = ((b - 1)/(a - 1))(t - 1) + 1$, where $a = f_{i-1}(x)$ and $b = (1 - \lambda_i(x)) f_{i-1}(x) + \lambda_i(x)(-1)$. Now define $\phi_i(x, t) = (x, s_x(t))$. Clearly, ϕ_i is continuous. Then define Φ_i: $g_{i-1}(N) \to N^+$ by

$$\Phi_i(x) = \begin{cases} H_i \phi_i H_i^{-1}(x) & \text{for } \; x \in g_{i-1}(N) \cap H_i(\bar{U} \times [-1, 1]), \\ x & \text{otherwise,} \end{cases}$$

and $g_i = \Phi_i g_{i-1}$. Clearly Φ_i and thus g_i is a well-defined embedding since $\phi_i \,|(\bar{U}_i - U_i) \times [-1, 1] \cup \bar{U}_i \times \{1\} = 1$, since ϕ_i is an embedding (because each s_x is), and since

$$\begin{aligned} & g_{i-1}(N) \cap H_i(\bar{U}_i \times [-1, 1]) \\ & \quad = H_i(\bar{U}_i \times [0, 1]) \cup \{(x, t) \mid t \geqslant f_{i-1}(x) \quad \text{and} \quad x \in \bar{U}_i\} \end{aligned}$$

by (c) for g_{i-1}. Note that (b) now defines $f_i(x)$, and that (a), (c), and (b) are satisfied by construction.

Remark 1.7.2 (The noncompact case). The same method of proof works in the noncompact case since it is possible to order the U_i, although infinite, so that every point in N^+ has an open neighborhood which moves only finitely often.

Remark 1.7.3 (The PL case). The theorem is still true if all manifolds and maps mentioned (including the definition of local collaring) are PL. The same proof goes through except that the particular definition of ϕ_i must be altered slightly. Namely, to make ϕ_i PL it is easiest to triangulate $\bar{U}_i \times [-1, 1]$ so that $\bar{V}_i \times [-1, 1]$ and $H_i^{-1}g_{i-1}(N)$ are subpolyhedra and the projection $\pi\colon \bar{U}_i \times [-1, 1] \to \bar{U}_i$ is simplicial. Then, it is easy to define a simplicial map ϕ_i so that it has the desired properties.

1.8. CELLULAR SETS AND APPLICATIONS

A set X in an n-dimensional manifold M is said to be **cellular** in M if $X = \bigcap_{i=1}^{\infty} D_i^n$, where each D_i^n is an n-cell such that Int $D_i^n \supset D_{i+1}^k$. Thus, one of the obvious necessary conditions that an embedding $f\colon I^k \to E^n$ be equivalent to the inclusion of I^k into E^n is that $f(I^k)$ be cellular in E^n. It is easy to see that the topologist's $\sin(1/x)$ curve (which includes the limit arc) pictured in Fig. **1.8.1** is cellular in E^2 and so cellular sets do not even have to be locally connected.

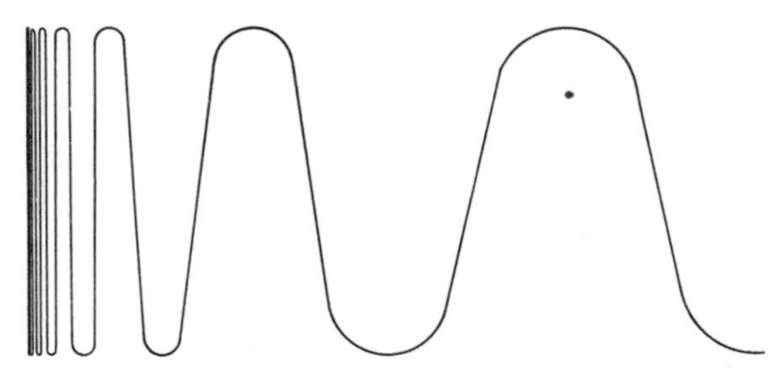

Figure 1.8.1

It is appropriate to begin our study of cellular sets by proving the following theorem, although with more work we could prove a better theorem. (It is easy to see that one improvement which can be made after one has access to the generalized Schoenflies theorem is the removal of the condition in parentheses.)

Decomposition Theorem 1.8.1. *Let $\{X_i\}$ be a collection of closed subsets of the interior of an n-manifold M for which there exists a collection of n-cells $\{D_i\}$ such that $X_i \subset \text{Int } D_i$ and for which there is an $\epsilon > 0$ such that* $\text{dist}(D_i, D_j) > \epsilon$ *whenever $i \neq j$. Then, there is a homeomorphism $h\colon M/\{X_i\} \twoheadrightarrow M$ such that $h \mid \text{Bd } M = 1$ (and $h(\tilde{D}_i/X_i)$ is contained in an open n-cell U_i in M for some n-cell $\tilde{D}_i$ such that $X_i \subset \text{Int } \tilde{D}_i$) if and only if every X_i is cellular.*

Corollary 1.8.1. *Let X be a compact subset of E^n. Then E^n/X is homeomorphic to E^n if and only if X is cellular.*

A subset X of a manifold M is **point-like** if $M - X$ is homeomorphic to $M - p$ for some $p \in M$.

Corollary 1.8.2. *A cellular subset X in the interior of a manifold is point-like.*

Corollary 1.8.3. *Let M be a topological manifold and let X be a cellular subset of* Int M. *Suppose that M/X is homeomorphic to a manifold N. Then, M is homeomorphic to N.*

EXERCISE 1.8.1. Give examples to show that it is possible to have a manifold M (with or without boundary) and a closed subset X of Int M such that M/X is a manifold but such that X is not cellular in M.

Before proving Theorem **1.8.1** we will establish a couple of lemmas.

Lemma 1.8.1. *Let D^n be an n-cell and let $\{X_i\}_{i=1}^r$ be a finite collection of disjoint closed subsets of* Int D. *Suppose that f is a homeomorphism of $D/\{X_i\}_{i=1}^r$ into an n-manifold M. Then, $f(D/\{X_i\}_{i=1}^r)$ is the union of $f(\text{Bd } D)$ and one of its complementary domains.*

PROOF. [We will first show that $f(D/\{X_i\}) \subset (f(\text{Bd } D) \cup C)$ where C is a complementary domain of $f(\text{Bd } D)$.] Since $f(\text{Int } D/\{X_i\})$ is connected and does not intersect $f(\text{Bd } D)$ it must be contained in one of the complementary domains C of $f(\text{Bd } D)$ in M. Thus, $f(D/\{X_i\}) \subset [f(\text{Bd } D) \cup C]$.

[We will complete the proof by showing that $(f(\text{Bd } D) \cup C) \subset f(D/\{X_i\})$.] First we want to see that $f(\text{Int } D - \bigcup_{i=1}^r X_i)$ is open. Well, certainly $\text{Int } D - \bigcup_{i=1}^r X_i$ is open since the X_i are closed. Thus, since $f \mid \text{Int } D - \bigcup_{i=1}^r X_i$ is one-to-one, it follows from the Invariance of Domain Theorem that $U = f(\text{Int } D - \bigcup_{i=1}^r X_i)$ is an open subset of C. Now consider the set $V = C - (U \cup f(\bigcup_{i=1}^r X_i)) = C - f(D/\{X_i\})$. Since D is compact, $D/\{X_i\}$ is compact and so $f(D/\{X_i\})$ is compact, hence closed in M. Thus V is open in C. Now we have that

$$C = U \cup V \cup f\left(\bigcup_{i=1}^{r} X_i\right),$$

where U and V are disjoint open subsets of C. This means that the finite set $f(\bigcup_{i=1}^r X_i)$ separates C which is impossible. Thus, $V = \emptyset$ and $C \subset f(D/\{X_i\})$. Hence, $(f(\text{Bd } D) \cup C) \subset f(D/\{X_i\})$ as desired.

Lemma 1.8.2. *Let D be an n-cell and let X be a closed subset of* Int D. *Suppose that there is a homeomorphism $f\colon D/X \to M$, where M is an*

n-manifold, such that $f(D/X)$ is contained in an open n-cell U in M. Then, X is cellular in D.

PROOF. It follows from Lemma **1.8.1** that $f(D/X) = f(\text{Bd } D) \cup C$ where C is a complementary domain of $f(\text{Bd } D)$ in M. Certainly it will follow that X is cellular if we can show that for any neighborhood V of X in D there is an n-cell contained in V which contains X in its interior. Let V be a neighborhood of X. Then, $f(V)$ is open in C and contains the point $f(X)$. It is easy to construct a homeomorphism h of U which carries $f(D/X)$ into $f(V)$ and which is fixed on an n-cell neighborhood W [which is contained in $f(V)$] of the point $f(X)$. Let g be the map of D into itself defined by

$$g(x) = \begin{cases} x, & x \in X, \\ f^{-1}hf(x), & x \notin X. \end{cases}$$

Since $f^{-1}hf = 1$ on $f^{-1}(W)$, g is a well-defined homeomorphism. Hence, $g(D)$ is the desired n-cell in V containing X in its interior.

EXERCISE 1.8.2. Let D and D_* ($D_* \subset \text{Int } D$) be n-cells. Then, given $\epsilon > 0$, show that there is a homeomorphism $h\colon D \twoheadrightarrow D$ such that $h(x) = x$ for all $x \in \text{Bd } D$ and diam $h(D_*) < \epsilon$.

Proof of Theorem 1.8.1

Necessity. This implication follows immediately from Lemma **1.8.2**, for by letting $(\tilde{D}_i, X_i, U_i, h \mid D_i/X_i)$ play the role of $(D, X, U, h \mid D/X)$ of the lemma it follows that X_i is cellular.

Sufficiency. Let $\{D_j^i\}_{j=1}^{\infty}$ be a sequence of n-cells in D_i whose intersection is X_i and such that $D_{j+1}^i \subset \text{Int } D_j^i$. By Exercise **1.8.2**, we can get a homeomorphism h_1^i of D_i onto itself which is fixed on Bd D_i and such that the diameter of $h_1^i(D_1^i)$ is less than 1. Let $h_2^i\colon D_i \twoheadrightarrow D_i$ be defined by

$$h_2^i(x) = \begin{cases} h_1^i(x), & x \in D_i - D_1^i, \\ g^i h_1^i(x), & x \in D_1^i, \end{cases}$$

where $g^i\colon h_1^i(D_1^i) \twoheadrightarrow h_1^i(D_1^i)$ is a map which is the identity on Bd $h_1^i(D_1^i)$ and takes $h_1^i(D_2^i)$ onto a set of diameter less than $\frac{1}{2}$. Inductively, let h_k^i be a homeomorphism of D_i onto itself such that $h_k^i = h_{k-1}^i$ on $D_i - D_{k-1}^i$ and the diameter of $h_k^i(D_k)$ is less than $1/k$. Since $| h_k^i(x) - h_{k+1}^i(x)| < 1/k$ for all $x \in D_i$, $h^i = \lim_k h_k^i$ [that is, $h^i(x)$ is defined to be the limit of the sequence $\{h_k^i(x)\}_{k=1}^{\infty}$] is a map of D_i onto itself which is the identity on Bd D_i. Let $h\colon M/\{X_i\} \twoheadrightarrow M$ be defined by $h(X_i) = h^i(x)$ for some $x \in X_i$, $h(x) = h^i(x)$ for $x \in D_i - X_i$, and $h(x) = x$ for $x \in M - \bigcup_i D_i$.

EXERCISE 1.8.3. Show that h is a well-defined homeomorphism of $M/\{X_i\}$ onto M such that $h \mid Bd\, M = 1$.

REMARK 1.8.1. Notice that by the above proof D_i is a cell, hence $U_i = \text{Int}\, D_i$ and $\tilde{D}_i = D_1^{\,i}$ satisfy the other condition of the conclusion of Theorem **1.8.1**.

The following theorem was proved about 1900.

Jordan Curve Theorem. *If Σ is a 1-sphere in S^2, then $S^2 - \Sigma$ consists of exactly two disjoint domains of which Σ is the common boundary.*

An embedding f: $S^k \to S^n$ $[E^n]$ is said to be **flat** if it is equivalent to the inclusion of S^k into S^n $[E^n]$. Shortly after the Jordan curve theorem appeared the following classical generalization was prove.

Schoenflies Theorem. *Every embedding f: $S^1 \to S^2$ is flat.*

The following analog in high dimensions of the Jordan curve theorem was also proved early in this century.

Jordan–Brouwer Separation Theorem. *If Σ is a topological $(n - 1)$-sphere in S^n, then it separates S^n into exactly two disjoint domains of which it is the common boundary.* (See p. 363 of [Hocking and Young, 1] or p. 63 of [Wilder, 2].)

Of course it was conjectured that the analog of the Schoenflies theorem also held in high dimensions, that is, that every embedding f: $S^{n-1} \to S^n$ is flat.

EXERCISE 1.8.4. Show that f: $S^{n-1} \to S^n$ is flat if and only if the closure of each of the two complementary domains of $f(S^{n-1})$ in S^n are n-cells.

In 1921 Alexander announced that he had proved this generalized Schoenflies theorem, however in 1923 he showed [Alexander, 2] his proof to be incorrect by exhibiting a 2-sphere (called the **Alexander horned sphere**) in S^3 such that one of the complementary domains was not an open cell. (We will discuss this example in Section **2.4**.) The conjecture was then modified by adding a niceness condition to the embedded $(n - 1)$-sphere in S^n. In particular, it was conjectured that every bicollared embedding f: $S^{n-1} \to S^n$ is flat.

In 1959 Mazur [2] gave an elegant proof of the generalized Schoenflies theorem modulo a "niceness" condition. Then, in 1960 Brown [5] gave an elementary proof. Also, Morse [1] showed that the niceness condition imposed by Mazur could easily be removed. In this section we will give a proof of the following generalized Schoenflies theorem which is essentially that of Brown. (In Section **3.3** we will consider Mazur's technique of proof.)

Generalized Schoenflies Theorem 1.8.2. *A locally flat embedding* h: $S^{n-1} \twoheadrightarrow S^n$ *is flat.*

REMARK 1.8.2. Notice that the Generalized Schoenflies Theorem answers a special case of the Main Problem of Topological Embeddings. In particular, it says that if $X = S^{n-1}$ and $Y = S^n$ then the set of all locally flat embeddings of X into Y are in the same equivalence class.

REMARK 1.8.3. The following important piecewise linear analog of the generalized Schoenflies theorem is still unknown:

PL Schoenflies Conjecture. *Let* Σ^{n-1} *be a* PL $(n-1)$*-sphere in* S^n. *Then, the closures of the complimentary domains of* Σ^{n-1} *are* PL n*-balls.*

However the following weaker result is known.

Alexander-Newman Theorem. *If* Σ^n *is a* PL n*-sphere and* B^n *is a* PL n*-ball which is a subpolyhedron of* Σ^n, *then* $\mathrm{Cl}(\Sigma^n - B^n)$ *is a* PL n*-ball.*

Proofs of this theorem have appeared in [Alexander, 5], [Newman, 3], [Glaser, 3], and [Zeeman, 1]. A particularly nice recent proof appears in [Cohen, 3].

Before proving Theorem **1.8.2** we will establish a lemma.

Lemma 1.8.3. *Let* X_1 *and* X_2 *be disjoint closed subsets of* S^n *such that there is a homeomorphism* h: $S^n/\{X_1, X_2\} \twoheadrightarrow S^n$ *Then, both* X_1 *and* X_2 *are cellular in* S^n.

PROOF. Since S^n is connected and X_1 and X_2 are disjoint closed subsets of S^n, there must be a point $x \in S^n - (X_1 \cup X_2)$. There is an $\epsilon > 0$ such that B, the n-cell of radius ϵ about x, misses $X_1 \cup X_2$. Then, the n-cell $B_* = \mathrm{Cl}(S^n - B)$ contains $X_1 \cup X_2$ in its interior. Let $x_1 = h(X_1)$ and $x_2 = h(X_2)$. It follows from Lemma **1.8.1** that $h(B_*) = h(\mathrm{Bd}\, B_*) \cup C$ where C is that complementary domain of $h(\mathrm{Bd}\, B_*)$ which contains $x_1 \cup x_2$. Let U be an open subset of C which contains x_1 but not x_2. Then, it is easy to construct a homeomorphism f of S^n onto itself such that $fh(B_*) \subset U$ and $h \mid W = 1$ where W is a small neighborhood of x_1. It is easy to see that g: $B_*/\{X_2\} \to S^n$ defined by

$$g(x) = \begin{cases} x, & x \in X_1, \\ h^{-1}fh(x), & x \in B_* - X_1, \end{cases}$$

is a well-defined homeomorphism. Then, it follows from Lemma **1.8.2** that X_2 is cellular. In a similar manner one can show that X_1 is cellular.

Proof of Theorem 1.8.2. Since our embedding $h\colon S^{n-1} \to S^n$ is locally flat, it follows from Theorem **1.7.5** and the Jordan–Brouwer separation theorem (or more simply from [Rushing, 6]) that there is an embedding $\bar{h}\colon S^{n-1} \times [-1, 1] \to S^n$ such that $\bar{h}(x, 0) = h(x)$ for all $x \in S^n$. We know by the Jordan–Brouwer separation theorem that $\bar{h}(S^{n-1} \times -1)$, $\bar{h}(S^{n-1} \times 0)$ and $\bar{h}(S^{n-1} \times 1)$ each separates S^n into two complementary domains having it as the common boundary. Let A be the closure of the complementary domain of $\bar{h}(S^{n-1} \times 1)$ which does not contain $\bar{h}(S^{n-1} \times -1)$ and let B be the closure of the complementary domain of $\bar{h}(S^{n-1} \times -1)$ which does not contain $\bar{h}(S^{n-1} \times 1)$. Let

$$f\colon S^{n-1} \times [-1, 1]/\{S^{n-1} \times 1, S^{n-1} \times -1\} \to S^n$$

be a homeomorphism such that $f(S^{n-1} \times 1)$ is the north pole and $f(S^{n-1} \times -1)$ is the south pole. Then,

$$f\bar{h}^{-1} \mid \bar{h}(S^{n-1} \times (-1, 1))\colon \bar{h}(S^{n-1} \times (-1, 1)) \to S^n$$

can be extended to a homeomorphism

$$f_*\colon S^n/\{A, B\} \twoheadrightarrow S^n$$

by defining $f_*(A) =$ north pole and $f_*(B) =$ south pole. Let D_A and D_B be the complementary domains of $\bar{h}(S^{n-1} \times 0)$ which contain A and B, respectively. By Lemma **1.8.3** we know that A and B are cellular. Thus, since

$$f_* \mid D_A/A\colon D_A/A \twoheadrightarrow S^n_+ \qquad \text{and} \qquad f_* \mid D_B/B\colon D_B/B \twoheadrightarrow S^n_-$$

are homeomorphisms, it follows from Corollary **1.8.3** that D_A and D_B are n-cells. Exercise **1.8.4** completes the proof of Theorem **1.8.2**.

Theorem 1.8.3 (Cantrell). *Let Σ^{n-1} be an $(n-1)$-sphere contained in S^n and let C be a complementary domain of Σ. If* $\mathrm{Cl}(C)$ *is a manifold, then it is an n-cell.*

Exercise 1.8.5. Prove Theorem **1.8.3**.

Theorem 1.8.4. *If M is a compact manifold such that $M = U \cup V$ where U and V are open n-cells, then M is an n-sphere.*

Proof. Let $h\colon \mathrm{Int}\, B^n \twoheadrightarrow U$ be a homeomorphism. It must be the case that $h^{-1}(V) \cup \mathrm{Bd}\, B^n$ is a neighborhood of $\mathrm{Bd}\, B^n$ in B^n; for, if not, we could get a sequence $\{x_i\} \to p$, $x_i \in \mathrm{Int}\, B^n - h^{-1}(V)$, $p \in \mathrm{Bd}\, B^n$, and the sequence $h(x_i)$ would have no convergent subsequence which would contradict the compactness of M. Thus, there is an ϵ between 0 and 1 such that the n-cell $B_*{}^n$ of radius ϵ and centered at the origin is such that Bd $B_*{}^n \subset h^{-1}(V)$. Now, $h(B_*{}^n)$ is a closed n-cell whose boundary is locally flat in $V \approx E^n$. By the generalized Schoenflies Theorem $\text{Cl}(M - h(B_*{}^n)) \subset V$ is an n-cell. Therefore,

$$M = h(B_*{}^n) \cup \text{Cl}(M - h(B_*{}^n))$$

is the union of two n-cells meeting in their common boundary and is thus an n-sphere.

Let S_1 and S_2 be nonintersecting, locally flat $(n - 1)$-spheres in S^n. Denote the closure of the complementary domain of S_i, $i = 1, 2$, containing S_j, $j = 2, 1$ by B_i, $i = 1, 2$. By the generalized Schoenflies theorem and Exercise **1.8.4**, B_i, $i = 1, 2$ are n-cells. We call $B_1 \cap B_2$ the **closure of the region between S_1 and S_2** and denote it by $[S_1, S_2]$.

Theorem 1.8.5. *Let S_1, S_2, B_1, B_2 and $[S_1, S_2]$ be as in the above definition. Then,*

(a) $[S_1, S_2] - S_2 \approx S^{n-1} \times [0, 1)$, *and*
(b) $[S_1, S_2] - (S_1 \cup S_2) \approx S^{n-1} \times (0, 1)$.

PROOF OF (a). Since S_2 is locally flat in S^n, it is bicollared by Theorem **1.7.5**, and so it follows easily from the generalized Schoenflies theorem that S^n — Int $_2$ is cellular in Int B_1. Thus, by Theorem **1.8.1** $B_1/(S^n - \text{Int } B_2)$ is an n-cell. Hence

$$[S_1, S_2] - S_2 = B_1 - (S^n - \text{Int } B_2) \approx I^n - \text{point} \approx S^{n-1} \times [0, 1)$$

as desired. Part (b) is proved similarly.

REMARK 1.8.4. The **n-dimensional annulus conjecture** is that $[S_1, S_2] \approx S^{n-1} \times [0, 1]$. It has only recently been verified for $n \geqslant 5$ [Kirby, Siebenmann, and Wall, 1], and is still unknown for $n = 4$. Even for $n = 2$, it is a good exercise.

CHAPTER 2

Wild Embeddings, Knotted Embeddings, and Related Topics

2.1. INTRODUCTORY DEFINITIONS

An embedding $f\colon P \to Q$ of a polyhedron P into a polyhedron Q is **tame** if there is a homeomorphism $h\colon Q \twoheadrightarrow Q$ such that hf is PL. The embedding $f\colon P \to Q$ is said to be **wild** if it is not tame. Here we have given a weak definition of tame so that our examples of wild embeddings will be stronger. The ambient space Q will be either E^n or S^n throughout this chapter, and the embedded polyhedron P will usually be I^k or S^k. (Here, you may think of S^k as the polyhedron ∂I^{k+1}.) An embedding f of I^k or S^k into E^n or S^n is said to be **flat** if there is a homeomorphism h of E^n or S^n, as the case may be, onto itself such that hf is the inclusion map. A tame embedding f of I^k or S^k into E^n or S^n is **knotted** if it is not flat.

EXERCISE 2.1.1. Show that every tame embedding f of I into E^n is flat. (Because of this, sometimes the term tame is substituted for flat for embeddings of I.)

2.2. THE GROUP OF A KNOT AND KNOTTED CODIMENSION TWO SPHERES

By a **knot** we mean an embedding $h\colon S^1 \to E^3$. We also refer to $h(S^1)$ as a **knot**, however our meaning will be clear from the context. All knots which we consider in this section will be tame. If $h\colon S^1 \to E^3$ is a knot then $\pi_1(E^3 - h(S^1))$ is called the **group of the knot** h.

Thus, the group of a knot is actually the fundamental group of the complement of the image of the knot. In this section we will indicate briefly how to compute the group of a knot. For more detailed discussions refer to [Fox, 1] or to [Crowell and Fox, 1]. (Our discussion here summarizes pp. 121 and 122 of the first reference, pp. 6, 7, and 88–91 of the second reference and includes a short discussion of group presentations.)

A knot K in E^3 is usually specified by a projection π into E^2. For instance, Fig. **2.2.1** represents the projection of a **trefoil knot** which we

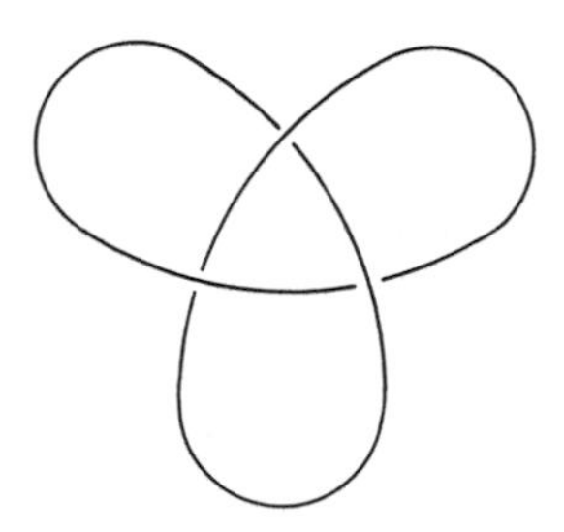

Figure 2.2.1

will consider later. The projection $P\colon E^3 \to E^2$ is defined by $P(x, y, z) = (x, y, 0)$ and $\pi\colon K \to E^2$ is defined to be $P \mid K$. A point $p \in \pi(K)$ is called a **multiple point** if $\pi^{-1}(p)$ contains more than one point. The **order** of $p \in \pi(K)$ is the cardinality of $\pi^{-1}(p)$. Thus, a **double point** is a multiple point of order two, a **triple point** is a point of order three, and so forth. Multiple points of infinite order can also occur. In general, the projection $\pi(K)$ may be quite complicated in the number and orders of multiple points present. Many times, however, a very complicated knot may be equivalent to another knot whose projection is relatively simple. A knot which is a polyhedron might be considered to be fairly simple if it satisfies the following definition.

A polyhedral knot K in E^3 is in **regular position** if (i) K has only a finite number of multiple points and they are all double points, and (ii) no double point is the image of any vertex of K. The second con-condition insures that every double point is a genuine crossing as in the left of Fig. **2.2.2** and that it is not the sort of double point shown in the right of Fig. **2.2.2**. Each double point of $\pi(K)$, where K is a knot in regular position, is the image of two points of K. The one with the larger z-coordinate is called an **overcrossing** and the other point is the corresponding **underscrossing**. The following theorem is proved on page 7 of [Crowell and Fox, 1].

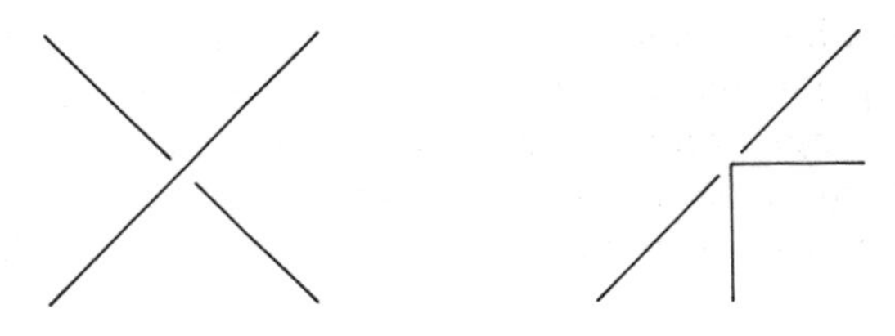

Figure 2.2.2

Theorem 2.2.1. *Any polyhedral knot in E^3 is equivalent under an arbitrarily small rotation of E^3 to a polyhedral knot in regular position.*

We will now digress to show how any group G can be represented by a presentation $[X: R]$. First of all, let us define the **free group** H generated by a set X. Consider the set $X_2 = X \times \{1, -1\}$. For each $x \in X$, let x^1 denote $(x, 1)$ and x^{-1} denote $(x, -1)$. A **word** made from X_2 is simply a finite formal product of elements of X_2. A word w is **reduced** if for every $x \in X$, x^1 never stands next to x^{-1} in w. The elements of the free group H generated by X are simply the set of all reduced words made from X_2 along with a symbol 1 which stands for the empty word. The group operation in H is defined as follows: Let w_1 and w_2 be arbitrary elements of H. If $w_1 = 1$, we define $w_1 w_2 = w_2$; if $w_2 = 1$, we define $w_1 w_2 = w_1$. Otherwise, w_1 and w_2 are both reduced words and so $w_1 w_2$ is a word. The word $w_1 w_2$ determines uniquely either the empty word 1 or a reduced word w by canceling from $w_1 w_2$ pairs of the form $x^{-1}x^1$ or $x^1 x^{-1}$ as far as possible. We define $w_1 w_2 \in H$ by taking $w_1 w_2 = 1$ or $w_1 w_2 = w$ accordingly. It is easy to check that this operation makes H a group with 1 the identity element.

Now we will show that any group G is isomorphic to a factor group of a free group. Let X be a subset of G which generates G. (You could take X to be G.) Consider the free group H generated by X. The inclusion function $i: X \to G$ extends to a homomorphism $h: H \twoheadrightarrow G$. Let O denote the kernel of h. Then by the fundamental theorem of homomorphisms G is isomorphic to H/O.

It is clear that $G \approx H/O$ is completely determined by $[X: R]$, where X is a set of generators for the free group H and R is a set of elements of H whose normal closure is O, that is, the smallest normal subgroup of H containing R is O. We call the elements of X the **generators** of G, the elements of R the **defining relations** of G and $[X: R]$ a **presentation** of G.

Given a polyhedral knot K in E^3, we will now describe an algorithm for reading from a regular projection of K a set of generators and defining relations for the group G of K. In a regular projection of K, the number n of double points is finite. Over each double point, K has an undercrossing point and an overcrossing point. The n undercrossing points divide K into n arcs. Let x_j denote the element of G represented by a loop that circles once around the jth arc in the direction of a left-handed screw and does nothing funny (see Fig. **2.2.3**). (In order for

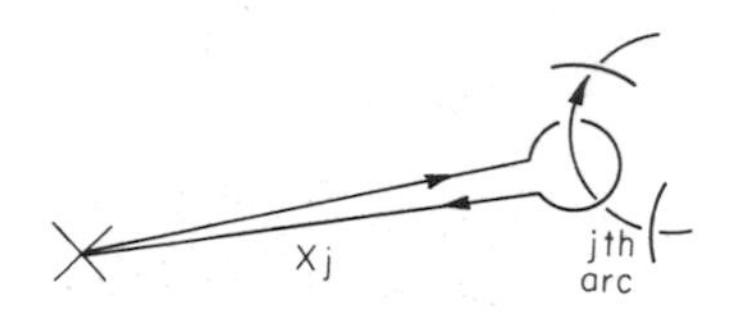

Figure 2.2.3

left-handed screw to mean anything we first give an orientation to K.) It is intuitively clear that $X = \{x_1, ..., x_n\}$ generates G, and it is even not too difficult to prove.

At each crossing a relation can be read. For instance, in Fig. **2.2.4** the relation would be $x_j x_i x_j^{-1} x_k^{-1}$. Fig. **2.2.5** shows why this is a true relation.

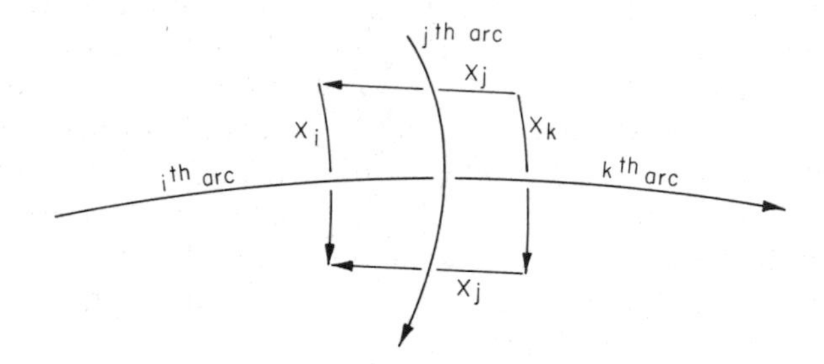

Figure 2.2.4

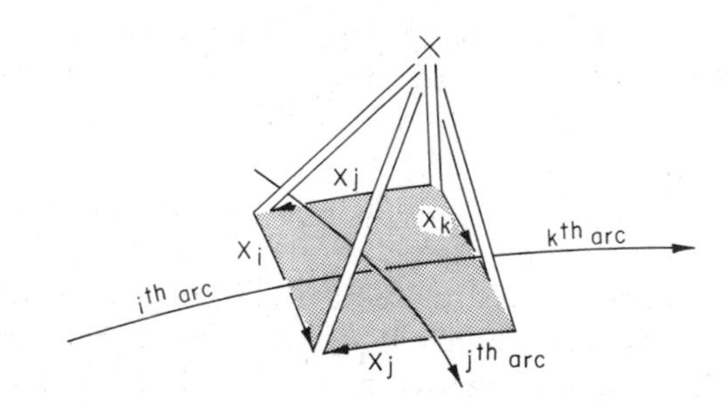

Figure 2.2.5

The n relations $r_1, ..., r_n$ obtained in this way form a complete system of defining relations for G. This may seem clear intuitively, but in fact, it is rather difficult to prove. Now we have obtained a presentation $[x_1, ..., x_n: r_1, ..., r_n]$ of G. It is easy to see that any one of the relations is a consequence of the others. Thus, we arrive at a presentation $[x_1, ..., x_n: r_1, ..., r_{n-1}]$ of G.

We will conclude this section by applying the above algorithm to show that the group of the trefoil knot K is non-Abelian. Denote the generators by x, y, and z. Then, by referring to Fig. **2.2.6**, we see that

$$\pi_1(E^3 - K) = [x, y, z: x^{-1}yzy^{-1}, y^{-1}zxz^{-1}, z^{-1}xyx^{-1}].$$

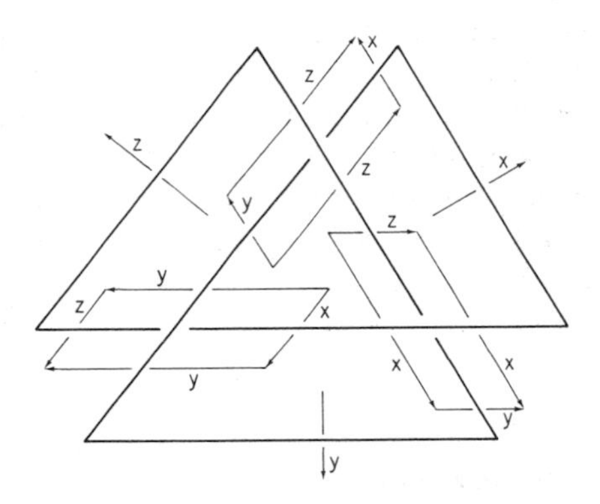

Figure 2.2.6

From the third relation it follows that $z = xyx^{-1}$, hence an equivalent presentation is

$$[x, y: x^{-1}yxyx^{-1}y^{-1}, y^{-1}xyx^{-1}xxy^{-1}x^{-1}, xy^{-1}x^{-1}xyx^{-1}]$$

$$\approx [x, y: x^{-1}yxyx^{-1}y^{-1}, y^{-1}xyxy^{-1}x^{-1}, 1]$$

$$\approx [x, y: yxy = xyx, xyx = yxy] \approx [x, y: xyx = yxy].$$

Consider the symmetric group S_3 of degree three, which is generated by the cycles (12) and (23). First observe that S_3 is non-Abelian, since

$$(12)(23) = (132) \neq (123) = (23)(12).$$

Let H denote the free group for which x and y are a free basis. Then, the homomorphism θ of H onto S_3 defined by

$$\theta(x) = (12), \qquad \theta(y) = (23)$$

induces a homomorphism of the knot group G onto S_3 provided $\theta(xyx) = \theta(yxy)$. But, this is the case since

$$\begin{aligned}\theta(xyx) &= \theta(x)\,\theta(y)\,\theta(x) = (12)(23)(12) = (13)\\ &= (23)(12)(23) = \theta(y)\,\theta(x)\,\theta(y) = \theta(yxy).\end{aligned}$$

Thus, the knot group G of the trefoil knot K can be mapped homomorphically onto a non-Abelian group and so G is non Abelian.

It follows immediately from the algorithm that $\pi_1(E^3 - S^1)$ is infinite cyclic. (This can also be proved by showing that $E^3 - S^1$ is of the same homotopy type as S^1 and showing that $\pi_1(S^1)$ is infinite cyclic.) Thus, the trefoil knot is knotted.

Let X be a compact topological space. Then, we define the **suspension**, $\mathscr{S}(X)$, of X to be the quotient space $(X \times I)/\{X \times -1, X \times 1\}$, where $I = [-1, 1]$. (It is easy to show that the suspension of a polyhedron as defined in Section **1.6** is topologically equivalent to its suspension as a topological space just defined.) If P^k is a k-polyhedron and Q^n is an n-polyhedron, $k < n$, then $n - k$ is called the **codimension** between P and Q.

Example 2.2.1. *There is a knotted codimension two sphere Σ^{n-2} in S^n for $n \geqslant 3$. Furthermore, $\pi_1(S^n - \Sigma^{n-2}) \not\approx Z$.*

Proof. We will prove this by induction on n. We have just seen that there is a 1-sphere Σ^1 in S^3 such that $\pi_1(S^3 - \Sigma^1) \not\approx Z$, where Z is the group of integers. Assume that there is an $(n-3)$-sphere Σ^{n-3} in S^{n-1} such that $\pi_1(S^{n-1} - \Sigma^{n-3}) \not\approx Z$. Then, it is easy to see that $\mathscr{S}(S^{n-1})$ is an n-sphere (which we consider to be S^n), $\mathscr{S}(\Sigma^{n-3})$ is an $(n-2)$-sphere and $\mathscr{S}(\Sigma^{n-3}) \subset \mathscr{S}(S^{n-1})$. Clearly, $(\mathscr{S}(S^{n-1}) - \mathscr{S}(\Sigma^{n-3})) \approx (S^{n-1} - \Sigma^{n-3}) \times (-1, 1)$. Since $(S^{n-1} - \Sigma^{n-3}) \times (-1, 1)$ deformation retracts onto $S^{n-1} - \Sigma^{n-3}$, it follows that $\pi_1(\mathscr{S}(S^{n-1}) - \mathscr{S}(\Sigma^{n-3})) \approx \pi_1(S^{n-1} - \Sigma^{n-3}) \not\approx Z$. Thus, $\mathscr{S}(\Sigma^{n-3})$ is knotted in S^n.

2.3. LOCAL HOMOTOPY GROUPS, WILD CODIMENSION TWO CELLS AND SPHERES, AND TAME NONLOCALLY FLAT CODIMENSION TWO CELLS AND SPHERES

Our considerations in this section will all be in codimension two. Many strange things can happen in codimension two. For instance, in the last section we saw that tame spheres can knot this codimension. Later we will show that spheres cannot knot in codimensions greater than two and Theorems **1.7.2** and **1.8.2** imply that (tame) spheres cannot (topologically) knot in codimension one. In this section, we define the local homotopy groups (as developed in [Tindell, 3], for example) in codimension two, and it turns out that they can be bad even for codimension two piecewise linear submanifolds. Because of this, we are easily able to exhibit wild codimension two cells and spheres. We shall see later that wild cells and spheres exist in other codimensions but their wildness is harder to establish. Also, because of the bad local homotopy groups we are able to give tame nonlocally flat codimension two cells and spheres. It follows from Theorem **1.7.2** that such cells and spheres cannot exist in other codimensions.

Let $M \subset Q$ be topological $n-2$ and n-manifolds, respectively. A **fundamental neighborhood sequence** at $p \in M$ is a sequence $\{V_i\}_{i=1}^{\infty}$ of neighborhoods of p in Q satisfying

(a) $V_1 \supset V_2 \supset \cdots$,

(b) $\bigcap_{i=1}^{\infty} V_i = p$, and

(c) each inclusion induced map $i_* : \pi_k(V_r - M) \to \pi_k(V_s - M)$ is an isomorphism onto for each k when $r \geqslant s$.

If $M \subset Q$ are topological $n-2$ and n-manifolds, respectively, and $p \in M$, then we say that the **local homotopy groups exist** at p if there is a fundamental neighborhood sequence $\{V_i\}$ at p. ($\pi_k(V_1 - M)$ is called the kth **local homotopy group** at p.)

Proposition 2.3.1. *The kth local homotopy group is well-defined; that is, if $\{V_i\}$ and $\{U_i\}$ are fundamental neighborhood sequences at $p \in M$, then $\pi_k(V_1 - M) \approx \pi_k(U_1 - M)$.*

PROOF. It follows from Conditions a and b of the definition of fundamental neighborhood sequence that there are integers r, s, and t such that $V_t \subset U_s \subset V_r \subset U_1$. Thus, we have the following inclusion induced commutative diagram where i_* and j_* are the onto isomorphisms assured by Condition c of the definition of fundamental neighborhood sequence.

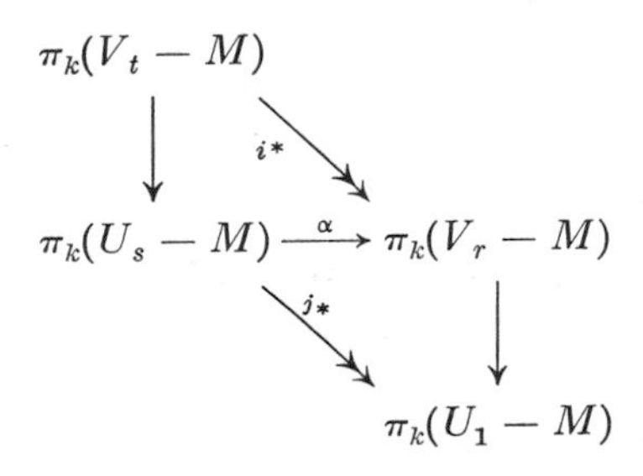

Since i_* is onto, α is onto and since j_* is one-to-one, α is one-to-one. Hence, $\pi_k(U_1 - M) \approx \pi_k(U_s - M) \approx \pi_k(V_r - M) \approx \pi_k(V_1 - M)$ as desired.

Although the local homotopy groups do not always exist, the next three propositions give important cases where they do exist.

Proposition 2.3.2. *If M is locally flat at $p \in \operatorname{Int} M$, then the local groups exist at p and are isomorphic to the corresponding homotopy groups of S^1.*

EXERCISE 2.3.1. Prove Proposition **2.3.2**.

Proposition 2.3.3. *Let Σ^{n-2} be a topological $(n-2)$-sphere in S^n, let M be the cone $v * \Sigma^{n-2}$ $[v \in \operatorname{Int} B^{n+1}]$, and let N be the cone $v * S^n = B^{n+1}$. Then, the local homotopy groups exist at the vertex v and are isomorphic to $\pi_k(S^n - \Sigma^{n-2})$.*

PROOF. Consider N to be $(S^n \times [0, 1])/(S^n \times 1)$ and let

$$V_i = (S^n \times [(i-1)/(i+1), 1])/(S^n \times 1).$$

(Notice that $V_1 = N = v * S^n$.) By using this fundamental neighborhood sequence $\{V_i\}$, it is easy to complete the proof.

Proposition 2.3.4. *If $M \subset N$ are PL $n-2$ and n-manifolds, respectively, then the local homotopy groups exist at every point of M.*

EXERCISE 2.3.2. Prove Proposition **2.3.4**.

The following example appears in [Doyle and Hocking, 1]. (Glaser [4] made use of this construction to show that for $n \geqslant 4$ there are uncountably many almost polyhedral wild $(n-2)$-cells in E^n. Correspondingly in E^3, Alford and Ball [1] constructed infinitely many almost polyhedral wild arcs so as to have an end-point as the "bad" point and Fox and Harrold [1] constructed uncountably many almost polyhedral wild arcs with an interior point as the "bad" point.)

Example 2.3.1 (Doyle and Hocking). *There is a wild $(n-2)$-cell D^{n-2} and a wild $(n-2)$-sphere Σ^{n-2} in S^n for $n \geqslant 4$.* (We will show that this is also the case for $n = 3$ later.)

PROOF. (We will construct a wild disk D and a wild 2-sphere Σ in S^4 and then it will be easy to see that essentially the same constructions yield a wild $(n-2)$-cell and a wild $(n-2)$-sphere in S^n for $n \geqslant 4$.) In S^3 let $\{\Delta_i^3\}$ be a sequence of 3-simplexes that converge to a point q

such that $\Delta_i^3 \cap \Delta_j^3 = \emptyset$ if $i \neq j$. Let $\{K_i\}$ be a sequence of trefoil knots such that $K_i \subset \Delta_i^3$. In $S_+^4 - S^3$, let $\{p_i\}$ be a sequence of points converging to q such that if Δ_i^4 is the simplex $\Delta_i^3 * p_i$, then $\{\Delta_i^4\}$ is a sequence of pairwise disjoint simplexes converging to q. If $\hat{\Delta}_i^4$ denotes

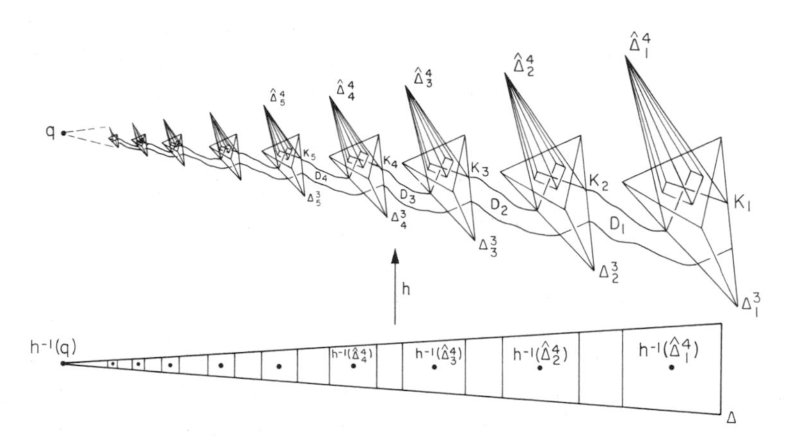

Figure 2.3.1

the barycenter of Δ_i^4, then the cone $\hat{\Delta}_i^4 * K_i$ is a polygonal disk in S^4. Now in S^3 join $\hat{\Delta}_1^4 * K_1$ and $\hat{\Delta}_2^4 * K_2$ by a polygonal disk D_1 so that $\hat{\Delta}_1^4 * K_1 \cup D_1 \cup \hat{\Delta}_2^4 * K_2$ is a polygonal disk disjoint from $(\bigcup_{i=3}^{\infty} \hat{\Delta}_i^4 * K_i) \cup q$. We next join $\hat{\Delta}_2^4 * K_2$ and $\hat{\Delta}_3^4 * K_3$ by a polygonal disk D_2 in S^3 so that $\hat{\Delta}_1^4 * K_1 \cup D_1 \cup \hat{\Delta}_2^4 * K_2 \cup D_2 \cup \hat{\Delta}_3^4 * K_3$ is a polygonal disk disjoint from $(\bigcup_{i=4}^{\infty} \hat{\Delta}_i^4 * K_i) \cup q$. This process is continued so that as $k \to \infty$ the diameter of D_k tends to 0 and we let D denote the disk $(\bigcup_{i=1}^{\infty}(\hat{\Delta}_i^4 * K_i \cup D_i)) \cup q$. As is indicated in Fig. **2.3.1**, it is easy to get a homeomorphism $h\colon \Delta \twoheadrightarrow D$ of a 2-simplex Δ onto D such that $h \mid (\Delta - h^{-1}(q))$ is PL.

Suppose that h is tame, that is, suppose there is a homeomorphism $f\colon S^4 \twoheadrightarrow S^4$ such that $fh\colon \Delta \to S^4$ is PL. Then, by Corollary **1.6.3**, $\{h(\hat{\Delta}_i^4)\}$ contains a point $h(\hat{\Delta}_j^4)$ that lies in the interior of a disk formed by the union of one or two 2-simplexes in a triangulation of $fh(\Delta)$. But, then the first local homotopy group at $h(\hat{\Delta}_j^4)$ must be infinite cyclic. However, this cannot be the case for by Proposition **2.3.3**, the first local homotopy group at $\hat{\Delta}_j^4$ is $\pi_1(\mathrm{Bd}\, \Delta_j^4 - K_j)$ which we showed to be non-Abelian in the last section.

To construct a wild 2-sphere Σ in S^4 simply add to the disk D constructed above, the cone over Bd D from a point in the interior of S_-^4.

Example 2.3.2. *There is a tame nonlocally flat $(n-2)$-cell D^{n-2} and* *a tame nonlocally flat* $(n - 2)$*-sphere* Σ^{n-2} *in* S^n, $n \geqslant 4$. (Notice that this theorem is false for $n = 3$ by Exercise **2.1.1**.)

PROOF. By Example **2.2.1**, there is a PL $(n - 3)$-sphere Σ^{n-3} in $\text{Bd}\, I^n \approx S^{n-1}$ such that $\pi_1(\text{Bd}\, I^n - \Sigma^{n-3}) \not\approx Z$. Let v be a point in $\text{Int}\, I^n$ and let $D^{n-2} = v * \Sigma^{n-3}$. By Proposition **2.3.3** the first local homotopy group of D^{n-2} at v is not Z, hence D^{n-2} cannot be locally flat. By using a suspension rather than a cone we could construct Σ^{n-2}.

2.4. WILD 1-CELLS, 1-SPHERES, 2-CELLS, AND 2-SPHERES IN S^3

Before beginning our constructions of wild embeddings we will discuss direct limit groups. A **relation** $<$ on a set λ is a subset of $\lambda \times \lambda$. If $(x, y) \in <$, we write (x, y) as $x < y$ and say that x is less than y. A relation $<$ is a **partial order relation** provided that (i) if $x < y$, then $y < x$ is false, and (ii) if $x < y$ and $y < z$ then $x < z$. A set λ is a **directed set** if λ is partially ordered by a relation $<$ such that (iii) for any pair of elements x, y in λ there is a $z \in \lambda$ such that $z > x$ and $z > y$. (Thus, the set of positive integers is directed.) Let $G = \{G_x\}_{x \in \lambda}$ be a collection of groups indexed by a directed set λ. For each x and y in λ such that $x < y$ or $x = y$ suppose that there is a homomorphism $H_x{}^y\colon G_x \to G_y$ of G_x into G_y. If H denotes the set of all such homomorphisms $H_x{}^y$, then (G, H) is said to be a **direct homomorphism system** if (i) $H_x{}^x = 1$ for each x in λ, and (ii) if $x < y < z$ then $H_y{}^z H_x{}^y = H_x{}^z$.

We will now define the **direct limit group** G' of a direct homomorphism system (G, H). Let G_* be the set of elements $\{g_x\}_{x \in \lambda'}$, $g_x \in G_x$, $\lambda' \subset \lambda$, such that if $x \in \lambda'$ and $H_x{}^y \in H$ then $y \in \lambda'$ and $g_y = H_x{}^y(g_x)$. Define an equivalence relation $\approx$ on G_* by letting $\{g_x\}_{x \in \lambda'} \approx \{g\}_{y \in \lambda''}$ if for some group of G their coordinates are defined and are equal. The elements of G' are the equivalence classes of $\approx$. Let g_1 and g_2 be two elements of G' and let $\{g_x\}_{x \in \lambda'}$ and $\{g_y\}_{y \in \lambda''}$ be representatives of g_1 and g_2, respectively. Let $\{g_x\}_{x \in \lambda'} + \{g_y\}_{y \in \lambda''}$ be the element of G_* obtained by adding the coordinates of $\{g_x\}_{x \in \lambda'}$ and $\{g_y\}_{y \in \lambda''}$ which lie in a common group of G. Then, $g_1 + g_2$ is defined to be the equivalence class of $\{g_x\}_{x \in \lambda'} + \{g_y\}_{y \in \lambda''}$ under $\approx$. It is easy to show that G' constitutes a group with addition so defined.

An alternate definition of the direct limit group G' of a direct homomorphism system (G, H) which is maybe a little more elegant than the above definition, although equivalent, will now be given. Let $\bigoplus_{x \in \lambda} G_x$

denote the weak direct sum of the groups in G, that is, all sets $\{g_x\}_{x\in\lambda}$, where $g_x \in G_x$ and all but a finite number of the g_x are the identity, and coordinatewise addition. Let the injection $i_y: G_y \to \bigoplus_{x\in\lambda} G_x$ be defined by $i_y(g_y) = \{g_x\}_{x\in\lambda}$, where $g_x = g_y$ if $x = y$ and $g_x = 1$ if $x \neq y$. Then, G' has the following presentation:

$$[i_y(g_y) \mid g_y \in G_y, y \in \lambda: i_y(H_x{}^y(g_x)) - i_x(g_x) \mid g_x \in G_x, x, y \in \lambda, x < y].$$

In the following material we will make use of the next two lemmas.

Lemma 2.4.1. *Let $M_1 \subset M_2 \subset \cdots$ be spaces such that each M_i is open in $M = \bigcup_{i=1}^{\infty} M_i$ and choose a base point in M_1. Then, $\pi_1(M)$ is the direct limit group of the direct homomorphism system $(\pi_1(M_i), i = 1, 2, \ldots, f_i, i = 1, 2, \ldots)$, where $f_i: \pi_1(M_i) \to \pi_1(M_{i+1})$ is the injection.* (Technically, we should say "inclusion induced homomorphism" rather than "injection"; however, we will use these terms synonymously.)

Exercise 2.4.1. Prove Lemma **2.4.1.**

Lemma 2.4.2. *If $(G_i, i = 1, 2, \ldots, f_i: G_i \to G_{i+1}, i = 1, 2, \ldots)$ is a direct homomorphism system where G_i has a presentation $[X_i: R_i]$, then the direct limit group G has the presentation*

$$\left[\bigcup_{j=1}^{\infty} X_j: \left(\bigcup_{j=1}^{\infty} R_j\right) \cup \left(\bigcup_{j=1}^{\infty}\left(\bigcup_{x_j\in X_j} x_j(f_j(x_j))^{-1}\right)\right)\right].$$

We are now ready to begin our construction of a wild arc in S^3. (This example, as well as the six examples which follow, is taken in substance from the classical paper [Fox and Artin, 1].) Consider I^3. Let $A_- = I^2 \times \{-1\}$ and $A_+ = I^2 \times \{1\}$. Then the points $r_- = (0, -\frac{1}{2}, -1)$, $s_- = (0, 0, -1)$ and $t_- = (0, \frac{1}{2}, -1)$ are on A_- and the points $r_+ = (0, -\frac{1}{2}, 1)$, $s_+ = (0, 0, 1)$ and $t_+ = (0, \frac{1}{2}, 1)$ are on A_+. In I^3 construct three nonintersecting oriented polygonal arcs K_- joining s_- to r_-, K_0 joining t_- to s_+, and K_+ joining r_+ to t_+. These arcs which have only their end-points in common with Bd I^3 are to be arranged as indicated in Fig. **2.4.1**. Denote the union of K_-, K_0, and K_+ by K.

Consider the suspension $\mathscr{S}(I^2)$ of I^2 from the points $p = (0, 0, -1)$ and $q = (0, 0, 1)$. Divide $\mathscr{S}(I^2)$ into an infinite number of sections by the family of parallel planes $z = \pm(i - 1)/i$, $i = 1, 2, \ldots$ (see Fig. **2.4.2**). For each positive integer n let D_n be the section

$$(n-1)/n \leqslant z \leqslant n/(n+1)$$

intersected with $S(I^n)$, for each negative integer n let D_n be the section $-n/(n-1) \leqslant z \leqslant -(n+1)/n$ intersected with $\mathscr{S}(I^n)$ and let $D_0 = \emptyset$.

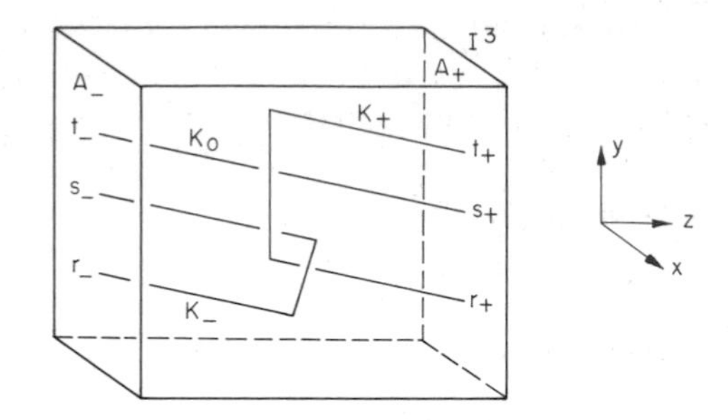

Figure 2.4.1

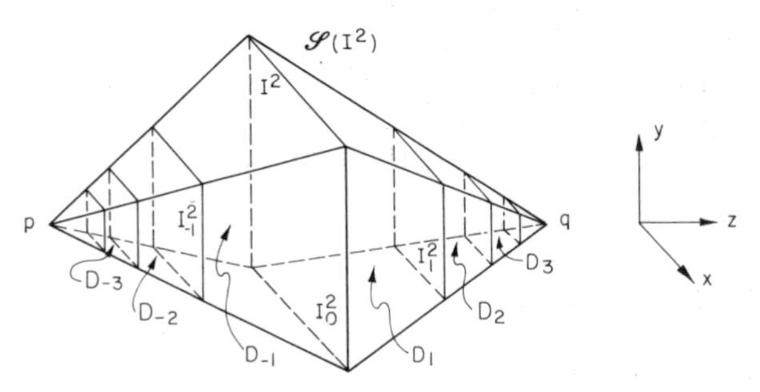

Figure 2.4.2

Let $\pi_3\colon I^3 \to I^2$ be defined by $\pi_3(x, y, z) = (x, y, 0)$. For $n = 0, 1, 2, \ldots$ let $I_n{}^2$ be the intersection of the plane $z = n/(n+1)$ with $\mathscr{S}(I^2)$ and for $n = -1, -2, \ldots$ let $I_n{}^2$ be the intersection of the plane $-n/(n-1)$ with $\mathscr{S}(I^2)$. For each integer n let $\Gamma_n\colon I^2 \twoheadrightarrow I_n{}^2$ be the projection of I^2 onto $I_n{}^2$ from p if $n < 0$ and from q if $n \geqslant 0$. Now for all integers $n \neq 0$ let $f_n\colon I^3 \twoheadrightarrow D_n$ be defined by letting

$$f_n \mid A_- = \begin{cases} \Gamma_{n-1}\pi_3 \mid A_-, & \text{if} \quad n \geqslant 1 \\ \Gamma_n\pi_3 \mid A_-, & \text{if} \quad n \leqslant -1 \end{cases}$$

$$f_n \mid A_+ = \begin{cases} \Gamma_n\pi_3 \mid A_+, & \text{if} \quad n \geqslant 1 \\ \Gamma_{n+1}\pi_3 \mid A_+, & \text{if} \quad n \leqslant -1 \end{cases}$$

and then extending linearly over segments in I^3 which connect a point in A_- with a point in A_+ having the same first two coordinates.

Let $X = p \cup (\bigcup_{n=1}^{\infty} (f_n(K) \cup f_{-n}(K))) \cup q$. Then, X has the regular projection into the yz-plane given in Fig. **2.4.3**.

Example 2.4.1 (Fox and Artin). *The complement of the arc X constructed above is not simply connected. (In fact, $\pi_1(S^3 - X)$ is non-Abelian.) Hence, X is wild.*

PROOF. By Lemma **2.4.1**, $\pi_1(S^3 - X)$ is the direct limit group of the direct homomorphism system $(\pi_1(M_i), i = 1, 2, ..., f_i, i = 1, 2, ...)$ where

$$M_i = S^3 - \left(X \cup \text{Cl}\left(\bigcup_{j=i+1}^{\infty} (D_j \cup D_{-j+1})\right)\right)$$

and $f_i: \pi_1(M_i) \to \pi_1(M_{i+1})$ is the injection homomorphism. We now obtain a set of generators and defining relations for $\pi_1(M_i)$ by employing a slight generalization of the algorithm discussed in Section **2.2**. Thus, $\pi_1(M_i)$ is generated by the elements $\bigcup_{j=-i+1}^{i}\{a_j, b_j, c_j\}$ indicated in Fig. **2.4.3** and has the following defining relations:

$$b_{-i+1}a_{-i+1}^{-1}c_{-i+1}^{-1} \qquad \left(\text{relation about } \bigcup_{j=-i}^{-\infty} D_j\right),$$

$$c_i a_i b_i^{-1} \qquad \left(\text{relation about } \bigcup_{j=i+1}^{\infty} D_j\right),$$

$$\left.\begin{aligned} a_{k+1} &= c_{k+1}^{-1}c_k c_{k+1} \\ b_k &= c_{k+1}^{-1}a_k c_{k+1} \\ c_{k+1} &= b_k^{-1}b_{k+1}b_k \end{aligned}\right\} \qquad -i+1 \leqslant k < i.$$

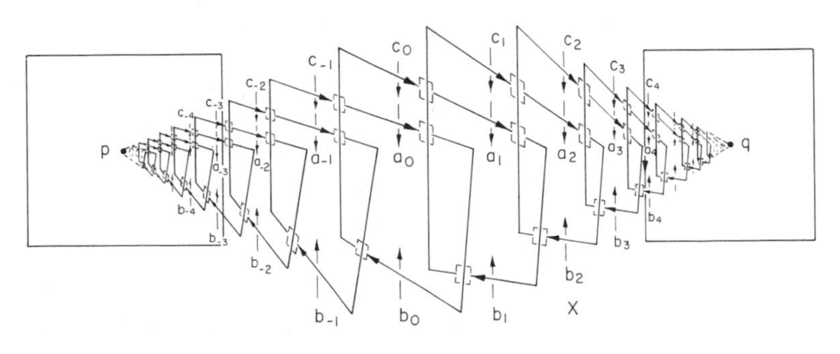

Figure 2.4.3

The injection homomorphism $f_i: \pi_1(M_i) \to \pi_1(M_{i+1})$ maps each generator of the set $\bigcup_{j=-i+1}^{i} \{a_j, b_j, c_j\}$ of generators of $\pi_1(M_i)$ into the same-named generator of $\pi_1(M_{i+1})$. Hence, it follows from Lemma **2.4.2** that $\pi_1(S^3 - X)$ has the following presentation:

$$\left[\bigcup_{\pm i=0}^{\infty}\{a_i, b_i, c_i\}: \bigcup_{\pm i=0}^{\infty}\{c_i a_i b_i^{-1}, a_{i+1} = c_{i+1}^{-1} c_i c_{i+1}, b_i = c_{i+1}^{-1} a_i c_{i+1}, c_{i+1} = b_i^{-1} b_{i+1} b_i\}\right]$$

$$\approx \left[\bigcup_{\pm i=0}^{\infty}\{b_i, c_i\}: \bigcup_{\pm i=0}^{\infty}\{c_i c_i^{-1} c_{i-1} c_i b_i^{-1}, b_i = c_{i+1}^{-1} c_i^{-1} c_{i-1} c_i c_{i+1}, c_{i+1} = b_i^{-1} b_{i+1} b_i\}\right]$$

$$\approx \left[\bigcup_{\pm i=0}^{\infty}\{b_i, c_i\}: \bigcup_{\pm i=0}^{\infty}\{b_i = c_{i-1} c_i, b_i = c_{i+1}^{-1} c_i^{-1} c_{i-1} c_i c_{i+1}, c_{i+1} = b_i^{-1} b_{i+1} b_i\}\right]$$

$$\approx \left[\bigcup_{\pm i=0}^{\infty}\{b_i, c_i\}: \bigcup_{\pm i=0}^{\infty}\{b_i = c_{i-1} c_i, b_i = c_{i+1}^{-1} c_i^{-1} c_{i-1} c_i c_{i+1}, \right.$$
$$\left. c_{i+1} = c_{i+1}^{-1} c_i^{-1} c_{i-1}^{-1} c_i c_{i+1} c_i c_{i+1} c_{i+1}^{-1} c_i^{-1} c_{i-1} c_i c_{i+1}\}\right]$$

$$\approx \left[\bigcup_{\pm i=0}^{\infty}\{c_i\}: \bigcup_{\pm i=0}^{\infty}\{c_{i-1} c_i = c_{i+1}^{-1} c_i^{-1} c_{i-1} c_i c_{i+1}, c_{i+1}^{-1} c_i^{-1} c_{i-1}^{-1} c_i c_{i+1} c_{i-1} c_i\}\right]$$

$$\approx \left[\bigcup_{\pm i=0}^{\infty}\{c_i\}: \bigcup_{\pm i=0}^{\infty}\{c_i c_{i+1} c_{i-1} c_i = c_{i-1} c_i c_{i+1}\}\right].$$

We will show that this group is nontrivial by giving a homomorphism h of it onto the nontrivial, non-Abelian subgroup of the symmetric group S_5 of degree five generated by the cycles (12345) and (14235). We define h by $h(c_i) = (12345)$ if i is odd and $h(c_i) = (14235)$ if i is even. It follows that h induces the desired homomorphism since

(a) for i even,

$$\begin{aligned} h(c_{i-1} c_i c_{i+1}) &= h(c_{i-1})\, h(c_i)\, h(c_{i+1}) = (12345)(14235)(12345) \\ &= (142) = (14235)(12345)(12345)(14235) \\ &= h(c_i)\, h(c_{i+1})\, h(c_{i-1})\, h(c_i) = h(c_i c_{i+1} c_{i-1} c_i), \end{aligned}$$

(b) for i odd,

$$\begin{aligned} h(c_{i-1} c_i c_{i+1}) &= h(c_{i-1})\, h(c_i)\, h(c_{i+1}) = (14235)(12345)(14235) \\ &= (345) = (12345)(14235)(14235)(12345) \\ &= h(c_i)\, h(c_{i+1})\, h(c_{i-1})\, h(-c_i) = h(c_i c_{i+1} c_{i-1} c_i). \end{aligned}$$

This completes the proof of Example **2.4.1**.

Lemma 2.4.3. *If q is an end-point of an arc Y which is tamely embedded in S^3 and $\{V_i\}$ is any sequence of closed neighborhoods of q such that $V_1 \supset V_2 \supset \cdots$ and $\bigcap_{i=1}^{\infty} V_i = q$, then there is an integer N such that the injection $\pi_1(V_N - Y) \to \pi_1(V_1 - Y)$ is trivial.*

Proof. By Exercise **2.1.1**, there is a homeomorphism of S^3 which takes Y onto $[0, 1]$ and takes q onto the origin 0, and so we may assume that Y is $[0, 1]$ and q is 0. Let $\{V_i\}$ be a sequence of closed neighborhoods of 0 such that $V_1 \supset V_2 \supset \cdots$ and $\bigcap_{i=1}^{\infty} V_i = 0$. Choose an $\epsilon > 0$ such the open ball B_ϵ about 0 of radius ϵ is contained in V_1 and choose an integer N such that $V_N \subset B_\epsilon$. Also choose a point in $V_N - [0, 1]$ to serve as the base point for the fundamental groups $\pi_1(V_N - [0, 1])$, $\pi_1(B_\epsilon - [0, 1])$ and $\pi_1(V_1 - [0, 1])$. Since the injection of $\pi_1(V_N - [0, 1])$ into $\pi_1(V_1 - [0, 1])$ is the composition of the injection of $\pi_1(V_N - [0, 1])$ into $\pi_1(B_\epsilon - [0, 1])$ and the injection of $\pi_1(B_\epsilon - [0, 1])$ into $\pi_1(V_1 - [0, 1])$ and since $\pi_1(B_\epsilon - [0, 1])$ is trivial, it follows that the injection of $\pi_1(V_N - [0, 1])$ into $\pi_1(V_1 - [0, 1])$ is trivial.

Consider the arc Y which is the set

$$f_{-1}(K_0) \cup f_{-1}(K_+) \cup \bigcup_{n=1}^{\infty} f_n(K) \cup q.$$

Then, Y has the regular projection into the yz-plane indicated in Fig. **2.4.4**.

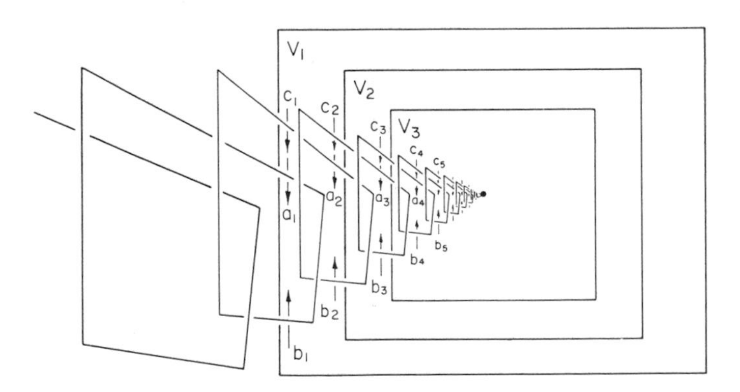

Figure 2.4.4

Example 2.4.2 (Fox and Artin). *The arc Y is wild and its complement is an open 3-cell. (Thus, Y is wild and cellular in S^3.)*

Proof. First we will show that the complement of Y is an open 3-cell. By fattening up Int Y, it is easy to construct a closed 3-cell D in

S^3 for which there is a homeomorphism $h: (I^3, [0, 1]) \twoheadrightarrow (D, Y)$. Let $g: I^3 \twoheadrightarrow I^3$ be a map such that $g \mid \mathrm{Bd}\, I^3 = 1$, $g([0, 1]) = (1, 0, 0)$, and $g \mid I^3 - [0, 1]$ is a homeomorphism. Now define a map $\bar{h}: S^3 \twoheadrightarrow S^3$ by $\bar{h} \mid S^3 - D = 1$ and $\bar{h} \mid D = hgh^{-1}$. Then $\bar{h} \mid S^3 - Y: S^3 - Y \twoheadrightarrow S^3 - q$ is a homeomorphism and so $S^3 - Y$ is an open 3-cell.

We will now show that Y is wild by showing that it does not satisfy Lemma **2.4.3**. Let $\{V_i\}$ be the sequence of closed 3-cell neighborhoods of q indicated in the above figure. This sequence clearly satisfies the hypotheses of Lemma **2.4.3**. By the proof of Example **2.4.1**, $\pi_1(V_N - Y)$ has the presentation $[\bigcup_{i=N}^{\infty}\{c_i\}: \bigcup_{i=N}^{\infty}\{c_i c_{i+1} c_{i-1} c_i = c_{i-1} c_i c_{i+1}\}]$. If the injection $\pi_1(V_N - Y) \to \pi_1(V_1 - Y)$ were trivial, then each element c_i, $i \geqslant N$, would be trivial in $\pi_1(V_1 - Y)$. However, by the proof of Example **2.4.1**, if we let $h(c_i) = (12345)$ if i is odd and let $h(c_i) = (14235)$ if i is even, then h induces a homomorphism of $\pi_1(V_1 - Y)$ into the symmetric group S_5. Since no c_i goes onto the identity permutation under h, it follows that no c_i is trivial in $\pi_1(V_1 - Y)$.

Let X be the arc of Example **2.4.1** and let X' be the corresponding arc pictured in Fig. **2.4.5**.

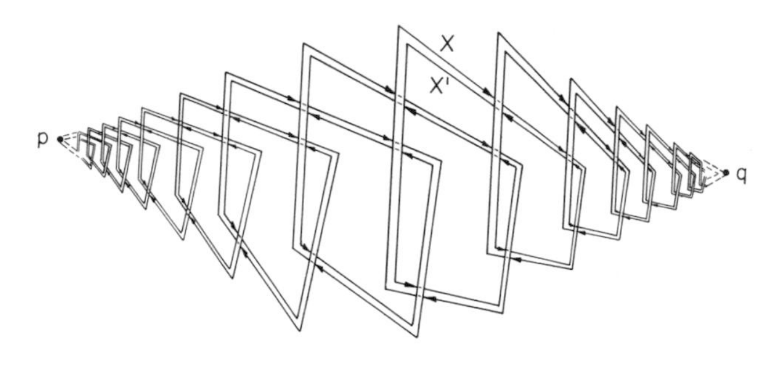

Figure 2.4.5

Example 2.4.3 (Fox and Artin). *Even though the simple closed curve $X \cup X'$ is nice in the sense that it obviously bounds a 2-cell D, it is still wild. In fact, $\pi_1(S^3 - (X \cup X'))$ is non-Abelian. (Of course, the 2-cell D is then also wild.)*

Proof. The fundamental group of the complement of $X \cup X'$ maps homomorphically by injection onto the fundamental group of the complement of X and is therefore non-Abelian by Example **2.4.1**.

Let Y be the arc of Example **2.4.2**, let Y' be the corresponding arc in Fig. **2.4.6** and let l be a segment connecting the left-hand end-points of Y and Y'.

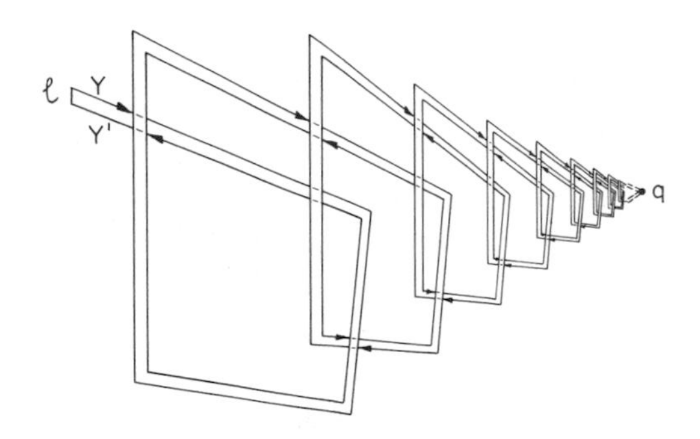

Figure 2.4.6

Example 2.4.4 (Fox and Artin). *The simple closed curve $Y \cup Y' \cup l$ is wild though the fundamental group of its complement is infinite cyclic and it obviously bounds a 2-cell D. (Of course, the 2-cell D is wild.)*

Proof. This simple closed curve is wild since by Example **2.4.2** the arc Y is wild. It is straightforward to calculate the fundamental group of its complement and check that it is infinite cyclic.

By fattening up Int X one can easily construct a homeomorphism $f: (I^3, I^1) \to S^3$ such that $f(I^1) = X$ where X is the arc of Example **2.4.1**. For instance, choose f so that $f(I^3)$ is as depicted in Fig. **2.4.7**.

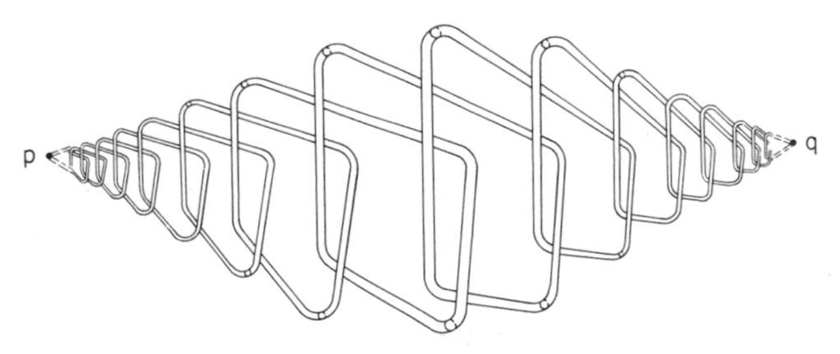

Figure 2.4.7

Example 2.4.5 (Fox and Artin). *The 2-sphere $f(\text{Bd}\, I^3)$ has a nonsimply connected complementary domain, hence is wild.*

Proof. Clearly $S^3 - X$ is of the same homotopy type as $S^3 - f(I^3)$ and since $\pi_1(S^3 - X)$ is nontrivial by Example **2.4.1**, it follows that $S^3 - f(I^3)$ is nonsimply connected.

As mentioned in the proof of Example **2.4.2**, it is easy to construct a homeomorphism $h: (I^3, [0, 1]) \to S^3$ such that $h([0, 1]) = Y$, where Y is the arc of Example **2.4.2**. Let $h(I^3)$ be as indicated in Fig. **2.4.8**.

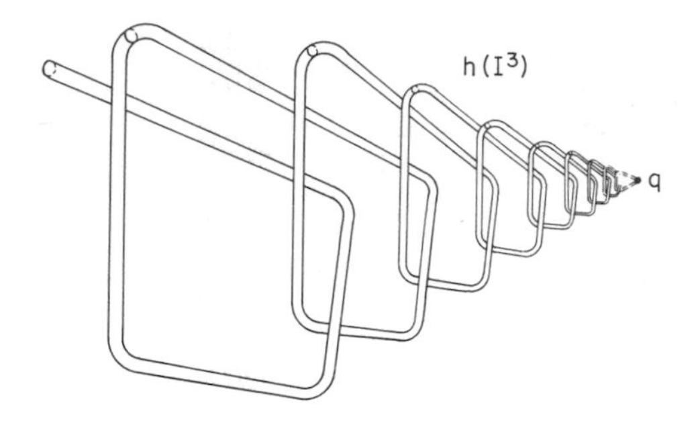

Figure 2.4.8

An $(n-1)$-sphere in S^n is said to be **weakly flat** if both of its complementary domains are open n-cells.

Example 2.4.6 (Fox and Artin). *The 2-sphere $h(\text{Bd}\, I^3)$ is weakly flat but wild (hence not flat).*

Example **2.4.6** can be established by making slight modifications in the proof of Example **2.4.2**.

Consider the arc $Z = Y \cup Y_r$ whose projection into the yz-plane is shown in Fig. **2.4.9**. (Think of Z as being polygonal modulo the endpoints, but having such a fine triangulation that you cannot distinguish the 1-simplexes.)

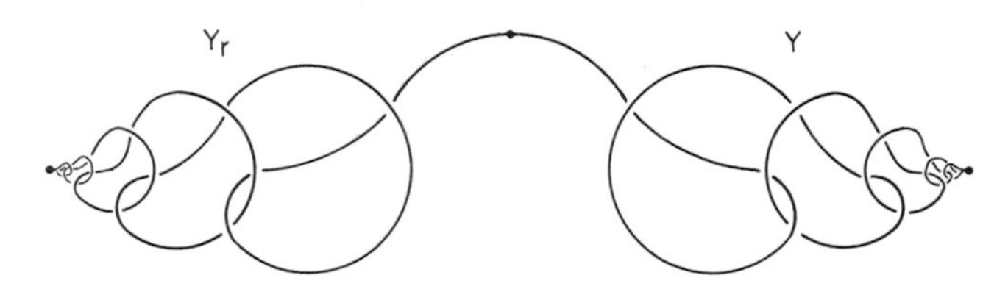

Figure 2.4.9

Example 2.4.7 (Fox and Artin). *The complement of the arc $Z = Y \cup Y_r$ is simply connected, but is not an open 3-cell. Hence Z is not cellular. However, the arcs Y and Y_r are cellular.*

The fact that Y and Y_r are cellular follows from Example **2.4.2**. The interested reader may refer to the paper of Fox and Artin [1] for the rest of the proof of this example.

The next example was given in [Harley, 1]. An example with similar properties for dimension four is given in [Glaser, 5] and for dimensions greater than four in [Glaser, 1]. Both Harley and Glaser show that the product of their examples with I is a cell. Other work related to these examples is mentioned at the end of Section **2.5**.

Example 2.4.8 (Harley). *There is a nonmanifold X such that*

(a) *X is the union of two 4-cells whose intersection is a 3-cell in the boundary of each, and*

(b) *X is the union of two 4-cells whose intersection is a 4-cell.*

Proof. Consider I^4 to be $I^3 \times I$. Put $D_1 = I^3 \times 0$. Then, Bd D_1 separates Bd I^4 into two 3-cells. Let $Y \cup Y_r$ be the arc of Example **2.4.7** embedded in Bd I^4 in such a way that one end-point lies in each complementary domain of Bd D_1 and $Y \cup Y_r$ intersects Bd D_1 in the common end-point of Y and Y_r (see Fig. **2.4.10**). Let $D_2 = I^3 \times [-1, 0]$ and

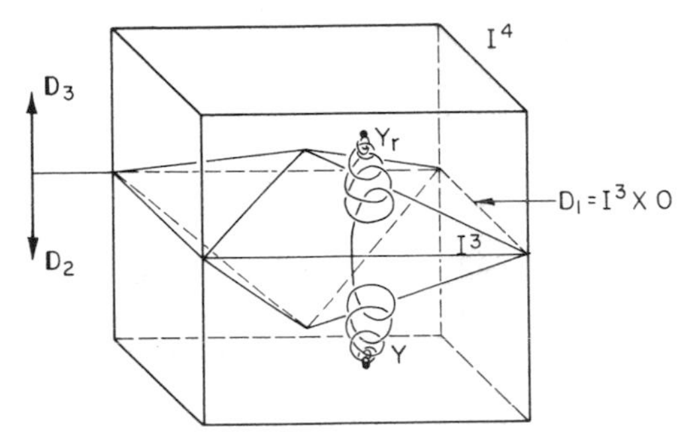

Figure 2.4.10

$D_3 = I^3 \times [0, 1]$ where the notation is chosen so that $Y \subset D_2$ and $Y_r \subset D_3$. Put $X = I^4/Y \cup Y_r$. Since $Y \cup Y_r$ is not cellular in Bd I^4, X is not a manifold. Since Y is cellular in Bd(D_2) and Y_r is cellular in Bd(D_3), both D_2/Y and D_3/Y_r are 4-cells. Hence (a) is satisfied. Let $\mathscr{S}(I^3)$ be the suspension of I^3 from the points $(0, 0, 0, \frac{1}{2})$ and $(0, 0, 0, -\frac{1}{2})$. Then, $(D_2 \cup \mathscr{S}(I^3))/Y$ and $(D_3 \cup \mathscr{S}(I^3))/Y_r$ are 4-cells satisfying (b).

Example 2.4.9 (Alexander). *There is a 2-sphere H* (**The Alexander horned sphere**) *in S^3 which has an open 3-cell for one complementary domain and which has a nonsimply connected 3-manifold for the other complementary domain. Also, this 2-sphere fails to be locally flat at exactly a Cantor set of points.*

The construction of Alexander's horned sphere H (originally given in [Alexander, 2]) is indicated in Fig. **2.4.11**. Perhaps the easiest way to show that the exterior complementary domain of H is not simply connected is to show that the simple closed curve J cannot be shrunk to a point in the exterior by employing Theorem 9 of [Bing, 8].

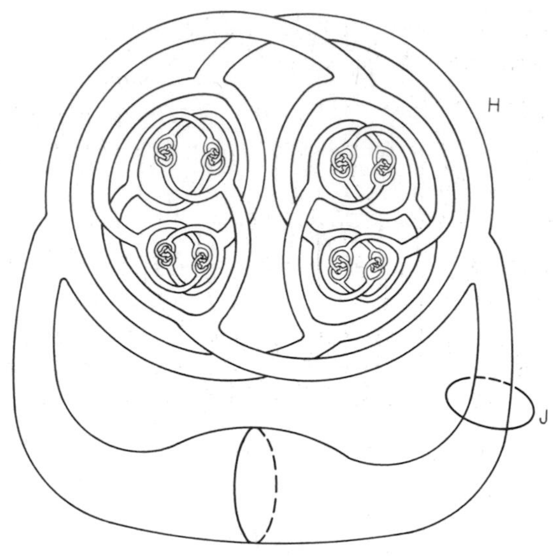

Figure 2.4.11

Another way to show that the exterior is nonsimply connected is to notice that it is homeomorphic to the complement of the pictured graph Γ (Fig. **2.4.12**) and to get a presentation for $\pi_1(S^3 - \Gamma)$ and

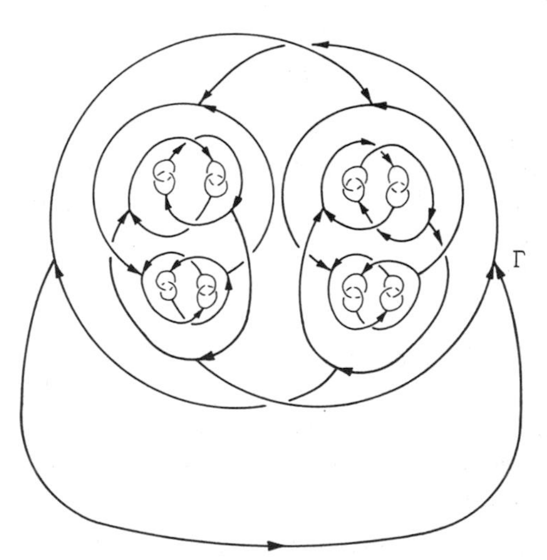

Figure 2.4.12

show the resulting group nontrivial. This is done in [Blankenship and Fox, 1].

A segment $\overline{axb}$ is said to **pierce** a sphere S at x if $S \cap \overline{axb} = x$ and a, b belong to different components of $E^3 - S$.

Example 2.4.10 (Fort). *There is a wild 2-sphere P (called the* **wild porcupine**) *which can be pieced at each point with a straight line segment.* (This example answered a question asked by Bing.)

The first such example was given in [Fort, 1]. The version of Example **2.4.10** given here is due to Bing [9]. This example is an "Alexander horned sphere," however, the construction is modified so that we will have the piercing property. The construction is indicated in Fig. **2.4.13.**

Figure 2.4.13

Example 2.4.11 (Antoine). *There is a Cantor set A in S^3 (called* **Antoine's necklace**) *which is not embedded equivalently to the standard Cantor set. In fact, $S^3 - A$ is not simply connected.*

The Cantor set we are about to construct first appeared in [Antoine, 1, 2]. Let T be a solid torus and let T_1, T_2, T_3, and T_4 be four solid tori embedded in T and linked as shown in Fig. **2.4.14.** (We could take any number of such tori but four makes a good picture.) In each T_i, let $T_1{}^i$, $T_2{}^i$, $T_3{}^i$, and $T_4{}^i$ be four solid tori embedded and linked in T_i as the T_i are embedded and linked in T. Continue inductively in this manner. At the ith stage, there are 4^i tori, whose union we call X_i. Then, **Antoine's necklace** A is $\bigcap_{i=1}^{\infty} X_i$. By the finite intersection property, A is nonempty. In fact, it can be shown that A is a totally disconnected,

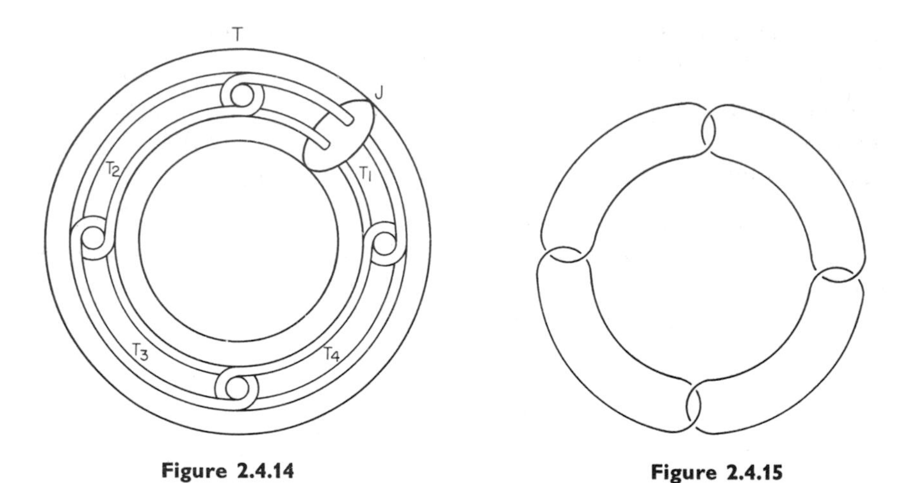

Figure 2.4.14 **Figure 2.4.15**

compact, perfect metric space, hence a Cantor set. It is shown in [Coelbo, 1] that the simple closed curve J cannot be shrunk to a point without hitting A. In [Blankenship and Fox, 1] a presentation of $\pi_1(T - A)$ is obtained by taking the direct limit of a sequence of graphs, the first of which is indicated in Fig. **2.4.15**. Then, a presentation for $\pi_1(S^3 - A)$ is easily obtained and shown to be nontrivial.

Example 2.4.12. *A 2-sphere which has the same properties of Example* **2.4.9** *can be constructed using Antoine's necklace.* (This was probably the first known wild 2-sphere.)

Spheres such as the one we are about to construct appeared in [Antoine, 2] and [Alexander, 3]. Let S_0 be a 2-sphere lying outside the torus T of Example **2.4.11**. Alter S_0 by removing a small disk and replacing the hole with a tube which runs to T and is capped off on Bd T. This yields a 2-sphere S_1. Cut four small holes in the disk $S_1 \cap \text{Bd } T$ and run tubes to T_1, T_2, T_3, and T_4 and cap them off in a similar manner forming S_2. By continuing in this fashion we construct a 2-sphere S $(= \lim S_i)$ which contains Antoine's necklace as indicated in Fig. **2.4.16**. The 2-sphere S is wild because since the simple closed curve J cannot be shrunk in the complement of A, it certainly cannot be shrunk in the complement of S.

Example 2.4.13. *There is a 2-sphere S (called* **Bing's Hooked Rug***) in S^3 and a simple closed curve J in $S^3 - S$ such that J cannot be shrunk to a point in $S^3 - S$, but for each disk Y in S, J can be shrunk to a point in $S^3 - Y$. Furthermore, each arc in S is tame.*

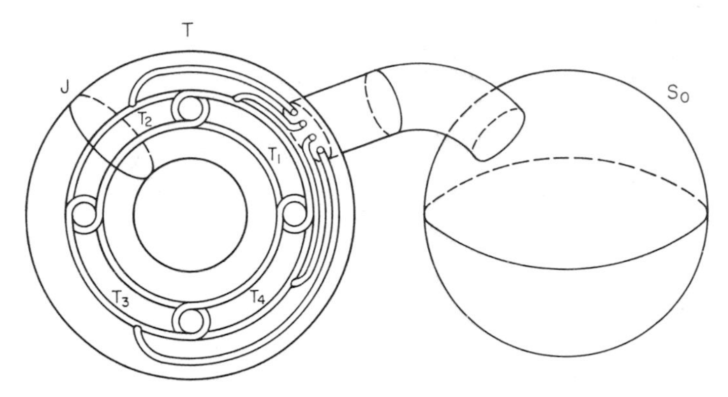

Figure 2.4.16

The construction of the hooked rug is suggested by Fig. **2.4.17**. Also indicated is the simple closed curve J satisfying the example. For a

Figure 2.4.17

complete proof the interested reader is referred to [Bing, 8]. (To read this paper is a nice way to satisfy one's geometrical appetite for the day.)

In [Alford, 1], Bing's construction is modified to produce uncountably many nonequivalent 2-spheres in S^3. Fig. **2.4.18** illustrates an Alford wild 2-sphere which has an arc of wild points.

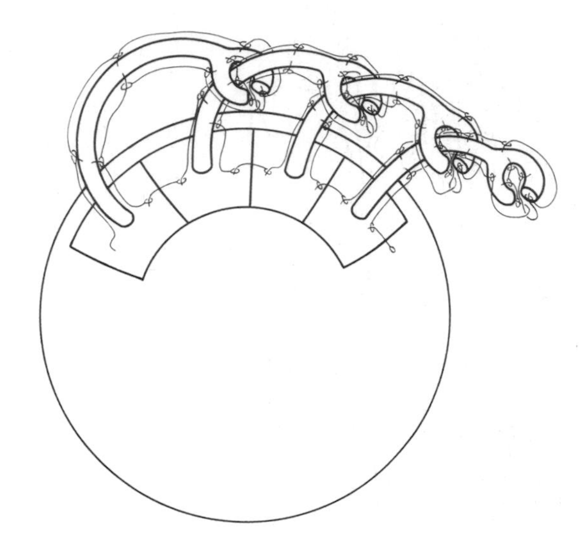

Figure 2.4.18

For a good survey paper on surfaces in E^3, the reader is referred to [Burgess and Cannon, 1].

2.5. E^n MODULO AN ARC CROSSED WITH E^1 IS E^{n+1}

Although the main objective of this chapter is to construct wild and knotted embeddings, we do not concern ourselves with such constructions in this section. However, we will use the main result of this section to construct wild cells and spheres in high dimensions in the next section. Furthermore, the results of this section are of independent interest. In 1958, Bing [12] gave an example of a nonmanifold (his **dogbone space**) whose product with E^1 is E^4. Andrews and Curtis [1] used Bing's technique of proof to show that E^n modulo an arc crossed with E^1 is E^{n+1}. First, we will establish a preliminary theorem that is a corollary to a result of Klee [1]. The rest of this section will be based on the papers of Bing and of Andrews and Curtis just mentioned.

Flattening Theorem 2.5.1 (Klee). *Every k-cell D in E^n is flat in*

$$E^{n+k} = E^n \times E^k.$$

PROOF. Since D is a k-cell, there is a homeomorphism f_* from the closed subset D of E^n onto $I^k \subset E^k$. Hence by Tietze's extension theorem

f_* extends to a map $f: E^n \to I^k \subset E^k$. Define $\phi_f: E^{n+k} \to E^{n+k}$ by $\phi_f(x, y) = (x, y + f(x))$ for $x \in E^n$, $y \in E^k$. Notice that ϕ_f is a homeomorphism of E^{n+k} onto itself which carries each "vertical" k-hyperplane onto itself. Since $D \subset E^n$, all points of D are of the form $(x, 0) \in E^{n+k}$ and so $\phi_f(x, 0)$ is of the form $(x, f(x))$. Thus, no two points of $\phi_f(D)$ have the same last k coordinates because f is one-to-one on D.

Let r be a retraction of E^k onto I^k and let $g: E^k \to E^n$ be defined by $g(y) = -f_*^{-1}r(y)$. Define $\psi_g: E^{n+k} \to E^{n+k}$ by $\psi_g(x, y) = (x + g(y), y)$ for $x \in E^n$, $y \in E^k$ (see Fig. 2.5.1). Notice that ψ_g is a homeomorphism

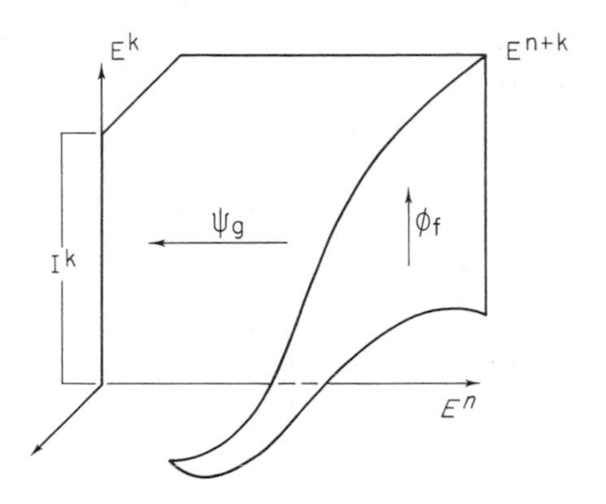

Figure 2.5.1

of E^{n+k} onto itself which takes each "horizontal" n-hyperplane onto itself. Now we will show that $\psi_g\phi_f: E^{n+k} \to E^{n+k}$ is such that $\psi_g\phi_f(D) = I^k \subset E^k \subset E^n \times E^k$. Suppose $(x, 0) \in D$. Then, $\psi_g\phi_f(x, 0) = \psi_g(x, f(x)) = (x + g(f(x)), f(x)) = (x + (-f_*^{-1}f(x)), f(x)) = (x - x, f(x)) = (0, f(x))$ where $f(x) \in I^k \subset E^k$. Since $f \mid D = f_*$ is a homeomorphism of D onto I^k, the proof is complete.

Factorization Theorem 2.5.2 (Bing). *Let $C = \bigcap_i T_i$ where each T_i is a compact neighborhood of C in E^n and $T_{i+1} \subset T_i$ for each i. Suppose that for each i and $\epsilon > 0$ there is an integer N and an isotopy μ_t of E^{n+1} onto E^{n+1} such that μ_0 = identity, μ_1 is uniformly continuous and*

(1) $\mu_t \mid (E^{n+1} - (T_i \times E^1)) = 1$,
(2) *μ_t changes $(n + 1)$st coordinates less than ϵ, and*
(3) *for each $w \in E^1$, diameter $\mu_1(T_N \times w) < \epsilon$.*

Then, $(E^n/C) \times E^1 \approx E^{n+1}$.

Proof. Consider the upper semicontinuous decomposition G' of E^{n+1} whose elements are of the form $g \times w$, where $g \in E^n/C$ and $w \in E^1$.

If the space corresponding to G' is B', then it is clear that $(E^n/C) \times E^1 \approx B'$. We will show that $(E^n/C) \times E^1$ is topologically E^{n+1} by showing that B' is topologically E^{n+1}. To do this it will suffice to show that there is a **pseudo-isotopy** f_t of E^{n+1} onto itself (that is, f_t is a homotopy such that f_t is a homeomorphism for $0 \leqslant t < 1$) such that $f_0 = 1$ and f_1 takes each element of G' onto a distinct point of E^{n+1}. We will obtain the pseudo-isotopy f_t by a sequence of applications of the isotopy μ_t in the hypothesis of Theorem **2.5.2**.

Let $\epsilon_1, \epsilon_2, \ldots$ be a sequence of positive numbers with a finite sum. We will define a monotone increasing sequence $\eta_1 = 1, \eta_2, \eta_3, \ldots$ of integers and a sequence of isotopies f_t^i, $i = 1, 2, \ldots$, $(i-1)/i \leqslant t \leqslant i/(i+1)$ of E^{n+1} such that

(1) $f_0^1 = 1$,
(2) $f_{i/(i+1)}^i = f_{i/(i+1)}^{i+1}$ and $f_{i/(i+1)}^i$ is uniformly continuous,
(3) $f_t^i \mid (E^{n+1} - (T_{\eta_i} \times E^1)) = f_{(i-1)/i}^i \mid (E^{n+1} - (T_{\eta_i} \times E^1))$,
(4) diameter $f_{i/(i+1)}^i(T_{\eta_{i+1}} \times w) < \epsilon_i$, $\quad w \in E^1$,
(5) f_t^i moves no point more than $2\epsilon_{i-1}$, and
(6) $f_{(i-1)/i}^i(E^n \times [w - \epsilon_i, w + \epsilon_i]) \supset f_{i/(i+1)}^i(E^n \times w)$, $\quad w \in E^1$.

Before constructing the sequence f_t^i, $i = 1, 2, \ldots$, we will show that the existence of such a sequence implies the existence of the pseudo-isotopy f_t that we seek.

Let f_t, $0 \leqslant t < 1$, be defined by $f_t(x) = f_t^i(x)$ if $(i-1)/i \leqslant t \leqslant i/(i+1)$. If we define f_1 by $f_1(x) = \lim_{t \to 1} f(x, t)$, then Condition 5 implies that f_1 is a map of E^{n+1} and so f_t, $0 \leqslant t \leqslant 1$, so defined is a pseudo-isotopy.

Condition 4 insures that $f_1(g)$ is a point for $g \in G'$.

Condition 3 implies that if $f_1(g_1) = f_1(g_2)$, $g_1, g_2 \in G'$, then one of g_1, g_2 is above the other in the w direction. This is the case since if g_1 and g_2 are in the same level then either g_1 or g_2 (say g_1) is a point of E^{n+1}, but then there is an integer i so large that $f_{(i-1)/i} \mid N = f_1 \mid N$ for some neighborhood N of g_1.

Finally, by employing Condition 6 one can show that no two points with different w coordinates go into the same point under f_1.

The existence of f_t^1 and η_2 follow directly from the hypothesis of Theorem **2.5.2**, and so we proceed inductively to define f_t^i and η_{i+1}. Let γ be a positive number so small that

$$\text{diameter } f_{(i-1)/i}^{i-1}(T_{\eta_i} \times [a, b]) < 2\epsilon_{i-1} \qquad \text{if} \quad |b - a| < \gamma.$$

The existence of such a γ follows from Condition 4 and the uniform continuity of $f^{i-1}_{(i-1)/i}$.

Let δ be a positive number so small that for each set S of diameter less than δ, diameter $f^{i-1}_{(i-1)/i}(S) < \epsilon_i$.

By the hypothesis of Theorem **2.5.2**, there is an isotopy μ_t, $(i-1)/i \leqslant t \leqslant i/(i+1)$, of E^{n+1} and an integer η_{i+1} such that

$$\mu_{(i-1)/i} = 1,$$

$$\mu_t \mid (E^{n+1} - (T_{\eta_i} \times E^1)) = 1,$$

$$\mu_t \text{ changes } (n+1)\text{st coordinates less than } \mathrm{Min}(\gamma, \epsilon_i),$$

$$\text{diameter } \mu_{i/(i+1)}(T_{\eta_{i+1}} \times w) < \delta, \qquad w \in E^1,$$

and

$$h_{i/(i+1)} \text{ is uniformly continuous.}$$

Now define, $f_t^i = f^{i-1}_{(i-1)/i}\, \mu_t$. Then, f_t^i clearly satisfies Conditions 1 and 2. It satisfies Condition 3 because $\mu_t \mid (E^{n+1} - (T_{\eta_i} \times E^1) = 1$. It satisfies Condition 4 because diameter $\mu_{i/(i+1)}(T_{\eta_{i+1}} \times w) < \delta$. It satisfies Condition 5 because

$$\text{diameter } f^{i-1}_{(i-1)/i}(T_{\eta_i} \times [a, b]) < 2\epsilon_{i-1} \qquad \text{if} \quad |\, b - a \,| < \gamma,$$

and μ_t changes $(n+1)$st coordinates less than γ. It satisfies Condition 6 because $\mu_{(i+1)/i}(E^n \times w) < E^n \times [w - \epsilon_i, w + \epsilon_i]$ and

$$f^i_{(i+1)/i}(E^n \times w) = f^{i-1}_{(i-1)/i}(\mu_{(i+1)/i}(E^n \times w)) \subset f^{i-1}_{(i-1)/i}(E^n \times [w - \epsilon_i, w + \epsilon_i]).$$

This completes the proof of Theorem **2.5.2**.

To show that (E^n modulo an arc α) $\times E^1$ is E^{n+1}, we will apply Theorem **2.5.2** where the set C is an arc. Thus, we must show how to construct the isotopies μ_t of the hypothesis. However, before doing this we will construct the neighborhoods T_i of α and some appropriate $(n+1)$-cells.

In this section let I denote $[0, 1] \subset E^1 \subset E^n \times E^1$. For $j = 4, 5, \ldots$ let P_j be the neighborhood of I consisting of a chain of $(n+1)$-cubes $P_j^1, P_j^2, \ldots, P_j^{j-1}$ of side length $(j+2)/j(j-1)$, P_j^1 having one face at $-1/j$ as in Fig. **2.5.2**. By Theorem **2.5.1**, there is a homeomorphism $\varphi\colon E^{n+1} \twoheadrightarrow E^{n+1}$ such that $\varphi(I) = \alpha$. Let $\varphi(P_j) = Q_j$ and $\varphi(P_j^\lambda) = Q_j^\lambda$. We denote the intersection of the Q's with the n-plane E^n by $Q_j \cap E^n = R_j$ and $Q_j^\lambda \cap E^n = R_j^\lambda$.

We can choose a subsequence $\{P_i\}$ of $\{P_j\}$, $j = 4, 5, \ldots$ such that

(a) For each i, diameter $Q_i^\lambda < 1/i$,

(b) For each i and each λ there is an s such that

$$Q_{i+1}^\lambda \subset (R_i^s \cup R_i^{s+1}) \times E^1,$$

(c) For each i and s, there is a λ such that $Q_{i+1}^\lambda \subset R_i^s \times E^1$, and if $\mu \leqslant \lambda$, then $Q_{i+1}^\mu \subset (R_i^1 \cup R_i^2 \cup \cdots \cup R_i^s) \times E^1$.

Conditions (a), (b), and the first part of (c) obviously may be satisfied. It is easy to show that the last part of (c) follows from the other conditions.

Let $T_i = R_{2i}$. Then $\{T_i\}$ will be the sequence of neighborhoods for α for which we will construct the isotopies required by Theorem **2.5.2**.

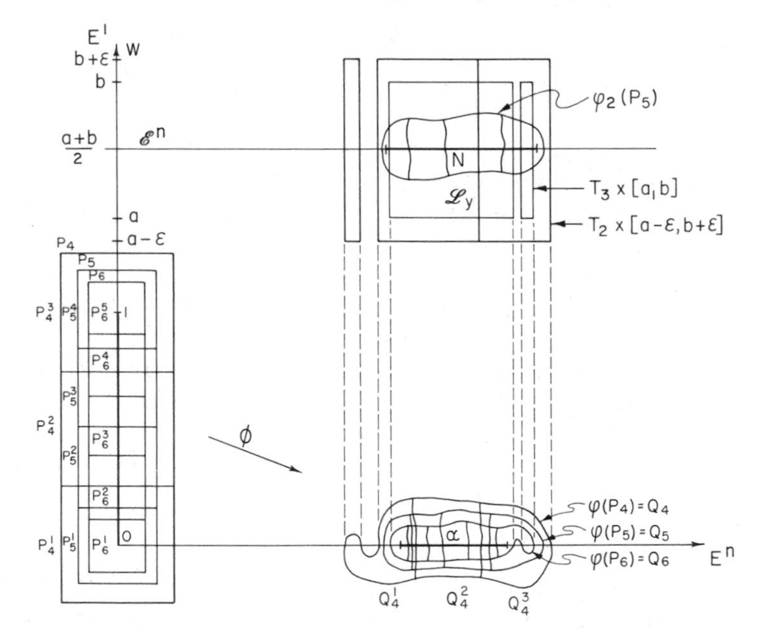

Figure 2.5.2

Lemma 2.5.1. *Given T_k and real numbers $\epsilon > 0$, $a < b$, there exists an $(n + 1)$-cell E such that*

$$T_k \times [a - \epsilon, b + \epsilon] \supset E \supset \operatorname{Int} E \supset T_{k+1} \times [a, b].$$

PROOF. Recall that $T_k = \varphi(P_{2k}) \cap E^n$ and $T_{k+1} = \varphi(P_{2k+2}) \cap E^n$. The $(n + 1)$-cell E will be obtained as a homeomorphic image of the $(n + 1)$-cell P_{2k+1}. Let w be the E^1 coordinate for $E^{n+1} = E^n \times E^1$.

Let φ_1 be obtained from φ by adding $(a + b)/2$ to the w-coordinate of each image point. Let φ_2 be obtained from φ_1 by compressing the w coordinate toward $w = (a + b)/2$ sufficiently so that

$$\varphi_2(P_{2k+1}) \subset T_k \times (a - \epsilon, b + \epsilon).$$

This is possible because of Condition (b) on the Q's and R's. Let $\mathscr{E}^n$ be the n-plane $w = (a + b)/2$, and let N be a nice neighborhood of $\varphi_2(T_{k+1})$ in $\mathscr{E}^n$ with N in $\varphi_2(P_{2k+1})$ (see Fig. **2.5.2**). For each $y \in T_{k+1}$, let $\mathscr{L}_y$ be the vertical line segment from $(y, a - \epsilon)$ to $(y, b + \epsilon)$.

Let d be the minimum vertical distance from $T_{k+1} \times ((a + b)/2)$ to the complement of $\varphi_2(P_{2k+1})$. Let θ stretch the part of $\mathscr{L}_y$ $(y \in T_{k+1})$ between $(y, ((a + b)/2) - d)$ and $(y, ((a + b/2) + d)$ to the part between (y, a) and (y, b) and compress the rest of $\mathscr{L}_y$, leaving the end-points fixed. Then, θ can be extended to an isotopy which moves points only vertically and is the identity outside of $N \times [a - \epsilon, b + \epsilon]$. The cell $E = \theta\varphi_2(P_{2k+1})$ satisfies the lemma.

Lemma 2.5.2. *Given T_k and any integer $m > 2$, there exists a sequence $E_1, E_2, \ldots, E_{m-2}$ of $(n + 1)$-cells such that*

$$(T_k \times [0, 2m - 3]) \supset E_1 \supset (T_{k+1} \times [1, 2m - 4]) \supset \cdots$$
$$\supset E_{m-2} \supset (T_{k+m-2} \times [m - 2, m - 1]).$$

Furthermore, each E_r may be written as the union of cells $U_1^r, U_2^r, \ldots, U_p^r$ such that $U_1^r \subset U_2^r \subset \cdots \subset U_p^r = E_r$ and each U_s^r

(a) *is the union of cells $\theta\varphi_2(P^{\mu}_{2(k+r)-1}$ where $\mu \leqslant \lambda_s^r$ and λ_s^r is such that*

$$Q^{\lambda_s^r}_{2(k+r)-1} \subset R^s_{2k} \times E^1$$

and

(b) $U_s^r \subset (R^1_{2k} \cup R^2_{2k} \cup \cdots \cup R^{s+1}_{2k}) \times E^1.$

Proof. The existence of the sequence $E_1, E_2, \ldots, E_{m-2}$ follows immediately from Lemma **2.5.1**. The cell E of Lemma **2.5.1** may be written as the union of cells $\theta\varphi_2(P^{\mu}_{2k+1})$. Also by Condition (c), we may write E as the union of cells $U_1, U_2, \ldots, U_p$ such that

$$U_1 \subset U_2 \subset \cdots \subset U_p = E$$

and each U_s is the union of cells $\theta\varphi_2(P^{\mu}_{2k+1})$ where $\mu \leqslant \lambda_s$ and λ_s is such that

$$Q^{\lambda_s}_{2k+1} \subset R^{s+1}_{2k} \times E^1 \subset T_k \times E^1.$$

Hence, we see that

$$U_s \subset (R^1_{2k} \cup R^2_{2k} \cup \cdots \cup R^{s+1}_{2k}) \times E^1,$$

because $\theta\varphi_2$ moves points only vertically after the homeomorphism φ. Thus, we can decompose E_r as indicated in Lemma **2.5.2**.

Lemma 2.5.3. *Let A be a closed n-cell which is the union of two n-cells A_1 and A_2 such that $A_1 \cap A_2 = \dot{A}_1 \cap \dot{A}_2$ is an $(n-1)$-cell. Let B be a closed subset of A such that $B \cap \dot{A} \subset A_2$. Then, there is an isotopy $\lambda\colon A \times I \to A$ such that λ_0 is the identity, $\lambda_t \mid \dot{A}$ is the identity, and $\lambda_1(B) \subset A_2$* (see Fig. **2.5.3**).

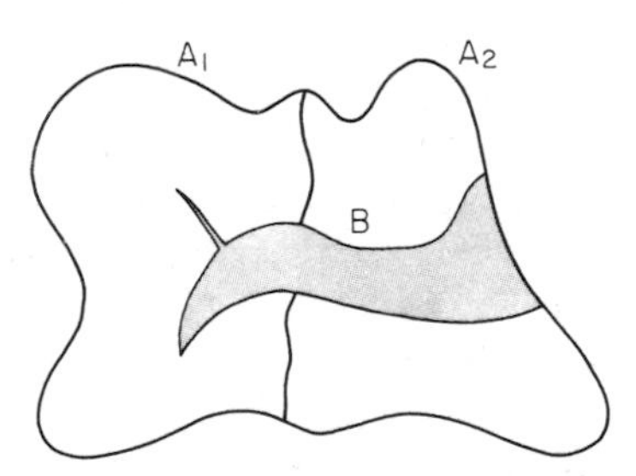

Figure 2.5.3

Proof. Let p be a point in $\dot{A}_2 - A_1$. Let h be a homeomorphism of A onto the closed unit ball B in E^n. Let γ be an isotopy stretching the rays toward $h(p)$ with γ_0 the identity, and $\gamma_1(h(B)) \subset h(A_2)$. Then, $h^{-1}\gamma h$ is the required isotopy.

Lemma 2.5.4. *Let*

$$L_0 = [0, 1] \cup [2m-4, 2m-3],$$
$$L_1 = [1, 2] \cup [2m-5, 2m-4], \ldots, L_{m-3} = [m-3, m].$$

Also,

$J_0 = [0, 2m-3]$, $J_1 = [1, 2m-4], \ldots, J_{m-3} = [m-3, m]$. *Let $\{T_i\}$ be the sequence of neighborhoods of α previously constructed and let $T_k \in \{T_i\}$, with $C_1, \ldots, C_m$ being the chain of R_{2k}^{λ}'s in T_k. Then, there is an isotopy of E^{n+1} starting with the identity and ending with a homeomorphism h of E^{n+1} onto itself such that*

$$h = 1 \text{ outside of } T_k \times J_0,$$
$$h = 1 \text{ on } (C_3 \cup C_4 \cup \cdots \cup C_m) \times L_0,$$
$$h = 1 \text{ on } (C_4 \cup C_5 \cup \cdots \cup C_m) \times L_1,$$
$$\vdots$$
$$h = 1 \text{ on } C_m \times L_{m-3},$$

and

$$h((T_k \cap (C_1 \cup C_2)) \times L_0) \subset (C_1 \cup C_2) \times J_0,$$
$$h((T_{k+1} \cap (C_1 \cup C_2 \cup C_3)) \times L_1) \subset (C_2 \cup C_3) \times J_0,$$
$$\vdots$$
$$h((T_{k+m-3} \cap (C_1 \cup C_2 \cup \cdots \cup C_{m-1})) \times L_{m-3}) \subset (C_{m-2} \cup C_{m-1}) \times J_0.$$

PROOF. We will just construct the homeomorphism h as

$$h = h_{m-3}h_{m-4} \cdots h_1,$$

and since each h_i will be obtainable by an isotopy starting at the identity, so will h.

Let E_1 be broken into $(n+1)$-cells $U_1{}^1, U_2{}^1, \ldots, U^1_{p_1}$ as in Lemma **2.5.2** (see Fig. **2.5.4**). Let $B = (T_{k+1} \times J_1) \cap U_1{}^1$, $A = U_1{}^1$,

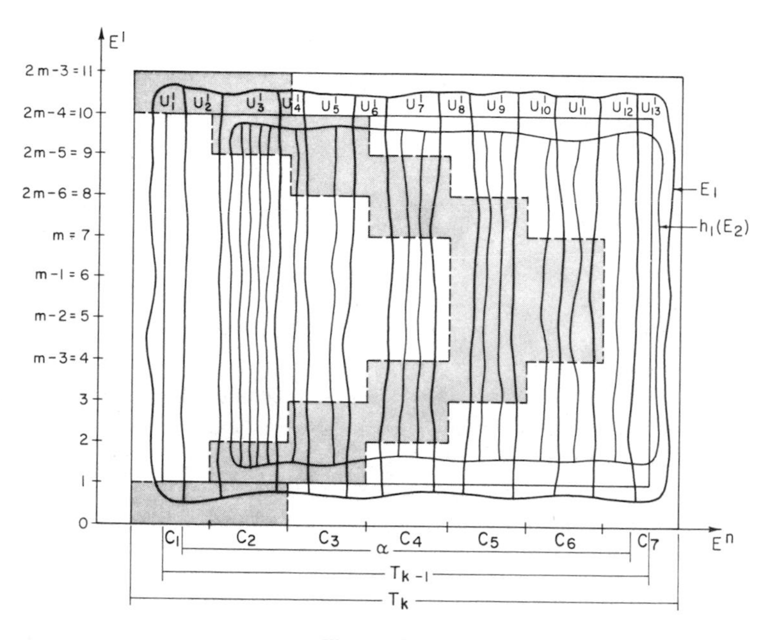

Figure 2.5.4

$A_2 = \theta\varphi_2(P^{\lambda_1{}^1}_{2k+1})$, and $A_1 = \mathrm{Cl}(A - A_2) = \bigcup_{\mu<\lambda_1{}^1} \theta\varphi_2(P^{\mu}_{2k+1})$. Take h_1 to be the homeomorphism given by Lemma **2.5.3**. We note that

$$h_1 = 1 \text{ outside } U_1{}^1 \subset (C_1 \cup C_2) \times J_0,$$
$$h_1 = 1 \text{ on } (C_3 \cup \cdots \cup C_m) \times J_1,$$

and

$$h_1((T_{k+1} \cap (C_1 \cup C_2)) \times J_1) \subset \theta\varphi_2(P^{\lambda_1^1}_{2k+1}) \subset C_2 \times J_0 .$$

Now let E_2 be broken into $(n + 1)$-cells $U_1^2, U_2^2, ..., U^2_{p_2}$ as in Lemma **2.5.2**. Then, $h_1(E_2)$ is broken into $(n + 1)$-cells

$$h_1(U_1^2), h_1(U_2^2), ..., h_1(U^2_{p_2}).$$

Again we apply Lemma **2.5.3** with $B = h_1((T_{k+2} \times J_2) \cap U_2^2)$, $A = h_1(U_2^2)$, $A_2 = h_1\theta\varphi_2(P^{\lambda_2^2}_{2k+3})$ and $A_1 = \mathrm{Cl}(A - A_2) = \bigcup_{\mu<\lambda_2^2} h_1\theta\varphi_2(P^{\mu}_{2k+2})$.

Hence we have a homeomorphism h_2 such that

$$h_2 = 1 \text{ outside of } h_1(U_2^2) \subset (C_2 \cup C_3) \times J_0 ,$$

$$h_2 = 1 \text{ on } (C_4 \cup \cdots \cup C_m) \times J_0 ,$$

$$h_2h_1((T_{k+2} \cap (C_1 \cup C_2 \cup C_3)) \times J_2) \subset C_3 \times J_0 .$$

Now $h_2h_1 = h_1$ outside $h_1^{-1}(U_2^2)$, and $h_1^{-1}(U_2^2) \cap (E^n \times L_0) = \emptyset$ by Lemma **2.5.3** so that we have

$$h_2h_1((T_k \cap (C_1 \cup C_2)) \times L_0) = h_1((T_k \cap (C_1 \cup C_2)) \times L_0) \subset (C_1 \cup C_2) \times J_0 .$$

The remaining h_j's are constructed similarly, and $h = h_{m-3}h_{m-4} \cdots h_1$ satisfies the theorem.

REMARK 2.5.1. Intuitively we may view the proof of Lemma **2.5.4** as follows: Suppose we have a collection of $m - 3$ worms, and suppose one of the worms swallows another worm tail first, and is then himself swallowed tail first by a third worm. Let the feast continue in this fashion until there is only one worm remaining. Now the fattest worm pulls in his tail a little way and in doing so he pulls in the tails of all of the worms inside him. Next, the second most fat worm pulls in his tail, consequently pulling in the tails of all of the worms inside him. This is continued until the innermost worm, although very cramped, pulls in his tail so as to give us h.

Lemma 2.5.5. *Let $T_k \in \{T_i\}$ have chambers $C_1, ..., C_m$ (that is, $C_\mu = R_{2k}$). Then, there is an isotopy of E^{n+1} starting with the identity and ending with a uniformly continuous homeomorphism φ of E^{n+1} onto itself such that*

(1) $\varphi = 1$ *outside* $T_k \times E^1$, and

(2) *For $w \in E^1$ there is an integer i such that*

$$\varphi(T_{k+m-3} \times w) \subset (C_i \cup C_{i+1} \cup C_{i+2} \cup C_{i+3}) \times [w - 2m + 3, w + 2m - 3].$$

PROOF. Figure **2.5.5** shows how to apply Lemma **2.5.4** to get Lemma **2.5.5**.

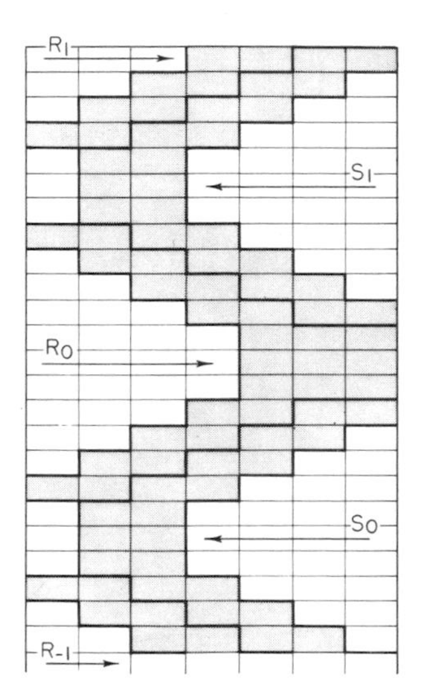

Figure 2.5.5

A point x of $T_{k+m-3} \times E^1$ represented in the regions R_j ($j = 0, 1, -1, 2, -2, \ldots$) is moved to the right so that its image is in $(C_i \cup C_{i+1}) \times E^1$ where C_i, C_{i+1} correspond to the columns containing the dotted rectangles in R_j on the same horizontal line with x. The w coordinate of x is not changed by more than $2m - 3$ in this adjustment. Similarly, points in the regions S_j ($j = 0, 1, -1, 2, -2, \ldots$) are moved to the left.

Factorization Theorem 2.5.3 (Andrews and Curtis). *Let α be an arc in E^n. Then, $(E^n/\alpha) \times E^1 \approx E^{n+1}$.*

PROOF. Let $\{T_i\}$ be the sequence of neighborhoods of α such that $\alpha = \bigcap_i T_i$ previously constructed. We will show that this sequence satisfies Theorem **2.5.2**. Thus, given $T_k \in \{T_i\}$ and $\epsilon > 0$ we must find an integer N and a homeomorphism φ such that φ changes w coordinates less than ϵ, φ is the identity outside $T_k \times E^1$, and, for each w, diameter $\varphi(T_N \times w) < \epsilon$. By changing the w scale of the homeomorphism φ of Lemma **2.5.5** so that $2m - 3 < \epsilon/2$, we satisfy the condition on the change of w coordinate. Finally, by choosing N so that no four consecutive chambers of T_{N-1} has diameter greater than $\epsilon/2$, we see that φ is the desired homeomorphism.

EXERCISE 2.5.1. Let α be an arc in S^n. Then, $\mathscr{S}(S^n/\alpha) \approx S^{n+1}$.

For the interested reader we will mention some results which are related to Theorem **2.5.3**. Bryant [1] has shown that the arc α of Theorem **2.5.3** can be replaced with certain cells. In [Bing, 2] it is shown that a set is a 3-cell if its product with I is a 4-cell. Poenaru [1] and Mazur [1] have given examples of PL 4-manifolds different from I^4 whose products with I are cells, and Curtis [1] and Glaser [1] have given similar examples for $n \geqslant 4$. Kwun and Raymond [1] have shown that the product of I^n modulo an arc in its interior and I^2 is I^{n+2}. Finally, in [Harley, 1] it is shown that if α is an arc in the boundary of I^n, then $(I^n/\alpha) \times I$ is I^{n+1}. (In particular, Harley shows that the product of Example **2.4.8** with I is a 5-cell.) Although we do not present an in-depth study of the decomposition theory of manifolds in this book, it is an important part of geometrical topology. Good references on the subject are [Armentrout, 1] and [Siebenmann, 2].

2.6. EVERYWHERE WILD CELLS AND SPHERES IN $E^{n \geqslant 3}$ OF ALL CODIMENSIONS

This section consists of some of the results of [Rushing, 1]. Wild cells of all possible codimensions in E^n, $n \geqslant 3$, were first constructed in [Blankenship, 1]. [Brown, 1] makes a nice application of Theorem **2.5.3** to construct wild cells and spheres of all possible codimensions in E^n. Although Brown did not go into detail on his suggested method for producing wild codimension two spheres, such spheres had previously been given in [Cantrell and Edwards, 1]. The work of Cantrell and Edwards was carried further in [Tindell, 1] to obtain some interesting wild embeddings in codimension two. Previous to the work of this section, everywhere wild arcs had been constructed in E^3 in [Bing, 1] and in [Alford, 1] and everywhere wild arcs had been constructed in E^n, $n \geqslant 4$ [Brown, 1]. (For applications of this section see [Seebeck, 2], [Sher, 2] and [Cantrell, Price, and Rushing, 1].) Our main result is the following theorem.

Theorem 2.6.1. *In E^n, $n \geqslant 3$, there are cellular, everywhere wild cells and everywhere wild spheres of all codimensions between* 0 *and* n.

Thus, it is immediate that there are also closed, everywhere wild strings and half-strings of all codimensions.

An **n-string (n-half-string)** is a set which is homeomorphic to E^n (E^n_+). A topological k-manifold M in E^n is **locally tame** at $x \in M$ if there is a neighborhood U of x in E^n and a homeomorphism of U onto E^n which carries $U \cap M$ onto a subpolyhedron of E^n. M is **wild** at $x \in M$ if it is not locally tame at x, and M is **everywhere wild** if it is wild at every point.

Let $A \subset X \subset Y$ be topological spaces. The set $Y - X$ is **projectively 1-connected** at A if each neighborhood U of A contains a neighborhood V of A such that each loop in $V - X$ is null homotopic in $U - X$. (Projective 1-connectivity of $Y - X$ at X is the same as the cellularity criterion for X discussed in Section **4.8**.) In the case that A is a point, we say that $Y - X$ is **1-LC** at A. Let $X \subset Y$ be topological spaces. Then, $Y - X$ is said to be **1-SS (1-short shrink)** at $x \in X$ if for every neighborhood U of x there is a neighborhood $V \subset U$ of x such that every loop in $V - X$ which is null homotopic in $Y - X$ is also null homotopic in $U - X$.

The following lemma originated in [Cantrell, Price, and Rushing, 1].

Lemma 2.6.1. (a) *Let $\Sigma^{n-2} \subset S^n$ (E^n) be an $(n-2)$-sphere which is locally flat at a point x. Then, $S^n - \Sigma^{n-2}$ ($E^n - \Sigma^{n-2}$) is* 1-SS *at x.*

(b) *Let $X^{n-2} \subset E^n$ be a closed $(n-2)$-string which is locally flat at a point x. Then, $E^n - X^{n-2}$ is* 1-SS *at x.*

PROOF. We will establish only the case $\Sigma^{n-2} \subset S^n$ is an $(n-2)$-sphere which is locally flat at x, because the proofs of the other cases are similar. Let U be any neighborhood of x and let $V \subset U$ be a flattening cell neighborhood for Σ^{n-2} at x, that is $(V, V \cap \Sigma^{n-2}) \approx (I^n, I^{n-2})$. Let l be a loop in $V - \Sigma$ which is null homotopic in $S^n - \Sigma$. By pushing radially away from x, we see that l is homotopic in $V - \Sigma$ to a loop l' in $\mathrm{Bd}\, V - \Sigma$ which is null homotopic in $S^n - (\mathrm{Int}\, V \cap \Sigma)$. The proof will be complete if we can show that l' is null homotopic in $\mathrm{Bd}\, V - \Sigma$. Since we know that l' is null homotopic in $S^n - (\mathrm{Int}\, V \cup \Sigma)$, it will suffice to show that the injection

$$\pi_1(\mathrm{Bd}\, V - \Sigma) \to \pi_1(S^n - (\mathrm{Int}\, V \cup \Sigma))$$

is a monomorphism. In order to do this consider the following Mayer-Vietoris sequence [Spanier, 1, pp. 186–190]:

$$\cdots \to H_2(S^n - (\Sigma - \mathrm{Int}\, V)) \to H_1(\mathrm{Bd}\, V - \Sigma)$$
$$\to H_1(S^n - (\mathrm{Int}\, V \cup \Sigma)) \oplus H_1(V - (\mathrm{Bd}\, V \cap \Sigma))$$
$$\to H_1(S^n - (\Sigma - \mathrm{Int}\, V)) \to \cdots.$$

Using Alexander duality [Spanier, 1, p. 296] on this sequence yields

$$\cdots \to 0 \to Z \to Z \oplus 0 \to 0 \to \cdots.$$

Hence, the inclusion of $\text{Bd } V - \Sigma$ into $S^n - (\text{Int } V \cup \Sigma)$ induces an isomorphism on first homology. But now any loop l in $\text{Bd } V - \Sigma$ which is null homotopic in $S^n - (\text{Int } V \cup \Sigma)$ is also null homologous in $S^n - (\text{Int } V \cup \Sigma)$, consequently null homologous in $\text{Bd } V - \Sigma$. Since $\pi_1(\text{Bd } V - \Sigma)$ is Abelian, it follows [Hu, 1, pp. 44–47] that l is null homotopic in $\text{Bd } V - \Sigma$ and so the injection

$$\pi_1(\text{Bd } V - \Sigma) \to \pi_1(S^n - (\text{Int } V \cup \Sigma))$$

is a monomorphism as desired.

Lemma 2.6.2. *Let $A \subset X \subset Y$ be topological spaces such that $Y - X$ is not 1-SS at any point of A. Suppose $W \subset Z \subset E^1$, where W is open in E^1, and suppose $R \subset Y \times (E^1 - W)$. Then, $(Y \times E^1) - ((X \times Z) \cup R)$ is not 1-SS at any point of $A \times W$.*

PROOF. Suppose that $(Y \times E^1) - ((X \times Z) \cup R)$ is 1-SS at some point $p = (a, t)$ of $A \times W$. Let U be any neighborhood of a in Y. Then, $U \times W$ is a neighborhood of p in $Y \times E^1$. Therefore, there is a neighborhood $V_* \subset U \times W$ of p such that every loop in $V_* - ((X \times Z) \cup R)$ which is null homotopic in $(Y \times E^1) - ((X \times Z) \cup R)$ is null homotopic in $(U \times W) - ((X \times Z) \cup R)$. Choose a neighborhood V of a in Y and an $\epsilon > 0$ such that

$$p = (a, t) \in V \times (t - \epsilon, t + \epsilon) \subset V_* .$$

By our construction, it is easy to see that every loop in $V - X$ which is null homotopic in $Y - X$ is also null homotopic in $U - X$ and so it follows that $Y - X$ is 1-SS at a which is a contradiction.

Although a manifold $M \subset E^n$ of codimension two may be locally tame at a point $x \in M$ and yet fail to be locally flat at x, the following lemma is easily established.

Lemma 2.6.3. *If a manifold $M \subset E^n$ fails to be locally flat at every point, then M is everywhere wild.*

Example 2.6.1. *For $n \geqslant 3$, there is an $(n-2)$-cell $F^{n-2} \subset E^n$ which lies on the boundary of an $(n-1)$-cell D^{n-1} such that $D^{n-1} - F^{n-2}$ is locally flat and such that $E^n - \text{Bd } D^{n-1}$ fails to be 1-SS at every point of $\text{Int } F^{n-2}$. (Hence, by Lemma* **2.6.1**, *F^{n-2} fails to be locally flat at every point and so is everywhere wild by Lemma* **2.6.3**.*) Furthermore, F^{n-2} is cellular in E^n.*

Following Example **2.4.13**, we indicated a construction of Alford (given more completely in [Alford, 1]) of a wild 2-sphere S in E^3 whose set of wild points is an arc F^1. Let D^2 be a 2-cell in S which has F^1 on its boundary. Then, $D^2 - F^1$ is locally flat. It follows from the construction of S that $E^3 - S$ is not 1-SS at any point of $\operatorname{Int} F^1$. In particular, there is a neighborhood U of each $x \in \operatorname{Int} F^1$ such that for any neighborhood $V \subset U$ of x there is a loop in $V - S$ which is null homotopic in $E^3 - S$, but not in $U - S$. For a fixed $x \in \operatorname{Int} F^1$, let $B \subset U$ be a flattening open disk neighborhood of $\operatorname{Bd} D^2$ in S at x, that is, $(B, B \cap \operatorname{Bd} D^2) \approx (E^2, E^1)$. Now let $U' \subset U$ be a neighborhood of x in E^3 such that $U' \cap S \subset B$ and suppose that $E^3 - \operatorname{Bd} D^2$ is 1-SS at x. Then, there is a neighborhood $V \subset U'$ such that each loop in $V - \operatorname{Bd} D^2$ which is null homotopic in $E^3 - \operatorname{Bd} D^2$ is also null homotopic in $U' - \operatorname{Bd} D^2$. Let l: $\operatorname{Bd} I^2 \to V - S$ be a loop which is null homotopic in $E^3 - S$, but not in $U - S$. By the above assumption, there is an extension f: $I^2 \to U' - \operatorname{Bd} D^2$ of l. Clearly, there are two closed disks D_+ and D_- in $B - \operatorname{Bd} D^2$ such that $f(I^2) \cap S \subset D_+ \cup D_-$. Let G denote the complementary domain of S in E^3 which contains $l(\operatorname{Bd} I^2)$. Let X denote the component of $f^{-1}(G)$ which contains $\operatorname{Bd} I^2$ and consider the components of $I^2 - X$. Let A_+ be the union of all of those components having frontiers whose images are contained in D_+ and let A_- be the union of all of those components having frontiers whose images are contained in D_-. [By unicoherence (see Theorem 5.19 on p. 60 of [Wilder, 2]) those frontiers are connected and so their images are contained in either D_+ or D_-.] Then, by Tietze's extension theorem $f \mid A_+ \cap f^{-1}(S)$ can be extended to a map f_+: $A_+ \to D_+$ and $f \mid A_- \cap f^{-1}(S)$ can be extended to a map f_-: $A_- \to D_-$. Redefine f to be f_+ on A_+ and f_- on A_-. By using a collar of D_+ and D_- in $\operatorname{Cl}(G \cap u)$ (which exist since D_+ and D_- are locally flat), we can "pull in" f to obtain f_*: $I^2 \to U - S$ and so l would be null homotopic in $U - S$ which is a contradiction. Hence, $E^3 - \operatorname{Bd} D^2$ is not 1-SS at x. Consequently, F^1 is not locally flat at x by Lemma **2.6.1** and so F^1 is everywhere wild by Lemma **2.6.3.** (In [Gillman, 1], it is shown that F^1 does not pierce any disk which also shows it everywhere wild. (It is easy to see that F^1 satisfies the cellularity criterion of [McMillan, 1] (which we shall discuss in Chapter **4**) and is thus cellular. (Gillman [1, 3] observed that F^1 is cellular and Alford [2] did the same.)

Inductively, assume that there is an $(n-2)$-cell $F^{n-2} \subset E^n$ which lies on the boundary of an $(n-1)$-cell D^{n-1} such that $D^{n-1} - F^{n-2}$ is locally flat and such that $E^n - \operatorname{Bd} D^{n-1}$ fails to be 1-*SS* at every point of Int F^{n-2}. Let $F^{n-1} = F^{n-2} \times [-1, 1]$, and $D^n = D^{n-1} \times [-1, 1]$. Then, $F^{n-1} \subset \operatorname{Bd} D^n \subset E^{n+1} = E^n \times E^1$. Since $D^{n-1} - F^{n-2}$ is locally flat in E^n, it follows that $D^n - F^{n-1}$ is locally flat in E^{n+1}. Since

$$\operatorname{Bd} D^n = (\operatorname{Bd} D^{n-1} \times [-1, 1]) \cup (D^{n-1} \times \{-1, 1\}),$$

Lemma **2.6.2** implies that $E^{n+1} - \text{Bd}\, D^n$ is not 1-SS at any point of $\text{Int}\, F^{n-1}$. It is a consequence of the next lemma that F^{n-1} is cellular in E^{n+1}.

Lemma 2.6.4. *If $X \times E^1 \approx E^{n+1}$ and $A \subset X$ is cellular in E^{n+1}, then $A \times [-1, 1]$ is also cellular in E^{n+1}.*

EXERCISE 2.6.1. Prove Lemma **2.6.4**.

Example 2.6.2. *For $n \geqslant 3$, there is a closed $(n-2)$-string X^{n-2} in E^n such that $E^n - X^{n-2}$ fails to be 1-SS at every point of X^{n-2}. (Hence, by Lemma **2.6.1** and Lemma **2.6.3**, X^{n-2} is everywhere wild.) Furthermore, X^{n-2} is the boundary of a closed $(n-1)$-half-string Y^{n-1} such that $Y^{n-1} - X^{n-2}$ is locally flat in E^n.*

Do Alford's construction a countable number of times on $E^2 \subset E^3$ so as to make each interval $[n, n+1]$, n an integer, the wild arc. (In carrying out the construction on $[n, n+1]$, run the hooks from n toward $n+1$.) The resulting 1-string X^1 obviously lies on a closed 2-string W^2 such that $E^3 - W^2$ fails to be 1-SS at every point of X^1. By an argument similar to that following Example **2.6.1**, we see that $E^3 - X^1$ fails to be 1-SS at every point of X^1. It is clear that X^1 bounds a closed 2-half-string Y^2 such that $Y^2 - X^1$ is locally flat in E^3. The fact that $X^{n-1} = X^{n-2} \times E^1$ and $Y^n = Y^{n-1} \times E^1$ satisfy Example **2.6.2** in dimension $n+1$, follows from Lemma **2.6.2**.

Example 2.6.3. *For $n \geqslant 3$, there is an everywhere wild $(n-2)$-sphere Σ^{n-2} in E^n. Furthermore, Σ^{n-2} lies on the boundary of an $(n-1)$-cell D^{n-1} such that $D^{n-1} - \Sigma^{n-2}$ is locally flat in E^n.*

To get Example **2.6.3**, simply one-point compactify the triple (X^{n-2}, Y^{n-1}, E^n) of Example **2.6.2** to obtain $(\Sigma^{n-2}, D^{n-1}, S^n)$ and then remove a point not on D^{n-1}.

Example 2.6.4. *For $n \geqslant 3$, there is an arc α_n in E^n such that $E^n - \alpha_n$ is not projectively 1-connected at α_n.*

The following construction is based on some work of [Brown, 1]. If α_3 is the arc of Example **2.4.1** then $S^3 - \alpha_3$ is neither 1-connected nor projectively 1-connected at α_3. Suppose that $\alpha_n \subset S^n$ is an arc such that $S^n - \alpha_n$ is neither 1-connected nor projectively 1-connected at α_n, and let $\alpha_{n+1} = \mathscr{S}(\alpha_n) \subset \mathscr{S}(S^n/\alpha_n)$. ($\mathscr{S}$ denotes suspension.) It follows from Exercise **2.5.1** that $\mathscr{S}(S^n/\alpha_n) \approx S^{n-1}$. Also, since $S^n - \alpha_n$ is not 1-connected, $\mathscr{S}(S^n/\alpha_n) - \mathscr{S}(\alpha_n)$ is not 1-connected. If S^{n+1} were projectively 1-connected at α_{n+1}, there would be a neighborhood

V of α_{n+1} in S^{n+1} such that every loop in $V - \alpha_{n+1}$ could be shrunk to a point in $S^{n+1} - \alpha_{n+1}$. Thus, any loop in $\mathscr{S}(S^n/\alpha_n) - \mathscr{S}(\alpha_n)$ could be pushed up the product structure into $V - \alpha_{n+1}$ and then be shrunk to a point, which would imply that $\mathscr{S}(S^n/\alpha_n) - \mathscr{S}(\alpha_n)$ is 1-connected. Hence, $S^{n+1} - \alpha_{n+1}$ is not projectively 1-connected at α_{n+1}. The arc α_n can be considered to be in E^n by removing a point of S^n not on α_n and so α_n satisfies Examples **2.6.4**.

REMARK 2.6.1. For arcs related to those of Example **2.6.4**, see Exercise **4.8.3**.

Lemma 2.6.5. *Let $A \subset X \subset Y$ be topological spaces such that $Y - X$ is not 1-LC at any point of A. Let $W \subset Z \subset E^1$ where W is open in E^1. Then, $(Y \times E^1) - (X \times Z)$ is not 1-LC at any point of $A \times W$.*

The proof of Lemma **2.6.5** is similar to that of Lemma **2.6.2** and so we omit it.

Example 2.6.5. *For $n \geqslant 4$, there is a closed 1-string X_n^1 in E^n such that $E^n - X_n^1$ fails to be 1-LC at every point of X_n^1. (Hence, X_n^1 fails to be locally flat at every point and is everywhere wild by Lemma* **2.6.3**.)

Let $\alpha_{n-1} \subset E^{n-1}$ be the arc of Example **2.6.4**. Then, $(E^{n-1}/\alpha_{n-1}) - \alpha_{n-1}$ fails to be 1-LC at α_{n-1}. Let $X_n^1 = \alpha_{n-1} \times E^1 \subset (E^{n-1}/\alpha_{n-1}) \times E^1$. By Theorem **2.5.3**, $(E^{n-1}/\alpha_{n-1}) \times E^1 \approx E^n$, and by Lemma **2.6.5**, $E^n - X_n^1$ fails to be 1-LC at every point of X_n^1.

Example 2.6.6. *For $n \geqslant 4$, there is an arc β_n in E^n such that $E^n - \beta_n$ fails to be 1-LC at every interior point of β_n. (Hence, β_n is everywhere wild.) Furthermore, β_n is cellular.*

In the notation of the preceding paragraph and by the same reasoning, $\beta_n = \alpha_{n-1} \times [-1, 1] \subset (E^{n-1}/\alpha_{n-1}) \times E^1 \approx E^n$ is the desired arc. It is cellular by Lemma **2.6.4**.

Example 2.6.7. *For integers n and k such that $n \geqslant 3$, $0 < k < n$, and $n - k \neq 2$, there is a k-cell D_n^k in E^n such that $E^n - D_n^k$ fails to be 1-LC at every interior point of D_n^k. (Hence, D_n^k is everywhere wild.) Furthermore, D_n^k is cellular.*

Gillman [2] modified Example **2.4.13** slightly to obtain an everywhere wild 2-sphere with certain surprising properties. Let D_3^2 be any 2-cell on that wild sphere. Then, $E^3 - D_3^2$ is not 1-LC at any interior point of D_3^2, however it is easy to see that D_3^2 satisfies the cellularity criterion of [McMillan, 1] (which we shall discuss in Chapter **4**) and so is cellular in E^3. We will now assume that Example **2.6.7** holds in dimension n

and show that it holds in dimension $n + 1$. The existence of D_{n+1}^1 follows from Example **2.6.6**. Let $k + 1$ be such that $1 < k + 1 < n + 1$ and $(n + 1) - (k + 1) \neq 2$. By assumption there is a k-cell $D_n{}^k \subset E^n$ which satisfies Example **2.6.7**. Let

$$D_{n+1}^{k+1} = D_n{}^k \times [-1, 1] \subset E^n \times E^1 = E^{n+1}.$$

It follows from Lemma **2.6.5** that $E^{n+1} - D_{n+1}^{k+1}$ fails to be 1-LC at every interior point of D_{n+1}^{k+1} and it follows from Lemma **2.6.4** that D_{n+1}^{k+1} is cellular.

Example 2.6.8. *For integers n and k such that $n \geqslant 3$, $0 < k < n$, and $n - k \neq 2$, there is a closed k-string $X_n{}^k$ in E^n such that $E^n - X_n{}^k$ fails to be* 1-LC *at every point of $X_n{}^k$. (Hence, $X_n{}^k$ is everywhere wild.)*

One can obtain $X_3{}^2$ by one-point compactifying E^3 and then removing a point of the 2-sphere constructed in Example **2.4.13**. The existence of X_{n+1} follows from Example **2.6.5**. By the same reasoning of the preceding paragraph the existence of appropriate X_{n+1}^{k+1} follows by letting

$$X_{n+1}^{k+1} = X_n{}^k \times E^1 \subset E^n \times E^1 = E^{n+1}$$

and applying Lemma **2.6.5**.

Example 2.6.9. *For integers n and k such that $n \geqslant 3$, $0 < k < n$, and $n - k \neq 2$, there is an everywhere wild k-sphere $S_n{}^k$ in E^n.*

To get $S_n{}^k$, one-point compactify the pair $(X_n{}^k, E^n)$ of Example **2.6.8** and remove a point not on the resulting k-sphere.

EXERCISE 2.6.2. Let $D^k \subset S^n$ be a k-cell. Show that

(*a*) If $D^k - S^n$ is not simply connected, then $\mathscr{S}(S^n) - \mathscr{S}(D^k)$ is not simply connected. (Hence, $\mathscr{S}(D^k)$ is not cellular in $\mathscr{S}(S^n) \approx S^{n+1}$.)

(*b*) If $n - k \neq 2$ $(n - k = 2)$ and $S^n - D^k$ is not 1-LC (1-SS) at any point of D^k, then $\mathscr{S}(S^n) - \mathscr{S}(D^k)$ is not 1-LC (1-SS) at any point of Int $\mathscr{S}(D^k)$. (Hence, $\mathscr{S}(D^k)$ is everywhere wild in $\mathscr{S}(S^n) \approx S^{n+1}$.)

Give examples of a 2-cell in S^3 and arcs in E^n, $n \geqslant 4$, which are appropriate to plug in this exercise.

EXERCISE 2.6.3 (Bierman). An arc α in S^n **pierces** an $(n - 1)$-sphere Σ^{n-1} at a point p if for some subarc β of α, $\beta \cap \Sigma^{n-1} = p$ and the endpoints of β are in different components of $S^n - \Sigma^{n-1}$. Show that for $n \geqslant 3$, there is an everywhere wild arc in S^n which can pierce a locally flat $(n - 1)$-sphere at one and only one point. (You may use the following fact which we will establish in Chapter **3**: If $\Sigma^{n-1} \subset E^n$, $n \geqslant 4$, is an $(n - 1)$-sphere which is locally flat except possibly at a single point, then Σ^{n-1} is locally flat.)

EXERCICE 2.6.4 (Daverman). Let X denote a Cantor set in B^k, where X is contained in ∂B^k in case $k = n$, and f an embedding of X onto a Cantor set in a connected n-manifold $M^n (k \leqslant n)$. Show that there exists an embedding F of B^k in M^n such that $F \mid X = f$ and $F(B^k)$ is locally flat modulo $F(X)$. Observe that this exercise can be used to construct wild cells.

REMARK 2.6.2. Exercise 2.6.4 is proved in [Daverman, 2]. By ingeniously defining the Cantor sets X and $F(X)$, Daverman exhibits certain wild cells containing a scarcity of tame disks. Construction of wild disks in E^n which lie on no 2-sphere are given in [Lacher, 6].

2.7. SOME WILD POLYHEDRA IN LOW CODIMENSIONS

In this section we show that under special dimension restrictions the union of two cells may be "badly" embedded in the n-sphere even though each of the cells is "nicely" embedded. The problem of determining whether the union of cells is nicely embedded in the n-sphere if each of the cells is nicely embedded is related to many topological embedding problems. For instance, the n-dimensional annulus conjecture (now known to be true for $n \neq 4$) is a special case. Cantrell and Lacher [2] have shown that an affirmative answer implies local flatness of certain submanifolds. Also, this problem is related to the conjecture that an embedding of a complex into the n-sphere which is locally flat on open simplexes is ϵ-tame in codimension three. The problem was first investigated by Doyle [2, 3] in the three-dimensional case and by Cantrell [7] in high dimensions and later by Lacher [2], Cantrell and Lacher [1, 2], Kirby [2], Černavskiĭ [4, 5] and Rushing [5]. Also, Sher [1] has generalized a construction of Debrunner and Fox [1] to obtain refined counterexamples in certain cases (see Exercise **2.7.6**). As mentioned above, in this section we will concentrate on giving certain "easy" counterexamples. We will obtain positive results on this problem later. (For further discussion, see Sections **3.9** and **5.2**.)

Let D_1 and D_2 be cells in E^n such that $D_1 \cap D_2 = \operatorname{Bd} D_1 \cap \operatorname{Bd} D_2$ is a cell. We say that $D_1 \cup D_2$ is a **flat pair** if there is a homeomorphism h of E^n such that $h(D_i)$ is a simplex and $h(D_1 \cap D_2)$ is a face of $h(D_i)$, $i = 1, 2$.

Let $\beta(n, m_1, m_2, k)$ stand for the following statement: If D_1 and D_2 are locally flat cells in E^n of dimensions m_1 and m_2, respectively, and if $D_1 \cap D_2 = \text{Bd } D_1 \cap \text{Bd } D_2 = D$ is a k-cell which is locally flat in $\text{Bd } D_1$, in $\text{Bd } D_2$ and in E^n, then $D_1 \cup D_2$ is a flat pair.

The following is the main result of this section.

Theorem 2.7.1. *$\beta(n, m_1, m_2, k)$ is false for $k = n - 3$ and $n \geqslant 3$.*

Lemma 2.7.1. *Let D_1 and D_2 be cells in E^n, $n \geqslant 3$, such that $D_1 \cup D_2$ is a flat pair and $D_1 \cap D_2 = \text{Bd } D_1 \cap \text{Bd } D_2 = D$ is an $(n-3)$-cell. Let $q \in \text{Int } D$ and let $\{V_i\}_{i=1}^{\infty}$ be a sequence of closed neighborhoods of q in E^n such that*

(a) $V_1 \supset V_2 \supset \cdots$,

(b) $\bigcap_{i=1}^{\infty} V_i = q$, *and*

(c) $V_i - (D_1 \cup D_2)$ *is arcwise connected.*

Then, there exist indices N and $m > N$ such that for $k \geqslant m$ the image group under the injection $i_{kN}\colon \pi_1(V_k - (D_1 \cup D_2)) \to \pi_1(V_N - (D_1 \cup D_2))$ is infinite cyclic.

Exercise 2.7.1. Prove Lemma **2.7.1**. (Hint: This proof is quite similar to the proof of Proposition **2.3.1**.)

Lemma 2.7.2. *Let $\alpha \subset E^n$ be an arc for which there is a straight line $L \subset E^n$ such that no line parallel to L intersects α in more than one point. Then, α is flat.*

Proof. We may suppose that L is parallel to the line $\{(0, ..., 0, t)\colon t \in E^1\}$, for if not we can rotate E^n so that this is the case. Since no line parallel to L intersects α in more than one point, the projection $\pi\colon \alpha \to E^{n-1}$, defined by $\pi(x_1, ..., x_n) = (x_1, ..., x_{n-1}, 0)$ where $(x_1, ..., x_n) \in \alpha$ is a homeomorphism. We now define a map $h\colon \pi(\alpha) \to E^1$ by $h(x_1, ..., x_{n-1}) = x_n$ where $\pi(x_1, ..., x_{n-1}, x_n) = (x_1, ..., x_{n-1}, 0)$. Certainly $h(\pi(\alpha))$ is compact and hence is contained in a closed interval $[a, b]$. By applying Tietze's extension theorem, we get a map $f\colon E^{n-1} \to E^1$ such that $f \mid \pi(\alpha) = h$.

Define a homeomorphism $\phi_f\colon E^n \to E^n$ by

$$\phi_f(x_1, ..., x_n) = (x_1, ..., x_{n-1}, x_n + f(x_1, ..., x_{n-1})).$$

Consider $\phi_f^{-1}\colon E^n \to E^n$. First, $\phi_f^{-1}(\alpha) = \pi(\alpha) \subset E^{n-1}$ since for $(x_1, ..., x_n) \in \alpha$ we have

$$\phi_f^{-1}(x_1, ..., x_n) = \phi_f^{-1}(x_1, ..., x_{n-1}, h(x_1, ..., x_{n-1}))$$
$$= \phi_f^{-1}(x_1, ..., x_{n-1}, 0 + f(x_1, ..., x_{n-1})) = (x_1, ..., x_{n-1}, 0) \in \pi(\alpha).$$

But by Theorem **2.5.1**, $\pi(\alpha)$ is flat. Hence, α is flat.

Example 2.7.1 (Fox and Artin). *$\beta(3, 1, 1, 0)$ is false, that is, there is a wild arc $H^\natural$ in E^3 which is the union of two flat arcs $H^\#$ and $H^\flat$ sharing a common end-point. Furthermore, the counterexample fails to satisfy Lemma* **2.7.1**

In the construction of this example we will use the notation developed following Lemma **2.4.2**. Denote by $K^\#$ and $K^\flat$ the two arcs situated in I_3 which joint t_- to t_+ and r_- to r_+, respectively, as shown in Fig. **2.7.1**. The two arcs $H^\# = \bigcup_{n=1}^{\infty} f_n(K^\#) \cup q$ and $H^\flat = \bigcup_{n=1}^{\infty} f_n(K^\flat) \cup q$ intersect in their common end-point q. Their union is the arc $H^\natural = H^\# \cup H^\flat$. These three arcs have the regular projection into the yz-plane pictured in Fig. **2.7.2**.

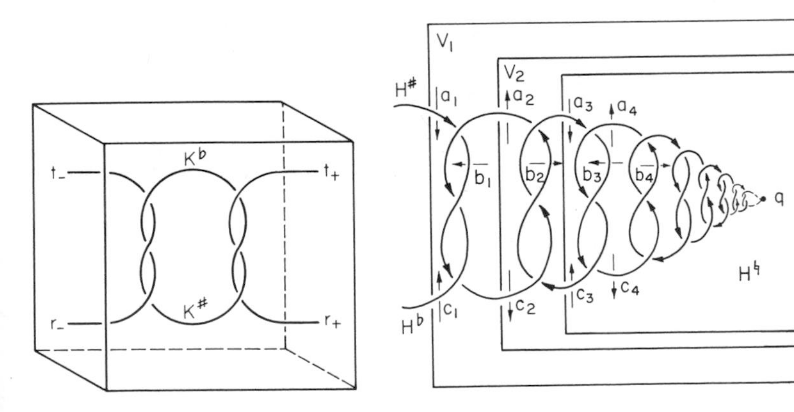

Figure 2.7.1 **Figure 2.7.2**

The fact that both $H^\#$ and $H^\flat$ are flat follows directly from Lemma **2.7.2**. We will show that $H^\natural$ is wild by showing that it does not satisfy Lemma **2.7.1**. Let $\{V_i\}$ be the sequence of closed 3-cell neighborhoods of q indicated in Fig. **2.7.2**. This sequence clearly satisfies the hypothesis of Lemma **2.7.1**. By the same technique used in Example **2.4.1**, we see that $\pi_1(V_m - H^\natural)$ has the following presentation:

$$\Bigg[\bigcup_{i=m}^{\infty} \{a_i , b_i , c_i\}: \Bigg(\bigcup_{\substack{i \geqslant m \\ i \text{ odd}}} \{a_i = c_i , a_{i+1} a_i a_{i+1}^{-1} b_i^{-1}, b_i a_{i+1} b_i^{-1} c_i^{-1}, b_i c_i^{-1} c_{i+1}^{-1} c_i\} \Bigg)$$

$$\cup \Bigg(\bigcup_{\substack{i \geqslant m \\ i \text{ even}}} \{a_i = c_i , a_{i+1}^{-1} a_i^{-1} a_{i+1} b_i , b_i^{-1} a_{i+1}^{-1} b_i c_i , b_i^{-1} c_i c_{i+1} c_i^{-1}\} \Bigg) \Bigg]$$

$$\approx \Bigg[\bigcup_{i=m}^{\infty} \{a_i , b_i\}: \Bigg(\bigcup_{\substack{i \geqslant m \\ i \text{ odd}}} \{b_i = a_{i+1} a_i a_{i+1}^{-1} , b_i a_{i+1} b_i^{-1} a_i^{-1}, b_i = a_i^{-1} a_{i+1} a_i\} \Bigg)$$

$$\cup \Bigg(\bigcup_{\substack{i \geqslant m \\ i \text{ even}}} (b_i^{-1} = a_{i+1}^{-1} a_i^{-1} a_{i+1} , b_i^{-1} a_{i+1}^{-1} b_i a_i , b_i^{-1} = a_i a_{i+1}^{-1} a_i^{-1}\} \Bigg) \Bigg]$$

$$\approx \Bigg[\bigcup_{i=m}^{\infty} \{a_i\}: \Bigg(\bigcup_{\substack{i \geqslant m \\ i \text{ odd}}} \{a_{i+1} a_i a_{i+1}^{-1} a_{i+1} a_i^{-1} a_{i+1}^{-1} a_i a_i^{-1}, a_{i+1} a_i a_{i+1}^{-1} = a_i^{-1} a_{i+1} a_i\} \Bigg)$$

$$\cup \Bigg(\bigcup_{\substack{i \geqslant m \\ i \text{ even}}} \{a_{i+1}^{-1} a_i^{-1} a_{i+1} a_{i+1}^{-1} a_i a_{i+1} a_i^{-1} a_i , a_{i+1}^{1} a_i^{-1} a_{i+1} = a_i a_{i+1}^{-1} a_i^{-1}\} \Bigg) \Bigg]$$

$$\approx \Bigg[\bigcup_{i=m}^{\infty} \{a_i\}: \bigcup_{i=m}^{\infty} \{a_i a_{i+1} a_i = a_{i+1} a_i a_{i+1}\} \Bigg].$$

We see from the last relation that if there exist indices N and $m > N$ such that the injection $i_{mN}\colon \pi_1(V_m - H^\natural) \to \pi_1(V_N - H^\natural)$ is Abelian then $a_m = a_{m+1} = \cdots$ in $\pi_1(V_N - H^\natural)$. This is not possible since we can define a homeomorphism h of $\pi_1(V_N - H^\natural)$ into the permutation group on three letters by $h(a_i) = (12)$ if i is odd and $h(a_i) = (13)$ if i is even. It follows that h induces the desired homeomorphism since it is easy to check that for any i,

$$h(a_{i+1})\, h(a_i)\, h(a_{i+1}) = h(a_i)\, h(a_{i+1})\, h(a_i) = (23).$$

Hence there is no N and $m > N$ such that the injection

$$i_{mN}\colon \pi_1(V_m - H^\natural) \to \pi_1(V_N - H^\natural)$$

is Abelian and so $H^\natural$ is wild by Lemma **2.7.1**.

Example 2.7.2. *$\beta(3, 2, 1, 0)$, $\beta(3, 2, 2, 0)$, $\beta(3, 3, 1, 0)$, $\beta(3, 3, 2, 0)$ and $\beta(3, 3, 3, 0)$ are all false, Furthermore, the counterexamples fail to satisfy Lemma* **2.7.1**.

Example **2.7.2** clearly follows from Example **2.7.1** by fattening up the arcs $H^{\#}$ and H^{b} into 2-cells and 3-cells as appropriate.

Lemma 2.7.3. *Suppose that* D_1 *and* D_2 *are cells in* E^n *such that* $D_1 \cap D_2 = \text{Bd}\, D_1 \cap \text{Bd}\, D_2 = D$ *is an* $(n - 3)$*-cell and suppose that* $D_1 \cup D_2$ *fails to satisfy the conclusion of Lemma* **2.7.1**. *Then,* $D_1{}^* = D_1 \times I$ *nd* $D_2{}^* = D_2 \times I$ *are cells in* E^{n+1} *such that*

$$D_1{}^* \cap D_2{}^* = \text{Bd}\, D_1{}^* \cap \text{Bd}\, D_2{}^* = D^*$$

is an $(n - 2)$*-cell and* $D_1{}^* \cup D_2{}^*$ *fails to satisfy the conclusion of Lemma* **2.7.1**.

Exercise 2.7.2. Prove Lemma **2.7.3**. (Hint: This proof is similar to the proof of Lemma **2.6.2**.)

Proof of Theorem 2.7.1. Theorem **2.7.1** follows immediately from Example **2.7.1**, Example **2.7.2**, and Lemma **2.7.3**.

Exercise 2.7.3. An arc is said to be **mildly wild** if it is wild and can be expressed as the union of two tame arcs. Thus, the arc of Example **2.7.1** is mildly wild. Show that the arc pictured in Fig. **2.7.3** is mildly wild. Apparently [Wilder, 1, p. 634, footnote ‡] was the first to consider such an arc. Such arcs are also studied in [Fox and Harrold, 1].

Figure 2.7.3

Exercise 2.7.4. An n-**frame** F_n is a union of n arcs, $F_n = \bigcup_{i=1}^{n} A_i$, with a distinguished point p such that, if $n = 1$, p is an end-point of A_1, and if $n > 1$, p is an end-point of each A_i and $A_i \cap A_j = p$, $i \neq j$. In E^k let B_i be the arc in the x_1x_2-plane defined in polar coordinates by $r \leqslant 1$, $\theta = \pi(1 - 1/i)$. For n a positive integer, the **standard** n-**frame** G_n is defined by $G_n = \bigcup_{i=1}^{n} B_i$. An n-frame F_n in E^k is said to be **flat** if there is a homeomorphism of E^k onto itself which carries F_n onto G_n. Otherwise, F_n is said to be **wild**. The exercise is to construct an n-frame in E^3 for all $n \geqslant 2$ such that every m-subframe, $1 < m \leqslant n$, is wild.

Exercise 2.7.5. An n-frame $F_n = \bigcup_{i=1}^{n} A_i \subset E^k$ is said to be **mildly wild** if it is wild and $F_n - (A_i - p)$ is flat for $i = 1, 2, \ldots, n$. Show that for $n \geqslant 1$

there are mildly wild n-frames in E^3. A general construction for $n \geqslant 2$ is suggested by Fig. **2.7.4**. It is clear that each proper subframe is flat from

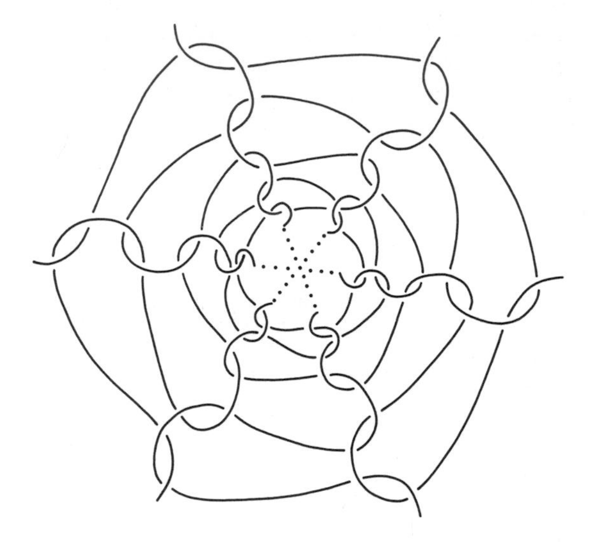

Figure 2.7.4

Fig. **2.7.5**, that is, the process of moving a proper subframe onto a standard frame resembles the unraveling of a piece of knitted goods in which a "run" has

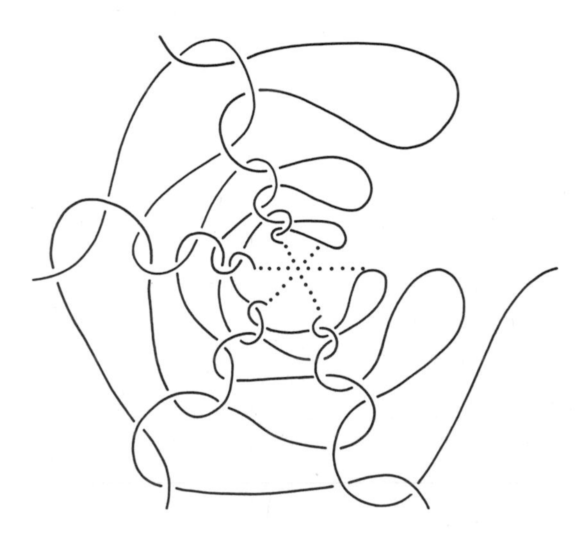

Figure 2.7.5

appeared. To show the suggested frame wild actually is a hard exercise. This is done in [Debrunner and Fox, 1] and their proof involves the construction of a complicated, but quite interesting, group. We also note that a mildly wild triod (3-frame) was attempted in [Doyle, 1], however the proof is incorrect because the presentation of the group G_p is wrong. Doyle's triod can in fact be shown to be flat.

Exercise 2.7.6. Given $1 \leqslant m < n$ show how to generalize the construction of the last example to construct an n-frame in E^3 such that every m-subframe is flat but every $(m + 1)$-subframe is wild. Such a construction is given in [Sher, 1].

Exercise 2.7.7. Use Exercises **2.7.3** through **2.7.6** and the techniques of this section to obtain other wild embeddings in high dimensions.

CHAPTER 3

Flattening, Unknotting, and Taming Special Embeddings

3.1. INTRODUCTION

Although some of the results covered in this chapter have been generalized, as will be apparent from the numerous references, many of the techniques of proof have become classical. (In Chapter **5** we will present some generalizations.) The serious student of topological embeddings, in the author's opinion, is well advised to begin his study of taming and unknotting theorems with these topics. One appeal of the included material for beginning students of the area is that none of the proofs use engulfing. (Engulfing is covered in the next chapter.) It is best to have some feeling for the theory of topological embeddings before attempting to master that valuable tool.

To get the idea of "straightening out bad-looking things" we begin with the easily obtained result that an almost polyhedral arc is flat. Next, we carry the generalized Schoenflies Theorem a step further by establishing the fact that any $(n-1)$-sphere in S^n, $n \geqslant 4$, which is locally flat modulo a point is flat. (This is not true for $n = 3$ as is shown by Example **2.4.6**.) This proof, which was developed by Cantrell and employs a technique of Mazur, uses the fact that an arc which is locally flat modulo an end-point is flat. It follows from the work three sections later that such an arc can be assumed to be polyhedral modulo the end-point and so by the first section it is flat. Next some theorems of Lacher which flatten half-strings and strings are presented, and the result of Cantrell just mentioned is used. An unknotting theorem for polyhedra follows which was first considered by Gugenheim and later refined by Bing and Kister. Our version of this theorem handles infinite polyhedra. We then employ this unknotting theorem to $\epsilon(x)$-tame infinite polyhedra in the trivial range. This is succeeded by a taming theorem of Bing and Kister for polyhedra which lie in hyperplanes. Our proof is not that of Bing and Kister, but an adaptation of some early work of Bryant which uses the preceding two sections. Next another result of Bryant is established which $\epsilon(x)$-tames embeddings that are locally tame modulo nice subsets. Finally, in the last section all of the results of the preceding sections of this chapter are used to prove a theorem of Kirby which generalizes Cantrell's theorem very nicely. More detailed discussions of these results as well as many related references will appear in the various sections.

3.2. ALMOST POLYHEDRAL ARCS ARE FLAT

Flattening Theorem 3.2.1. *Let $\alpha \subset E^n$, $n \geqslant 4$, be an arc which is locally polyhedral modulo a countable set $X \subset \alpha$, that is, if $p \in \alpha - X$, then p has a closed neighborhood in α which is a polygonal arc. Then, α is flat in E^n.*

PROOF. Put α in a countable number of lines $\{L_i\}_{i=1}^{\infty}$ by extending all of the one-simplexes of α and running a line through each point of X. Consider the countable collection (L_i, L_j), $i \neq j$, of pairs of these lines. Each such pair determines a hyperplane H_{ij} of dimension at most three. Translate each hyperplane H_{ij} to a hyperplane H_{ij}^t which passes through the origin. Since each of the H_{ij}^t is nowhere dense in E^n, by the Baire category theorem there is a point p of E^n not in any of the H_{ij}^t.

Let l be the line which passes through p and the origin. Certainly l hits each H_{ij}^t only in the origin, because if it hit some H_{ij}^t somewhere else then it would be contained in that H_{ij}^t which contradicts the fact that $p \in l$ is not in H_{ij}^t. Also, it is easy to see that any line l' parallel to l hits each H_{ij}^t in at most one point since a translation of E^n which takes a point $x \in l' \cap H_{ij}^t$ to the origin also takes l' onto l and H_{ij}^t onto itself. Similarly, any line l' parallel to l can hit each H_{ij} in at most one point. It now follows easily that each line l' parallel to l can hit $\bigcup_{i=1}^{\infty} L_i$ in at most one point. Hence no line parallel to l can intersect α in more than one point and so α is flat by Lemma **2.7.2**.

Some early results related to the theorem of this section were obtained in [Cantrell and Edwards, 2]. Related higher-dimensional results are contained in [Seebeck, 1] and [Cantrell and Rushing, 1, Theorem 2].

3.3. AN $(n-1)$-SPHERE IN $S^{n\geq 4}$ WHICH IS LOCALLY FLAT MODULO A POINT IS FLAT

Mazur [2] developed an elegant technique in the process of proving the generalized Schoenflies theorem with a "niceness" condition added to the hypothesis. (It follows from a short argument in [Morse, 1] that this "niceness" condition can be removed.) Mazur's technique has many applications, some of which may be found in [Stallings, 1]. (For instance, Stallings showed that an invertible cobordism (defined in Section **4.13**) minus one end is the product of the other end with a half-open interval. For related work see Section **4.13**.) Since Brown's proof of the generalized Schoenflies theorem (Theorem **1.8.2**) is included in this book, we will not give the Mazur–Morse proof. However, Mazur's technique will be fully developed in this section in the presentation of an application due to Cantrell. Cantrell's result, the statement of which follows, was announced in [1] and his proof can be obtained by combining [2] and [3].

Flattening Theorem 3.3.1 (Cantrell). *If Σ is an $(n-1)$-sphere in S^n $(n \geqslant 4)$, $p \in \Sigma$, and Σ is locally flat at each point of $\Sigma - p$, then Σ is flat in S^n.*

In this section we will actually establish the following stronger result.

Theorem 3.3.2 (Cantrell). *If Σ is an $(n-1)$-sphere in S^n, $n \geqslant 4$, $p \in \Sigma$, G a complementary domain of Σ, and $\Sigma - p$ is locally collared in* Cl G, *then* Cl G *is an n-cell.*

Stallings, in proving that locally flat spheres in S^n of codimension at least three are flat, got the analog of Theorem **3.3.1** in codimensions at least three for free. (We will give Stalling's proof in Section **4.5**.) For some time after the appearance of Cantrell's result, it was wondered whether a codimension one sphere which is locally flat modulo two points is flat. In connection with this, it was shown in [Cantrell and Edwards, 1, Theorem 3.5] that if no embedding of the closed manifold M into the interior of the manifold N fails to be locally flat at precisely one or two points, then every embedding which is locally flat modulo a countable number of points is locally flat. The codimension one problem was solved by Kirby [1], Černavskiĭ [1], [4], and Hutchinson [1]. In particular, it follows from [Kirby, 1] that an embedding of an unbounded $(n-1)$-manifold into an unbounded n-manifold, $n \geqslant 4$, which is locally flat modulo a countable set of points, is locally flat.

We will prove this in Section **3.9** (see Corollary **3.9.1**). Some of Černavskiĭ's work on this problem will be presented in Chapter **5** (see Theorem **5.2.2** and Remark **5.2.3**).

The rest of this section will be devoted to establishing the results of Cantrell mentioned above and will be based on [Cantrell, 1–5].

For $t > 0$, let B_t be the n-ball in E^n with center at the origin and radius t. For $t > -1$, let

$$C_t = \{(x_1, x_2, \ldots, x_n) \in E^n \mid x_1^2 + x_2^2 + \cdots + x_{n-1}^2 + (x_n + t)^2 \leqslant (1+t)^2\}.$$

Lemma 3.3.1. *Let Σ be an $(n-1)$-sphere in S^n, let p be a point of Σ, and let G and H be the components of $S^n - \Sigma$. Suppose that $\Sigma - p$ is locally flat in S^n, that Σ is locally collared in* Cl H *at p and that h is a homeomorphism from* Bd B_1 *onto Σ such that $h((0, 0, \ldots, 0, 1)) = p$. Then, h can be extended to a homeomorphism g from* $\mathrm{Cl}(C_1 - B_{1/4})$ *into S^n.*

Lemma **3.3.1** follows easily from Theorem **1.7.4** and Theorem **1.7.6**.

Flattening Theorem 3.3.3. *Let Σ be an $(n-1)$-sphere in S^n, $n \geqslant 4$, let p be a point of Σ, and let G and H be the components of $S^n - \Sigma$. If $\Sigma - p$ is locally flat in S^n and if Σ is locally collared in* Cl H *at p, then Σ is flat.*

Proof. Cl H is an n-cell by Theorem **1.8.3**, hence by Exercise **1.8.4** we need only show that Cl G is an n-cell. Let g be the homeomorphism from $\mathrm{Cl}(C_1 - B_{1/4})$ into S^n given by Lemma **3.3.1**. Also, let A be the line segment in E^n from $(0, 0, \ldots, 0, \frac{1}{2})$ to $(0, 0, \ldots, 0, 1)$ and let K be the component of $S^n - g(\mathrm{Bd}\, B_{1/2})$ that contains G (see Fig. **3.3.1**). Note that Cl K is an n-cell. Let ϕ be a continuous map of $\mathrm{Cl}(C_1 - B_{1/2})$ onto

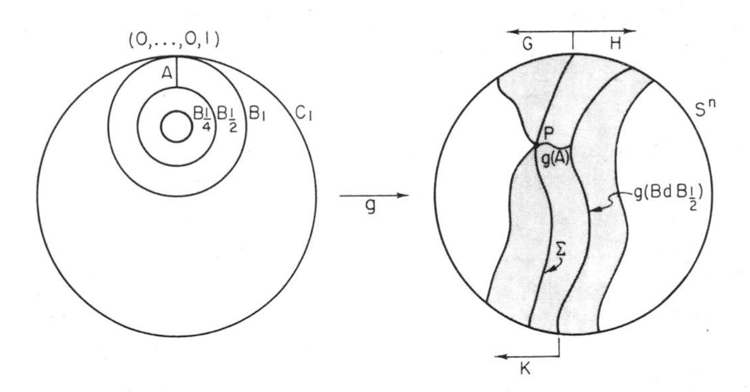

Figure 3.3.1

$\mathrm{Cl}(C_1 - B_1)$ such that (1) ϕ is fixed on Bd C_1, (2) $\phi(\mathrm{Bd}\, B_{1/2}) = \mathrm{Bd}\, B_1$, and A is the only inverse set under ϕ. The mapping f of Cl K onto Cl G defined by

$$f(x) = \begin{cases} x, & x \notin g(\mathrm{Cl}(C_1 - B_{1/2})), \\ g\phi g^{-1}(x), & x \in g(\mathrm{Cl}(C_1 - B_{1/2})) \end{cases}$$

is a continuous map of Cl K onto Cl G and $g(A)$ is the only inverse set.

We will next construct a map k of Cl K onto Cl K such that $g(A)$ is the only inverse set. Then, kf^{-1} will be a homeomorphism from Cl G onto the n-cell Cl K. It follows from a result proved later in this chapter (Corollary **3.6.1**) that $g(A)$ may be assumed to be locally polyhedral modulo p. (In [Rushing, 10, Remark 5], it is shown how to avoid using the corollary to Theorem **3.6.1**, by applying a technique of Černavskiĭ instead.) Thus, it follows from Theorem **3.2.1** that $g(A)$ is flat. Using this fact it is easy to construct an embedding $j\colon B_1 \to \mathrm{Cl}\, K$ such that $j(A) = g(A)$. Let ψ be a map of B_1 onto itself such that $\psi \mid \mathrm{Bd}\, B_1 = 1$ and A is the only inverse set, and let k be given by

$$k(x) = \begin{cases} j\psi j^{-1}(x), & x \in j(B_1), \\ x, & x \in \mathrm{Cl}\, K - j(B_1). \end{cases}$$

EXERCISE 3.3.1. If D is an n-cell in E^n, $n \geqslant 4$, $p \in \mathrm{Bd}\, D$, and $D - p$ is locally flat, then D is flat.

EXERCISE 3.3.2. If M is an n-manifold in the interior of an n-manifold N, $n \geqslant 4$, and E is the set of points at which M fails to be locally flat, then E is perfect or $E = \emptyset$.

EXERCISE 3.3.3. If D^n is an n-cell in E^n and $D^k \subseteq \mathrm{Bd}\, D^n$ is a k-cell, $0 \leqslant k < n$, such that $D^n - D^k$ is locally flat in E^n and D^k is locally flat in Bd D^n, then D^n is flat if $\lambda(D^k \times I)$ is locally flat in E^n for some collaring of Bd D^n in D^n. (For a more general result see Theorem 3.1 of [Cantrell and Lacher, 2].)

Proof of Theorem 3.3.2. Let h be a homeomorphism from Bd B_1 onto Σ such that $h((0, 0, \ldots, 0, 1)) = p$ and let G and H denote the complementary domains of Σ in S^n. Then, by Theorem **1.7.6**, h can be extended to an embedding g from $\mathrm{Cl}(C_1 - B_1)$ into Cl G. Let G_1 and G_2, respectively, be the components of $S^n - g(\mathrm{Bd}\, C_{1/2})$ and $S^n - g(\mathrm{Bd}\, C_1)$ that are contained in G (see Fig. **3.3.2**). We now observe that Cl G_1 is homeomorphic to Cl G. For, if h_* is a homeomorphism of E^n onto itself which is fixed on Bd C_1 and carries Bd $C_{1/2}$ onto Bd B_1, then

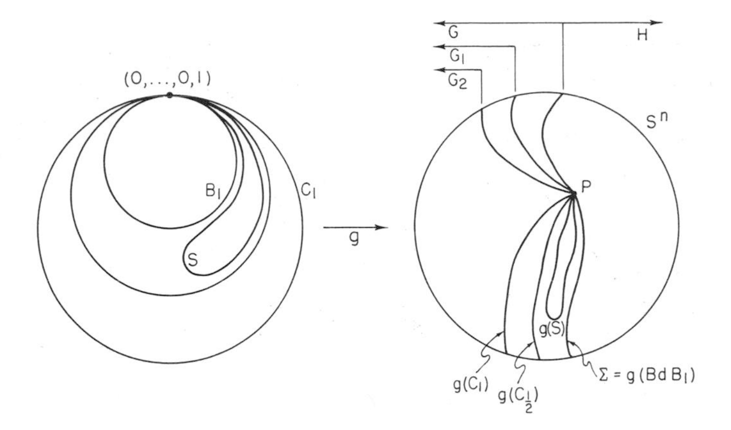

Figure 3.3.2

the mapping f defined by

$$f(x) = \begin{cases} x, & x \in G_2 \\ gh_* g^{-1}(x), & x \in \text{Cl } G_1 - G_2 \end{cases}$$

carries Cl G_1 homeomorphically onto Cl G. This gives the following fact: if one attaches a copy of Cl G_1 to $\text{Cl}(C_{1/2} - B_1)$ along Bd $C_{1/2}$ with g^{-1}, the space thus obtained is homeomorphic to Cl G_1. (It is simply Cl G.) This will be used to show that Cl G_1 is an n-cell and hence that Cl G is an n-cell.

Let S be a flat $(n-1)$-sphere in $\text{Cl}(C_{1/2} - B_1)$ such that

$$S \cap (\text{Bd } C_{1/2} \cup \text{Bd } B_1) = (0, 0, ..., 0, 1).$$

If K is the component of $S^n - g(S)$ which contains G_1, then by Theorem **3.3.3**, Cl K is an n-cell. Let Int S denote the complementary domain of S which does not intersect $g^{-1}(\Sigma) = \text{Bd } B_1$ and notice that Cl K can be realized by taking $P = \text{Cl}(C_{1/2} - B_1) - \text{Int } S$ and attaching Cl H to P along Bd B_1 by $g^{-1} \mid \Sigma$ and attaching Cl G_1 to P along Bd $C_{1/2}$ with $g^{-1} \mid \text{Bd}(G_1)$.

The set P is a closed n-cell (the closure of the exterior of S) with the interiors of two n-cells, sharing a common boundary point with Bd S, removed. The cell obtained from P by attaching Cl G_1 and Cl H to the interior boundary spheres of P with g^{-1} will be denoted by $\bar{P}$ (see Fig. **3.3.3**).

Let F be the part of the solid unit ball in E^n centered at $(0, 0, ..., 1, 0)$, determined by $x_n \geqslant 0$. Let $\{q_i\}_{i=0}^{\infty}$ be a sequence of points in the

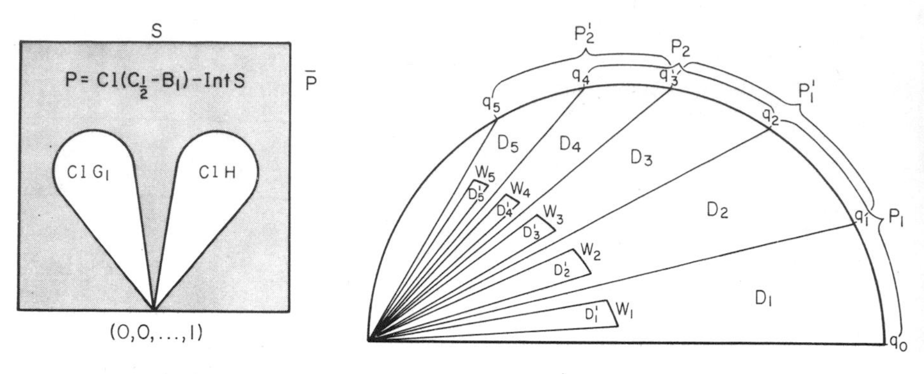

Figure 3.3.3

Figure 3.3.4

intersection of the plane $x_1 = x_2 = \cdots = x_{n-2} = 0$ and $\mathrm{Bd}\, F$ such that if $q_i = (0, 0, \ldots, a_{(n-1)i}, a_{ni})$, then $a_{(n-1)0} = 2$, $a_{n0} = 0$, the $a_{(n-1)i}$ converge monotonically to 0, and $a_{ni} > 0$ if $i > 0$. We then section F into a countable number of n-cells by projecting the $(n-2)$-plane $x_n = x_{n-1} = 0$ onto each of the q_i. The section determined by q_{i-1} and q_i is denoted by D_i. We then delete from D_i the interior of a cell D_i', similar in shape to D_i and except for the boundary point $(0, 0, \ldots, 0)$, contained in the interior of D_i. Any two adjacent sections then form a copy of P, and are labeled P_i, P_i' as in Fig. **3.3.4**. Notice that P_i and P_i' have $w_{2i} = \mathrm{Bd}\, D_{2i}'$ in common, and P_i', P_{i+1} have $w_{2i+1} = \mathrm{Bd}\, D_{2i+1}'$ in common.

Let ϕ_i be a homeomorphism of P_i onto P_i' which leaves w_{2i} fixed and carries w_{2i-1} onto w_{2i+1}. (Of course, here we need $n \geqslant 3$.) Let ψ_i be a homeomorphism of P_i' onto P_{i+1} which leaves w_{2i+1} fixed and carries w_{2i} onto w_{2i+2}. We identify (P_1, w_1, w_2) with $(P, \mathrm{Bd}\, C_{1/2}, \mathrm{Bd}\, B_1)$. The sets $\mathrm{Cl}\, G_1$ and $\mathrm{Cl}\, H$ are sewn to P_1 along w_1 and w_2, respectively, with g^{-1}. The resulting n-cell is denoted by $\bar{P}_1$. Then, $\mathrm{Cl}\, G_1$ and $\mathrm{Cl}\, H$ are sewn into alternate holes bounded by w_{2i+1}, w_{2i+2} with attaching homeomorphisms

$$\phi_i \cdots \phi_2 \phi_1 g^{-1}\colon \mathrm{Bd}\, G_1 \to w_{2i+1},$$

$$\psi_i \cdots \psi_2 \psi_1 g^{-1}\colon \mathrm{Bd}\, H \to w_{2i+2}.$$

The sets thus obtained from P_i and P_i' are denoted by $\bar{P}_i$ and $\bar{P}_i'$ and we set $F_1 = \bigcup_{i=1}^{\infty} \bar{P}_i$.

Since ϕ_1 is the identity on w_2, we can extend ϕ_1 to a homeomorphism of $\bar{P}_1$ onto $\bar{P}_1'$ and conclude that $\bar{P}_1'$ is also an n-cell. In a similar manner we extend ψ_i to a homeomorphism of $\bar{P}_i'$ onto $\bar{P}_{i+1}$ and extend ϕ_i to a homeomorphism of $\bar{P}_i$ onto $\bar{P}_i'$. It then follows that each $\bar{P}_i$ and each $\bar{P}_i'$ is an n-cell.

We now observe that F_1 is an n-cell. We map the boundary of

$D_{2i-1} \cup D_{2i}$ onto the boundary of $\bar{P}_i$ with the identity homeomorphism. Since $D_{2i-1} \cup D_{2i}$ and $\bar{P}_i$ are n-cells, this homeomorphism between their boundaries can be extended to a homeomorphism between the cells. These extensions for $i = 1, 2, \ldots$ yield a homeomorphism of F onto F_1.

We next observe that F_1 is a copy of $C_{1/2} - B_1$ with Cl G_1 sewn along one boundary sphere. This is established by showing that F_1, with G_1 removed from $\bar{P}_1$, is homeomorphic to F, with Int D_1' removed. Let i be the identity mapping on $D_1 -$ Int D_1' and on $\text{Bd}(D_{2i} \cup D_{2i+1})$, $i = 1, 2, \ldots$. Since $D_{2i} \cup D_{2i+1}$ and $\bar{P}_i'$ are closed n-cells and i restricts to the identity on their boundaries, i can be extended over their interiors. These extensions over each of the $D_{2i} \cup D_{2i+1}$ yield the desired homeomorphism.

We have seen that F_1 can first be viewed as a closed n-cell, and secondly as Cl G_1 sewn into a boundary sphere of $\text{Cl}(C_{1/2} - B_1)$. We have previously observed that a space of the second type is equivalent to G_1. Hence Cl G_1, or equivalently Cl G, is an n-cell.

Exercise 3.3.4. Use Theorem **3.3.1** to show that if X is a closed, locally flat $(n-1)$-string in E^n, $n \geqslant 4$, then there is a homeomorphism of E^n onto itself that takes X onto E^{n-1}.

Exercise 3.3.5. (We suggest in this exercise an elegant alternate proof of Corollary 6 of [Doyle and Hocking, 2].) Let Σ^2 be a 2-sphere in E^3 and suppose that p is a point of Σ^2 such that $\Sigma^2 - p$ is locally flat. Use the technique of this section to show that if there is a flat arc α in Σ^2 through p, then Σ^2 is flat.

Remark 3.3.1. By employing Exercise **3.3.3** and the $\gamma(n, n-1, n-2)$ and $\beta(n, n-1, n-2)$ statements, to be discussed later, one can prove an analog of Exercise **3.3.5** in high dimension where p is replaced by an $(n-3)$-cell and α is replaced by an $(n-2)$-cell.

3.4. FLATTENING CELLS, HALF-STRINGS, AND STRINGS

The material of this section is based on results developed by Lacher in [1]. Similar results were independently developed by Černavskiĭ in Section 2 of [2]. We begin this section by showing that locally flat cells are flat, a result which has been in the folklore of topological embeddings for some time.

Flattening Theorem 3.4.1. *Let D be a locally flat k-cell in an n-manifold M. Then, D has a neighborhood U in M such that $(U, D) \approx (E^n, I^k)$* ($\approx$ means is homeomorphic to).

PROOF. (We will prove the theorem for the case that D is an arc and leave the generalization of the argument to cells as an exercise.) Since D is locally flat, there are open sets $U_1, U_2, \ldots, U_s$ in M such that D is covered by the U_i and such that $(U_i, U_i \cap D)$ is homeomorphic either to (E^n, E^1_+) or to (E^n, E^1) for each i. Let $\{D_1, \ldots, D_r\}$ be a covering of D with subarcs satisfying the following properties:

(1) each D_i is contained in some U_j,
(2) $E_i = D_i \cup \cdots \cup D_r$ is an arc for each i, and
(3) $D_i \cap E_{i+1}$ is an end-point of both D_i and E_{i+1}.

Suppose that for $i = 1, 2, \ldots, r-1$, we can construct a homeomorphism h_i of M onto itself such that $h_i(E_i) = E_{i+1}$. Then, $h = h_{r-1} \cdots h_2 h_1$ would be a homeomorphism of M which maps $D = E_1$ onto $D_r = E_r$. Thus, the proof would be complete since D_r has a neighborhood of the desired type.

Let us now construct h_1. It follows from the above three conditions that there is a U_j and a homeomorphism $\phi_1: U_j \twoheadrightarrow E^n$ such that $\phi_1(U_j \cap D) = E^1_+$ and $\phi_1(D_1) = I^1_+$. Let ψ_1 be a homeomorphism of E^n which is the identity outside some compact set and which takes E^1_+ onto $\mathrm{Cl}(E^1_+ - I^1_+)$. Define h_1 on M by

$$h_1 = \begin{cases} \phi_1^{-1}\psi_1\phi_1 & \text{on } U_j, \\ \text{identity} & \text{on } M - U_j. \end{cases}$$

Clearly h_1 takes E_1 onto E_2. One may construct $h_2, \ldots, h_{r-1}$ similarly.

EXERCISE 3.4.1. Generalize the above argument so that it will handle cells of all dimensions.

Corollary 3.4.1. *Every locally flat embedding of I^k into E^n or S^n, $k \leqslant n$, is flat.*

PROOF. The corollary follows by applying Theorem **1.8.2** (the generalized Schoenflies theorem).

Lemma 3.4.1. *Suppose that K is a compact subset of a locally flat k-half-string X in the n-manifold M, $k \leqslant n$. Then, there is an open set U in M such that $K \subset U$ and $(U, U \cap X) \approx (E^n, E^k_+)$.*

EXERCISE 3.4.2. Establish Lemma **3.4.1**. (Hint: Use Theorem **3.4.1**.)

Next we will prove a simple-minded engulfing lemma (see Fig. **3.4.1**).

Lemma 3.4.2. *Let X be a k-half-string in the n-manifold M, $k \leqslant n$. Let U, V be open sets in M such that $(U, U \cap X) \approx (E^n, E^k_+) \approx (V, V \cap X)$*

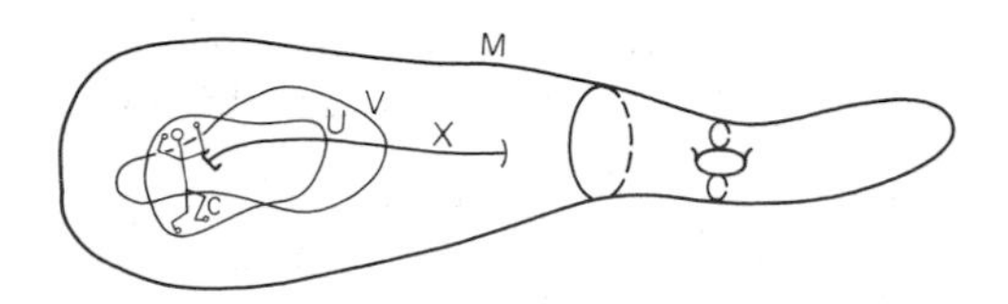

Figure 3.4.1

and $(U \cap X) \subset (V \cap X)$. If C is a compact subset of U, then there is a homeomorphism g of M onto itself such that

(1) $g \mid X =$ *identity,*
(2) $g \mid M - U =$ *identity, and*
(3) $g(V)$ *contains* C.

PROOF. Let $W = \phi(U \cap V)$, where $\phi\colon (U, U \cap X) \twoheadrightarrow (E^n, E_+^k)$ is a homeomorphism. Let f be a homeomorphism of E^n onto itself which is the identity on E_+^k and outside some compact set and such that $f(W)$ contains $\phi(C)$. Then, define g on M by $g = \phi^{-1}f\phi$ on U and $g =$ identity otherwise.

Lemma 3.4.3. *Let X be a locally flat k-half-string in the n-manifold M. Then, there exists a sequence $V_1, V_2, \ldots$ of open subsets of M such that*

(1) $(V_i, V_i \cap X) \approx (E^n, E_+^k)$ *for each* i,
(2) $\bar{V}_i \subset V_{i+1}$ *for each* i, *and*
(3) $X \subset \bigcup_{i=1}^{\infty} V_i$.

PROOF. Let $K_1, K_2, \ldots$ be a sequence of compact sets whose union is X, and let $U_1, U_2, \ldots$ be a sequence of open subsets of M such that $(U_i, U_i \cap X) \approx (E^n, E_+^k)$, $K_i \subset U_i$, and $(U_i \cap X) \subset (U_{i+1} \cap X)$ for each i. Let $h_i\colon (U_i, U_i \cap X) \twoheadrightarrow (E^n, E_+^k)$ be a particular homeomorphism. The existence of U_i and h_i follows directly from Lemma **3.4.1.** We shall apply Lemma **3.4.2** recursively on the U_i.

Step 1. Apply Lemma **3.4.2** with $U = U_1$, $V = U_2$, and $C = C_1 = h_1^{-1}(B_1) \cup K_1$, where B_1 is a (round) ball in E^n centered at the origin. Let $g = g_2$ be given by Lemma **3.4.2,** $\tilde{U}_2 = g_2(U_2)$, and $\tilde{h}_2 = h_2 g_2^{-1} \mid \tilde{U}_2$. Note that $\tilde{h}_2$ maps the pair $(\tilde{U}_2, \tilde{U}_2 \cap X)$ homeomorphically onto (E^n, E_+^k) and $\tilde{U}_2$ contains $h_1^{-1}(B_1) \cup K_1$.

Step 2. Apply Lemma **3.4.2** again with $U = \tilde{U}_2$, $V = U_3$, and $C = C_2 = \tilde{h}_2^{-1}(B_2) \cup K_2$, where B_2 is a ball in E^n centered at the origin so that $\tilde{h}_2(C_1) \subset \operatorname{Int}(B_2)$. Let $g = g_3$ be given by Lemma **3.4.2,** $\tilde{U}_3 = g_3(U_3)$, and $\tilde{h}_3 = h_3 g_3^{-1}$. Then, $\tilde{h}_3$ maps $(\tilde{U}_3, \tilde{U}_3 \cap X)$ homeomorphically onto (E^n, E_+^k) and $\tilde{U}_3$ contains $\tilde{h}_2^{-1}(B_2) \cup K_2$.

Continuing this process, we get a sequence $\tilde{U}_1, \tilde{U}_2, \ldots$ of open sets in M and a sequence $\tilde{h}_1, \tilde{h}_2, \ldots$ of homeomorphisms such that

(a) $\tilde{h}_i: (\tilde{U}_i, \tilde{U}_i \cap X) \twoheadrightarrow (E^n, E^k_+)$ for all i,
(b) $\tilde{h}_{i+1}^{-1}(\text{Int } B_{i+1}) \supset \tilde{h}_i^{-1}(B_i)$ for all i, and
(c) $\tilde{h}_{i+1}^{-1}(B_{i+1}) \supset K_i$ for all i.

Define $V_i = \tilde{h}_i^{-1}(\text{Int } B_i)$, $i = 1, 2, \ldots$. Since B_i is an open (round) ball centered at the origin in E^n, condition 1 follows from (a). Conditions 2 and 3 follow from (b) and (c), respectively, and the lemma is proved.

The following theorem is proved by "the technique of meshing a straight structure and a wiggly structure" developed by Brown [2] in the process of proving that the monotone union of open n-cells is an open n-cell. Later we will make other uses of this technique when considering work of Stallings (Theorem **4.5.3**) and Černavskiĭ (Lemma **5.2.3**).

Flattening Theorem 3.4.2 (Lacher). *Let X be a locally flat k-half-string in the n-manifold M, $k \leqslant n$. Then, there is an open set U in M containing X such that $(U, X) \approx (E^n, E^k_+)$.*

PROOF. Let $V_1, V_2, \ldots$ be a sequence of open subsets of M as given by Lemma **3.4.3**. We may assume that each V_i has compact closure. Let $h_i: (V_i, V_i \cap X) \twoheadrightarrow (E^n, E^k_+)$ be a homeomorphism for $i = 1, 2, \ldots$.

Sequences $Q_1, Q_2, \ldots$ of n-cells in M and $g_1, g_2, \ldots$ of embeddings $g_i: (Q_i, Q_i \cap X) \to (E^n, E^k_+)$ will be constructed so that

$$\text{(a)}\quad Q_i \subset \text{Int } Q_{i+1}, \qquad \text{(b)}\quad g_{i+1} \mid Q_i = g_i, \qquad \text{(c)}\quad \bigcup_{i=1}^{\infty} Q_i = \bigcup_{i=1}^{\infty} V_i,$$

and

$$\text{(d)}\quad \bigcup_{i=1}^{\infty} g_i(Q_i) = E^n, \qquad \text{(e)}\quad \bigcup_{i=1}^{\infty} g_i(Q_i \cap X) = E^k_+ .$$

Then, we may define U to be $\bigcup_{i=1}^{\infty} Q_i$ and a homeomorphism $g: (U, X) \twoheadrightarrow (E^n, E^k_+)$ by $g \mid Q_i = g_i$.

For each $t > 0$, B_t denotes the ball with center 0 and radius t in E^n. Let $Q_1 = h_1^{-1}(B_1)$, $g_1 = h_1 \mid Q_1$, and consider $h_2 h_1^{-1}(B_2) \subset E^n$. Note that $h_2(\bar{V}_1)$ is a compact subset of E^n, and so there is a homeomorphism $\phi_2: E^n \twoheadrightarrow E^n$ such that $\phi_2 =$ identity on $h_2 h_1^{-1}(B_1)$, $\phi_2 =$ identity outside some compact set in E^n, $\phi_2(E^k_+) = E^k_+$, and $\phi_2(h_2 h_1^{-1}(B_2))$ contains $h_2(\bar{V}_1)$. (See the next paragraph for a construction of ϕ_2.) Then define Q_2 to be $h_2^{-1}\phi_2 h_2 h_1^{-1}(B_2)$ and g_2 to be $h_1 h_2^{-1} \phi_2^{-1} h_2 \mid Q_2$. Notice that

$g_2 \mid Q_1 = h_1 h_2^{-1} \phi_2^{-1} h_2 \mid Q_1 = h_1 \mid Q_1 = g_1$, $Q_1 \subset \operatorname{Int} Q_2$, $V_1 \subset Q_2$, and $g_2(Q_2, Q_2 \cap X) = (B_2, B_2 \cap E_+^k)$.

The homeomorphism ϕ_2 may be obtained as follows: Let A be a ball with center $h_2 h_1^{-1}(0)$ such that $A \subset h_2 h_1^{-1}(\operatorname{Int} B_1)$. Then there is a homeomorphism $\psi: E^n \twoheadrightarrow E^n$ such that ψ = identity outside of B_2, $\psi(E_+^k) = E_+^k$, and $\psi(B_1) \subset h_1 h_2^{-1}(\operatorname{Int} A)$. Also, there is a homeomorphism $\tilde{\psi}: E^n \twoheadrightarrow E^n$ such that $\tilde{\psi}$ = identity outside of some compact set, $\tilde{\psi}$ = identity on A, $\tilde{\psi}(E_+^k) = E_+^k$, and $\tilde{\psi}(h_2 h_1^{-1}(B_2))$ contains $h_2(\overline{V}_1)$. Both ψ and $\tilde{\psi}$ are homeomorphisms which map each half-ray emanating from the origin onto itself. Then, ϕ_2 may be defined by

$$\phi_2 = \begin{cases} h_2 h_1^{-1} \psi^{-1} h_1 h_2^{-1} \tilde{\psi} h_2 h_1^{-1} \psi h_1 h_2^{-1} & \text{on } h_2 h_1^{-1}(B_2), \\ \tilde{\psi} \quad \text{outside of } h_2 h_1^{-1}(B_2). \end{cases}$$

Continuing in this way, the sequences $Q_1, Q_2, \ldots$ and $g_1, g_2, \ldots$ may be constructed, and the proof is complete.

Remark 3.4.1. Although the above proof is precisely formulated, in order to get an intuitive feeling for the technique, one should regard ψ and $\tilde{\psi}$ as acting in M on the radial structure of V_1 and V_2, respectively.

Exercise 3.4.3. Use the technique of proof of Theorem **3.4.2** to establish the result of Brown mentioned above which states that the monotone union of open n-cells is an open n-cell.

Flattening Theorem 3.4.3 (Lacher). *Let X be a closed, locally flat k-half-string in E^n, $k \leqslant n$, $n \geqslant 4$. Then, $(E^n, X) \approx (E^n, E_+^k)$.*

Proof. Let X be a closed, locally flat k-half-string in E^n, $n \geqslant 4$. Theorem **3.4.2** supplies a neighborhood U of X in E^n and a homeomorphism $h: (U, X) \twoheadrightarrow (E^n, E_+^k)$. Suppose that Y is a locally flat, closed $(n-1)$-string in E^n such that $Y \cap E_+^k = \emptyset$ and such that $h^{-1}(Y)$ is closed in E^n. Exercise **3.3.3** states that locally flat, closed $(n-1)$-strings in E^n are trivially embedded if $n \geqslant 4$; thus, if we can find Y, and if V is the complementary domain of Y which contains E_+^k, the homeomorphism $g = h \mid h^{-1}(\overline{V})$ can be extended to a homeomorphism of (E^n, X) onto (E^n, E_+^k) as desired.

To complete the proof, Y must be constructed. If B_t is the ball of radius t and center 0 in E^n, $h^{-1} \mid B_t$ is uniformly continuous for each t. Choose $\epsilon_i > 0$ so that any set of diameter less than ϵ_i in B_i is mapped by h^{-1} onto a set of diameter less than 1. Let Y be a closed, locally flat $(n-1)$-string in E^n such that $Y \cap E_+^k = \emptyset$ and such that $\operatorname{dist}(y, E_+^k \cap (B_i - \operatorname{Int} B_{i-1})) < \epsilon_i$ for each $y \in Y \cap (B_i - \operatorname{Int} B_{i-1})$,

$i = 1, 2, \ldots$. It is then clear that a sequence $\{y_j\} \subset Y$ tends to infinity if and only if $\{h^{-1}(y_j)\}$ tends to infinity, so that $h^{-1}(Y)$ is closed in E^n. This completes the proof of Theorem **3.4.3**.

Corollary 3.4.2. *Let D be a k-cell in S^n, $k \leqslant n$, $n \geqslant 4$, and suppose that $D - p$ is locally flat, where $p \in \mathrm{Bd}\, D$. Then, $(S^n, D) \approx (S^n, I^k)$.*

Corollary 3.4.3. *Let M be a k-manifold with boundary contained in the interior of an n-manifold N, $k \leqslant n$, $n \geqslant 4$. Suppose that* Int M *is locally flat in N, and denote by E the set of points of* Bd M *at which M fails to be locally flat. Then, E does not contain an isolated point, and hence E is either empty or uncountable.*

PROOF. It follows from Corollary **3.4.2** that a k-cell in E^n, $n \geqslant 4$, may not fail to be locally flat at precisely one point if that point is a boundary point, and so Corollary **3.4.3** follows by applying this result locally to a supposed isolated point of E.

The next theorem is the analog of Theorem **3.4.2** for strings. Techniques of the nature of those employed in this section do not seem to be strong enough to establish an analog of Theorem **3.4.3** for strings. Later, we will present some work of Stallings [2] which uses engulfing and from which such an analog follows for strings of codimension at least three when $n \geqslant 5$.

Flattening Theorem 3.4.4 (Lacher). *Let X be a locally flat k-string in the n-manifold M, $k < n$. Then, there is an open set U in M containing X such that $(U, X) \approx (E^n, E^k)$.*

PROOF. Three lemmas which are analogs of Lemmas **3.4.1**, **3.4.2**, and **3.4.3** can be stated and proved by letting X be a locally flat k-string in M, $k < n$, instead of a locally flat k-half-string, and replacing E_+^k with E^k both in the statements and proofs of these three lemmas. Theorem **3.4.4** follows from the three new lemmas in exactly the same way that Theorem **3.4.2** followed from the old lemmas.

3.5. PL UNKNOTTING INFINITE POLYHEDRA IN THE TRIVIAL RANGE

A polyhedron P is said to **PL unknot** in the PL manifold Q if for any two PL embeddings f and g of P into Q which are homotopic in Q, there is a PL homeomorphism $h: Q \twoheadrightarrow Q$ such that $hf = g$.

If h can be realized by an ambient isotopy, then P is said to **isotopically PL unknot** in Q. (In the case $Q = E^n$, any two such embedding f and g are always homotopic.)

Gugenheim in his classical paper [1], written under the direction of Whitehead, proved, among other things, the following unknotting theorem (Theorem 5 of that paper).

Gugenheim's Unknotting Theorem. *A k-polyhedron* PL *unknots in E^n if $n \geqslant 2k + 2$.*

EXERCISE 3.5.1. Show that Gugenheim's unknotting theorem is a "best possible" result by exhibiting a p-sphere and a q-sphere which are linked in S^n (E^n) where $n = p + q + 1$ (see Exercise **1.6.14**).

Gugenheim's unknotting theorem was generalized in Theorem 5.5 of [Bing and Kister, 1] to an epsilontic type of isotopy unknotting theorem. A generalization (which will be applied in the next section) of the Bing–Kister unknotting theorem to infinite polyhedra is the main result of this section (Theorem **3.5.1**). In order to state this theorem in modern terminology we make the following definition: Let A be a subset of the topological manifold M which has metric d. If $\epsilon \geqslant 0$, then an **ϵ-push** of (M, A) is an isotopy h_t of M onto itself such that

(1) $h_0 = 1$,
(2) h_t is an ϵ-homeomorphism for all t,
(3) $h_t \mid M - N_\epsilon(A) = 1$,

If h_t is PL, then it is called a **PL ϵ-push** of (M, A).

Unknotting Theorem 3.5.1. *Suppose that P is a (possibly infinite) k-polyhedron, $n \geqslant 2k + 2$, that L is a finite subpolyhedron of P and that f and g are two* PL *embeddings of P into E^n such that* $\text{dist}(f, g) < \epsilon$ ($\epsilon \geqslant 0$) *and $f \mid P - L = g \mid P - L$. Then, there is a* PL *$\epsilon$-push h_t of $(E^n, f(L))$ which is fixed on $f(\text{Cl}(P - L))$ such that $h_1 f = g$. (Furthermore, no point of E^n moves along a path of length as much as ϵ.)*

If K is a finite k-complex, a **prismatic triangulation** of $K \times I$ is a triangulation for which there is an ordering of the vertices of K, $v_1, v_2, \ldots, v_p$ such that the triangulation includes all faces of s-simplexes of $K \times I$, $1 \leqslant s \leqslant k + 1$, having vertices

$$(v_{i_1}, 1), (v_{i_2}, 1), \ldots, (v_{i_j}, 1), (v_{i_j}, 0), \ldots, (v_{i_s}, 0)$$

for some choice of integers $i_1 < i_2 < \cdots < i_s$, where $v_{i_1}, v_{i_2}, \ldots, v_{i_s}$ span a simplex in K, and for some j, $1 \leqslant j \leqslant s$.

EXERCISE 3.5.2. Show that every ordering of the vertices of a complex K determines a prismatic triangulation of $K \times I$.

The following lemma is included only to motivate the proof of Lemma **3.5.2** and may be skipped if desired.

Lemma 3.5.1. *Let K be a (possibly infinite) k-complex and let L be a finite subcomplex of K. Let $K \times I$ have a triangulation which contains a prismatic triangulation of $L \times I$ as a subcomplex. Let X be the set of vertices in $|K| - \mathrm{Cl}(|K| - |L|)$ and let $R = \bigcup_{v \in X} \mathrm{St}(v \times I, K \times I)$. Suppose $\epsilon > 0$ and f is a map of $K \times I$ into E^n such that $f \mid (K \times 0) \cup R$ is an embedding, $f(x, t) = f(x, 0)$ for each vertex x of K not in X, f is linear on each simplex in $K \times I$ and $f(v \times I)$ has length less than ϵ for each vertex v in X. Then there is a PL ϵ-push h_t of $(E^n, f(L \times 0))$ which is fixed on $f(\mathrm{Cl}(K - L) \times I)$ such that $h_1 f(x, 0) = f(x, 1)$ for each $x \in K$. (Furthermore, no point of E^n moves along a path of length as much as ϵ.)* (See Fig. **3.5.1**.)

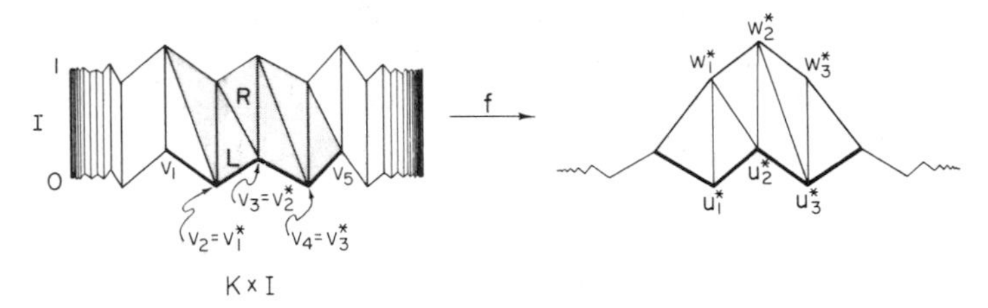

Figure 3.5.1

PROOF. Let $v_1^*, v_2^*, \ldots, v_m^*$ be the ordering of the vertices in X induced by the ordering of the vertices of L which determines the given prismatic triangulation of $L \times I$. For convenience denote $f(v_i^*, 0)$ and $f(v_i^*, 1)$ by u_i^* and w_i^*, respectively, and let $K_0 = f(K \times 0)$. The isotopy h_t will be constructed by taking the u_i^* onto the w_i^* one at a time.

The isotopy is broken into m pieces—one for each vertex of X. Each h_t will be linear on each simplex of K_0. In general $h_{i/m}(u_j^*)$ is w_j^* or u_j^* depending on whether $j \leqslant i$ or $j > i$. Hence, $h_{1/m}$ moves u_1^* to w_1^*, $h_{2/m}$ moves u_2^* to w_2^*, and so forth. Since each h_t is linear on each simplex of K_0, this completely describes each $h_{i/m}(K_0)$. We denote $h_{i/m}(K_0)$ by $K_{i/m}$.

Denote by $\mathscr{L}_i$ the complex $h_{(i-1)/m} f(Lk(v_i^*, K) \times 0)$. Then $\mathscr{L}_i$ is the link of u_i^* in $K_{(i-1)/m}$. For each simplex σ in $\mathscr{L}_1$, the join $[u_1^*, w_1^*] * \sigma$ intersects K_0 only in $u_1^* * \sigma$ since $[u_1^*, w_1^*] * \sigma \subset R$, $f \mid R$ is an embedding, $f(R) \cap f((K - L) \times I) = \emptyset$; and f is linear on each simplex in $K \times I$. Then, $[u_1{}^*, w_1{}^*] * \mathscr{L}_1$ intersects K_0 only in $u_1{}^* * \mathscr{L}_1$. In general $[u_i{}^*, w_i{}^*] * \mathscr{L}_i$ intersects $K_{(i-1)/m}$ only in $u_i{}^* * \mathscr{L}_i$.

Construction of h_t *for* $0 \leqslant t \leqslant 1/m$: It is not difficult to obtain a PL n-ball $B_1{}^n$ such that

(1) $([u_1{}^*, w_1{}^*] * \mathscr{L}_1) - \mathscr{L}_1 \subset \operatorname{Int} B_1{}^n$,
(2) $B_1{}^n \cap K_0 = u_1{}^* * \mathscr{L}_1$.
(3) $B_1{}^n$ is starlike with respect to each point of $[u_1{}^*, w_1{}^*]$ (that is, given $y \in [u_1{}^*, w_1{}^*]$, $B^n = y * \operatorname{Bd} B^n$).
(4) $B_1{}^n$ lies in a small neighborhood of $[u_1{}^*, w_1{}^*] * \mathscr{L}_1$.

As t moves from 0 to $1/m$, let h_t move $u_1{}^*$ linearly to $w_1{}^*$. For each t, h_t $(0 \leqslant t \leqslant 1/m)$ is fixed on $E^n - B^n$ and is linear on each segment from u_1 to $\operatorname{Bd} B^n$. Note that if h_t $(0 \leqslant t \leqslant 1/m)$ moves a point at all, it moves it in a straight path parallel to the vector from $u_1{}^*$ to $w_1{}^*$. Also note that h_t $(0 \leqslant t \leqslant 1/m)$ does not move any points of K_0 except possibly those in $u_1{}^* * \mathscr{L}_1$. If p is a point in a simplex of K_0 in $u_1{}^* * \mathscr{L}_1$ and p has barycentric coordinates $(x_1, x_2, ..., x_j)$ with $\Sigma x_i = 1$ and x_1 associated with the vertex $u_1{}^*$, the length of the path through which p is moved under h_t $(0 \leqslant t \leqslant 1/m)$ is x_1 times the distance from $u_1{}^*$ to $w_1{}^*$.

Construction of h_t *for* $i/m \leqslant t \leqslant (i+1)/m$, $i = 1, 2, ..., m-1$: In a similar fashion, the other $u_i{}^*$'s are pushed one at a time over to the corresponding $w_i{}^*$'s. In general, let $B_i{}^n$ be a PL n-ball which satisfies the properties corresponding to $B_1{}^n$ above. Choose g_t $[1/m \leqslant t \leqslant (i+1)/m]$ to be an isotopy such that $g_{i/m} = 1$, $g_t \mid E^n - B^n = 1 \mid E^n - B^n$, g_t moves u_{i+1}^* straight to w_{i+1}^*, and for each t, g_t is linear on each segment from u_{i+1}^* to $\operatorname{Bd} B^n$.

The ϵ-push h_t of $(E^n, f(L \times 0))$ is defined to be $g_t h_{i/m}$ for $i/m \leqslant t \leqslant (i+1)/m$.

Exercise 3.5.3. Verify that, by choosing $B_i{}^n$ to lie in a small enough neighborhood of $[u_i{}^*, w_i{}^*] * \mathscr{L}_i$, h_t is indeed an ϵ-push of $(E^n, f(L \times 0))$ by showing that no point of E^n moves along a path of length as much as ϵ. (Hint: First consider points of K_0 and then the remaining points of E^n.)

Exercise 3.5.4. Prove Theorem **3.5.1** for the case $n \geqslant 2k+3$ by using Lemma **3.5.1**.

Lemma 3.5.2. *Let K be a (possibly infinite) k-complex and let L be a finite subcomplex of K. Let $K \times I$ have a triangulation which contains a prismatic triangulation of $L \times I$ as a subcomplex. Let X be the set of vertices in* $|K| - \operatorname{Cl}(|K| - |L|)$ *and let* $R = \bigcup_{v \in X} \operatorname{St}(v \times I, K \times I)$.

Suppose that f is a map of $K \times I$ into E^n which is linear on each simplex of $K \times I$ and is such that $f(R) \cap f((K - L) \times I) = \emptyset$ and $f(x, t) = f(x, 0)$ for $(x, t) \in (K \times I) - R$. Suppose further that a and b are numbers, with $0 \leqslant a < b \leqslant 1$, and $\epsilon > 0$ so that $f \mid R^*$, where $R^* = (K \times [a, b]) \cap R$, is an embedding and $f(v_i \times [a, b])$ has length less than ϵ for each $v_i \in X$. Then, there is a PL *ϵ-push h_t of $(E^n, f(L \times a))$ which is fixed on $f(\mathrm{Cl}(K - L) \times I)$ such that $h_1 f(x, a) = f(x, b)$ for each $x \in K$. (Furthermore, no point of E^n moves along a path of length as much as ϵ.)* (See Fig. **3.5.2**.)

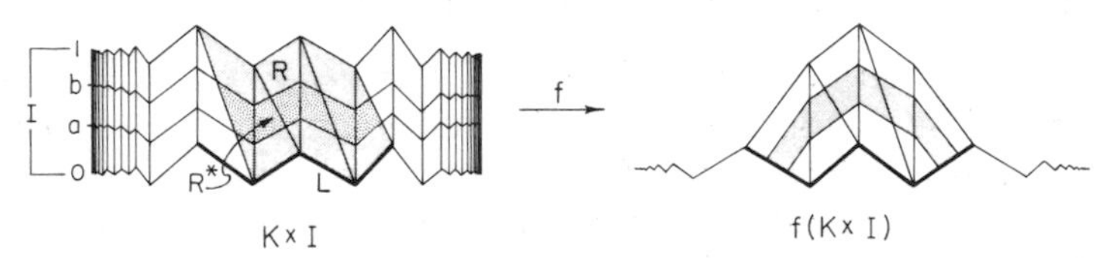

Figure 3.5.2

PROOF. (In proving this lemma one is tempted to seek a certain kind of triangulation of $K \times [a, b]$ that will enable one to apply the preceding lemma. However, in the case $k = 2$, $n = 3$, there is no triangulation of $K \times [a, b]$ such that each simplex lies in a simplex of $K \times I$ and has a vertical edge. Hence, we must modify the technique developed in the proof of Lemma **3.5.1**.)

Let $v_1^*, v_2^*, \ldots, v_m^*$ be the ordering of the vertices in X induced by the ordering of the vertices of L which determines the given prismatic triangulation of $L \times I$. Let $w_i^* = f(v_i^*, a)$, $u_i^* = f(v_i^*, b)$, $K_a = f(K \times a)$, and T be the cellular subdivision of $f(R^*)$ such that C is an element of T if and only if C is the image under f of the intersection of R^* with a simplex in $K \times I$.

Let C_i be the union of all elements of T which have $[w_i^*, u_i^*]$ as an edge. Let A_i be a collection of all maximal segments in C_i parallel to $[w_i^*, u_i^*]$. We regard these segments as oriented in the direction from w_i^* to u_i^* and refer to the sum of all lower ends of all such segments as the bottom of C_i and the sum of all upper ends as the top of C_i. Our isotopy is broken into m pieces, one for each C_i.

Construction of h_t for $0 \leqslant t \leqslant 1/m$: The only points of K_a that move under this part of the isotopy are those on the bottom of C_1 and $h_{1/m}$ carries these bottom points of C_1 to the corresponding top points of C_1. Any point of E^n that moves at all, moves parallel to $[w_1^*, u_1^*]$ and in the direction from w_1^* to u_1^*. Only points "close" to C_1 move. To get such an isotopy one might first enlarge the collection A_1 in a PL fashion to a continuous collection B_1 such that each element of B_1 is either a point or a segment parallel to $[w_1^*, u_1^*]$, each line parallel to $[w_1^*, u_1^*]$ contains one and only one element of B_1, no nondegenerate element of B_1 intersects K_a except those in A_1, and each point of a nondegenerate element of B_1 lies close to C_1. Let $\epsilon(b)$ be a continuous PL map which takes on small nonnegative real numbers, such that $\epsilon(b) = 0$ if and only if b is a degenerate element of B_1 and if b^+ denotes the segment obtained by extending b a distance of $\epsilon(b)$ in both directions, then $b^+ - b$ misses K_a. The isotopy is fixed except on the b^+'s and moves points along them so that the lower end of each b is moved to its top.

Construction of h_t *for* $i/m \leqslant t \leqslant (i+1)/m$, $i = 1, 2, ..., m-1$: Define $h_t = g_t h_{1/m}(1/m \leqslant t \leqslant 2/m)$ where g_t is as follows: g_0 is the identity; the only parts of $h_{1/m}(K_a)$ that move under g_t are those on the bottom of C_2 and $g_{2/m}$ carries these bottom points to the corresponding top points of C_2; any point of E^n that moves at all, moves parallel to $[w_2^*, u_2^*]$ and in the direction from w_2^* to u_2^*; only points close to C_2 are moved.

We continue defining h_t $(0 \leqslant t \leqslant 1)$ in this fashion until K_a has been moved onto $f(K \times b)$. In general, $h_t = g_t h_{(i-1)/m}((i-1)/m \leqslant t \leqslant i/m)$ where $g_{(i-1)/m}$ is the identity, $g_{i/m}$ pushes the bottom of C_i to the top, and g_t otherwise behaves in a decent fashion as suggested above.

The fact that no point of E^n moves along a path of length as much as ϵ, follows in a fashion suggested by Exercise **3.5.2** by insisting that nondegenerate elements of B_1 lie close to C_1.

Let σ and τ be two joinable simplexes in E^n. A point p in $\sigma * \tau$ **divides** σ, τ **in the ratio** a **to** b if p divides the segment from σ to τ on which it lies in the ratio a to b.

In E^n, let X be a finite set of points, Y be a possibly infinite set of points, and Y^* be a finite subset of Y. Then, X is in **ratio changing general position** with respect to the pair (Y, Y^*) if the following two conditions are satisfied.

1. No point p of X lies in an r-hyperplane $(r \leqslant n-1)$ determined by points of $X \cup Y - \{p\}$.
2. If σ_1, σ_2 are disjoint simplexes with vertices in X and τ_1, τ_2 are disjoint simplexes with vertices in Y^* where

$$\dim \sigma_1 + \dim \sigma_2 + \dim \tau_1 + \dim \tau_2 \leqslant n-2,$$

then either $(\sigma_1 * \tau_1) \cap (\sigma_2 * \tau_2) = \emptyset$ or the point of intersection of these two joins divides σ_1, τ_1 in a different ratio from the ratio in which it divides σ_2, τ_2.

Lemma 3.5.3. *Suppose that in E^n, X is a finite set of points, Y is a possibly infinite set of points and Y^* is a finite subset of Y such that X is in ratio changing general position with respect to the pair (Y, Y^*). Then, the set*

$$X_+ = \{p \mid X \cup \{p\} \text{ is in ratio changing general position with respect to } (Y, Y^*)\}$$

is dense in E^n.

PROOF. It follows from the Baire category theorem that the set A of points which are not in the countable number of r-hyperplanes ($r \leqslant n - 1$) determined by the various tuples of points of $X \cup Y$ is dense in E^n. Thus, A satisfies the first restriction imposed by ratio changing general position.

We will be through if we can show that each open set N_0 contains an open set N such that for each point p of $N \cap A$, $X \cup \{p\}$ is in ratio changing general position with respect to (Y, Y^*). In order to do this, consider disjoint simplexes σ_1, σ_2 with vertices in X and disjoint simplexes τ_1, τ_2 with vertices in Y^* such that

$$\dim \sigma_1 + \dim \sigma_2 + \dim \tau_1 + \dim \tau_2 \leqslant n - 3.$$

Let p_1 be any point in $N_0 \cap A$ and let σ_1' be the simplex $p_1 * \sigma_1$.

If $\sigma_1' * \tau_1$ misses $\sigma_2 * \tau_2$, choose N_1 to be a neighborhood of p_1 so small that $N_1 \subset N_0$ and for each point p in $N_1 \cap A$, $(p * \sigma_1) * \tau_1$ misses $\sigma_2 * \tau_2$. If $\sigma_1' * \tau_1$ intersects $\sigma_2 * \tau_2$ in a point q but does not divide σ_1', τ_1 in the same ratio that it divides σ_2, τ_2, then we also pick N_1 to be a small neighborhood of p_1. If q divides σ_1', τ_1 in the same ratio that it divides σ_2, τ_2, consider the simplex τ_1' consisting of all points p such that q divides some segment from p to a point of τ_1 in the same ratio that q divides σ_2, τ_2. Let p_2 be a point of N_0 which is in the r-plane determined by $\sigma_1' * \tau_1$, but which is not in the plane determined by σ_1 and τ_1'. Then, p_2 serves the role of p_1 in the preceding two cases and we can get a small N_1 containing it.

A finite number of applications of this method to all possible σ_i in X and τ_i in Y^*, $i = 1, 2$, where each N_i is chosen to lie in the preceding, leads to an N satisfying the conditions of the lemma.

Lemma 3.5.4. *Let K be a (possibly infinite) k-complex in E^n and let L be a finite subcomplex of K. Let Y be the vertices of K, let Y^* be the vertices of L and let $X = \{v_1^*, v_2^*, \ldots, v_m^*\}$ be the vertices of K in $|K| - \mathrm{Cl}(|K| - |L|)$. Then, given $\epsilon > 0$, there is a* PL *ϵ-push h_t of $(E^n, |L|)$ which is fixed on $\mathrm{Cl}(|K| - |L|)$ such that*

$\{h_1(v_1^*), h_1(v_2^*), ..., h_1(v_m^*)\}$ *is in ratio changing general position with respect to the pair* (Y, Y^*) *and* h_1 *is linear on each simplex of* K. *(Furthermore, each point of* E^n *moves along a polygonal path of length less than* ϵ.)

Proof. Our isotopy is broken into pieces, one for each v_i^*. For convenience in notation in describing the ith piece, we suppose that $v_1^*, v_2^*, ..., v_{i-1}^*$ are already in ratio changing general position with respect to (Y, Y^*) and $h_{(i-1)/m}$ is the identity.

Let D be a ball of small radius centered at v_i^*. Then, it is not difficult to obtain a PL n-ball B^n such that

(1) $(v_i^* * L) - L \subset \text{Int } B^n$, where L is the link of v_i in K,

(2) $B^n \cap K = v_i^* * L$,

(3) B^n is starlike from each point of D, (that is, given $y \in D$, $B^n = y * \text{Bd } B^n$).

(4) B^n lies in a small neighborhood of $v_i^* * L$.

Lemma **3.5.3** implies that there is a point v_i' in D very close to v_i^* so that $\{v_1^*, v_2^*, ..., v_{i-1}^*, v_i'\}$ is in ratio changing general position with respect to Y. Then, h_t $((i-1)/m \leqslant t \leqslant i/m)$ is the identity outside B^n, h_t moves v_i^* linearly to v_i' as t moves from $(i-1)/m$ to i/m and for each t, h_t is linear on each segment from v_i to $\text{Bd } B^n$. Note that paths swept out are short (no more than $\text{dist}(v_i^*, v_i')$).

Proof of Theorem 3.5.1. Regard K as a triangulation of P that contains a triangulation of L, which we also call L, such that f and g are linear on each simplex of K. Let Y denote the image of the vertices of K under f, let $Y^* = \{f(v_1), f(v_2), ..., f(v_p)\}$ where $v_1, v_2, ..., v_p$ are the vertices of L, and let $X = \{v_1^*, v_2^*, ..., v_m^*\}$ be the set of vertices of $|K| - \text{Cl}(|K| - |L|)$ with the ordering induced by the above ordering of the vertices of L. By using Lemma **3.5.4**, we may assume that $\{g(v_1^*), g(v_2^*), ..., g(v_m^*)\}$ is in ratio changing general position with respect to (Y, Y^*).

Let $K \times I$ be triangulated to contain a prismatic triangulation of $L \times I$ relative to the ordering $v_1, v_2, ..., v_p$ as a subcomplex. Let $F: K \times I \to E^n$ be the map defined by $F(v, t) = f(v)$ for all $(v, t) \in (K - L) \times [0, 1]$, $F(v_i, 0) = f(v_i)$ and $F(v_i, 1) = g(v_i)$ for the vertices $v_1, v_2, ..., v_p$ of L and F is linear on each simplex of $K \times I$. Let $R = \bigcup_{v \in X} \text{St}(v \times I, K \times I)$. It is not necessarily true that $F | R$ is a homeomorphism since there may be two simplexes whose images intersect. However, since $\{g(v_1^*), g(v_2^*), ..., g(v_m^*)\}$ is in ratio changing general position with respect to (Y, Y^*), there is a sequence $0 = t_0 < t_1 < \cdots < t_j = 1$ such that F is a homeomorphism on each

$(K \times [t_{i-1}, t_i]) \cap R$. Hence, the theorem follows from j applications of Lemma **3.5.2**.

REMARK 3.5.1. In proving Theorem **3.5.1**, one is tempted to use the methods of Lemma **3.5.1** to move the $f(v_i{}^*)$ to the $g(v_i{}^*)$. If this method had worked, we could have used ordinary general position arguments and not resorted to the more complicated ratio changing general position. However, this method fails. Suppose σ_1, σ_2 are two disjoint simplexes of $f(L)$ having v_1, v_2, repectively, as vertices in X. If $g(v_1) * f(\sigma_1)$ intersects $g(v_2) * f(\sigma_2)$ in a point q near both $g(v_1)$ and $f(v_2)$, then the first piece of an isotopy like that used in Lemma **3.5.1** would move a point near $f(\sigma_1)$ to q and the second piece would move q near $g(v_2)$, thereby sweeping out a path of length perhaps longer than ϵ.

Theorem **3.5.1** was stated in a form for application in the next section, however the techniques of this section suffice to prove the following theorem.

Unknotting Theorem 3.5.2. *Suppose that K is a (possibly infinite) polyhedron and that L is a subpolyhedron of K. Suppose that $\epsilon(x)$ is a positive real-valued function defined on K. Then, there exists a positive real-valued function $\delta(x)$ defined on K such that for any two* PL *embeddings f, g: $K \to E^n$ such that* $\text{dist}(f(x), g(x)) \leqslant \delta(x)$ *for all $x \in K$ and $f \mid L = g \mid L$, there is an $\epsilon(x)$-push* (defined analogously to ϵ-push) *h_t of $(E^n, f(K-L))$ which is fixed on $f(L)$ such that $h_1 f = g$. Furthermore, h_t is* PL *modulo* $E^n - (\text{Cl}(f(K)) - f(K))$.

3.6. ε(x)-TAMING LOCALLY TAME EMBEDDINGS OF INFINITE POLYHEDRA IN THE TRIVIAL RANGE

About 1962 a major breakthrough was made in the taming of general objects in [Gluck, 1, 2] where it was shown that, by using techniques of [Homma, 1] and of [Bing and Kister, 1] (developed in the last section), a locally tame embedding of a polyhedron into a PL manifold is ϵ-tame in the trivial range. (Much of the credit for this result is due to the creativity of Homma, since in [1] he developed the key techniques of the proof. Gluck observed that the "global" tameness condition of Homma can be replaced by a "local" tameness condition and Gluck also observed that Homma's techniques allow one to tame by isotopy rather than homeomorphism. Gluck's exposition of Homma's techniques is more readable than Homma's exposition.) It also followed that a locally flat embedding of a closed PL manifold into a PL manifold is ϵ-tame in the trivial range. Greathouse obtained similar results in [1]. These results were carried further in [Bryant, 2] and [Dancis, 1] where, by using similar techniques, it was shown that if the set of points on which an embedding is not known to be locally tame is mapped into a tame polyhedron in the trivial range with respect to the ambient PL manifold, then the embedding must be tame, in the trivial range. The results of Bryant and Dancis will be presented later. Other modifications of the Homma–Gluck techniques have been employed often in the literature to obtain results related to the ones mentioned above. (For instance, see [Bryant and Seebeck, 1] and [Dancis, 2].) Later, in Section **5.5**, we will discuss recent generalizations to lower codimensions of the results of this section as well as prove some such generalizations which were developed in [Rushing, 2–4]. Precise statements of the main results of this section follow which generalize the first results mentioned above to infinite polyhedra. (It should be mentioned that both Homma and Gluck failed to establish the solvability condition in the generality that they applied in their proof. However, enough care is taken in this section and the preceding section to eliminate this difficulty.)

Taming Theorem 3.6.1. *Let R be a subpolyhedron of the possibly infinite polyhedron P^k. Let $f\colon P^k \to M^n$ be a locally tame embedding of P^k into the interior of the* PL *n-manifold M^n, $2k+2 \leqslant n$, such that $f \mid R$ is* PL. *Let $\tilde{P}^k$ be another polyhedron and let $H\colon P^k \twoheadrightarrow \tilde{P}^k$ be a topological homeomorphism such that $h \mid R$ is* PL. *Then, fh^{-1} is $\epsilon(x)$-tame keeping $fh^{-1}(h(R)) = f(R)$ fixed.*

Corollary 3.6.1. *Let f be a locally flat embedding of the (possibly noncompact and possibly with boundary)* PL *manifold M^k into the interior of the* PL *n-manifold M^n, $2k+2 \leqslant n$, such that $f \mid R$ is* PL *where R is a subpolyhedron of M^k. Then, f is $\epsilon(x)$-tame keeping $f(R)$ fixed.*

The short proof of the next theorem (which is included as a sidelight) will involve a folklore result the proof of which will not be presented.

Unknotting Theorem 3.6.2. *Let f and f' be locally tame embeddings of the polyhedron P^k into the unbounded* PL *manifold M^n. If $2k+2 \leqslant n$ and f is homotopic to f', then there is a homeomorphism h of M^n onto itself which is isotopic to the identity such that $hf = f'$.*

Exercise 3.6.1. Show that any PL manifold M^n can be remetrized so that M is complete in the new metric and so that sets of sufficiently small diameter can be enclosed in an n-cell lying in M. (Whenever we have occasion to use a metric in this section, we will assume that it has these two properties.)

Let M be a manifold and let $x \in M$, then we define

$$N_\epsilon(x) = \{y \in M \mid \text{dist}(y, x) < \epsilon\}.$$

Let $\epsilon(x)$ be a positive continuous function defined on $X \subset M$. By the **$\epsilon(x)$-neighborhood** of X, $N_{\epsilon(x)}(X)$, we mean $\bigcup_{x \in X} N_{\epsilon(x)}(x)$. If $\epsilon(x)$ is defined on all of M, we define an **$\epsilon(x)$-isotopy** of M to be any isotopy $e_t\colon M \to M$ such that $e_0 = 1$ and $\text{dist}(x, e_t(x)) \leqslant \epsilon(x)$ for all $x \in M$ and $t \in I$. An $\epsilon(x)$-isotopy of M is called an **$\epsilon(x)$-push** of (M, X) if $e_t =$ identity outside $N_{\epsilon(x)}(X)$ for each $t \in I$. An embedding f of a (possibly infinite) polyhedron P into the PL manifold M is **$\epsilon(x)$-tame** if for each positive, continuous function $\epsilon(x)$ defined on M, there is an $\epsilon(x)$-push e_t of $(M, f(P))$ such that $e_1 f$ is PL. If R is a subpolyhedron of P such that $f \mid R$ is PL, then f is said to be **$\epsilon(x)$-tame keeping $f(R)$ fixed** if for each $\epsilon(x)$ there is an $\epsilon(x)$-push e_t of $(M, f(P - R))$ such that $e_t \mid f(R) = 1 \mid f(R)$ and $e_1 f$ is PL. An integer k is said to be in the **trivial range** with respect to an integer n if $2k + 2 \leqslant n$.

Let $f\colon P^k \to M^n$ be an embedding of the (possibly infinite) k-polyhedron P^k into the interior of the topological n-manifold M^n. If there is a triangulation of P^k such that for each point $x \in P^k$ there is an open neighborhood U of $f(x)$ in M^n and a homeomorphism $h_U\colon U \twoheadrightarrow E^n$ such that $h_U f$ is PL on some neighborhood of x, then we say that f is a **locally tame** embedding.

REMARK 3.6.1. Note that once a triangulation is chosen for P^k, the same triangulation must be used for deciding whether f is locally tame at each point $x \in P^k$. However, if another embedding f' is given, an entirely different triangulation of P^k may be used to decide whether f' is locally tame. Gluck in his definition of locally tame and throughout his papers [1, 2] uses the idea of abstract triangulations, that is, topological images of rectilinear triangulations. In order to make sense of abstract triangulations and PL maps between them, one must always go back to rectilinear triangulations, hence we will not use the notion of abstract triangulation (see the Remark **3.6.2**). However, it often aids in the intuitive understanding of a concept to visualize the "wiggly structure" induced by a topological homeomorphism from a rectilinear complex. Although our definition of locally tame is not the same as that of Gluck, is it easily seen to be equivalent. (It is quite instructive to consider this equivalence and to notice that the *hauptvermutung* for open cells is not necessary to establish it.)

Let X denote a metric space and A a subset whose closure is compact.

Let M denote a topological manifold with a complete metric d, and let $h: X \to M$ be a fixed embedding. Then, $\text{Hom}_h(X, A; M)$ denotes the set of all embeddings of X into M which agree with h on $X - A$. If $f, g \in \text{Hom}_h(X, A; M)$, we define a distance function

$$d(f, g) = \text{LUB}_{x \in X}\, d(f(x), g(x)) = \text{LUB}_{x \in \bar{A}}\, d(f(x), g(x)),$$

which exists because $\bar{A}$ is compact. This distance function makes $\text{Hom}_f(X, A; M)$ into a metric space.

Let F be a subset of $\text{Hom}_f(X, A; M)$ with the property that, for each $g \in \text{Hom}_f(X, A; M)$ and each $\epsilon > 0$, there is an $f \in F$ with $d(f, g) < \epsilon$. Then, we say that F is **dense** in $\text{Hom}_f(X, A; M)$.

Let F be a subset of $\text{Hom}_f(X, A; M)$ with the following property: for each $\epsilon > 0$ there is a $\delta > 0$ such that if $f', f'' \in F$ and $d(f', f'') < \delta$, then there is an ϵ-push h of $(M, f'(A))$ keeping $X - A$ fixed such that $hf' = f''$. Then, we say that F is **solvable**.

The next theorem is fundamental to the proofs of the main results of this section.

Theorem 3.6.3. *The union of two dense, solvable subsets of* $\text{Hom}_f(X, A; M)$ *is dense and solvable.*

Before proving Theorem **3.6.3** we will establish four preliminary lemmas. In these lemmas, M will denote a topological manifold with complete metric d and A will denote a subset of M with compact closure $\bar{A}$.

Lemma 3.6.1. *If* h_t *is an* ϵ*-push of* (M, A)*, then* h_t^{-1} *is a* 2ϵ*-push of* $(M, h_1(A))$.

PROOF. Since h_t is an ϵ-isotopy, so is h_t^{-1}. But h_t^{-1} may not restrict to the identity on $M - N_\epsilon(h_1(A))$. However, since $N_\epsilon(A) \subset N_{2\epsilon}(h_1(A))$, h_t^{-1} does restrict to the identity on $M - N_{2\epsilon}(h_1(A))$. The isotopy h_t is invertible by Exercise **1.3.8**, and so h_t^{-1} is a 2ϵ-push of $(M, h_1(A))$.

Lemma 3.6.2. *Let* $h_t^1, h_t^2, \ldots, h_t^r$ *be isotopies of* M *such that* h_t^1 *is an* ϵ_1*-push of* (M, A) *and* h_t^i *is an* ϵ_i*-push of* $(M, h_1^{i-1} \cdots h_1^1(A))$ *for* $i = 2, \ldots, r$. *Let* $h_1^0 = 1$ *and* $g_1^i = h_1^i h_1^{i-1} \cdots h_1^0$. *Then, the isotopy* $g_t = h_{(rt-i)}^{i+1} g_1^i$, $t \in [i/r, (i+1)/r]$, $i = 0, 1, \ldots, r$ *is an* $(\epsilon_1 + \epsilon_2 + \cdots + \epsilon_r)$*-push of* (M, A).

PROOF. Certainly for $t \in [i/r, (i+1)/r]$, g_t is a Σ_{i+1}-homeomorphism, where $\Sigma_j = \epsilon_1 + \cdots + \epsilon_j$. Furthermore, if $d(x, A) \geqslant \Sigma_r \geqslant \epsilon_1$, then $h_t^1(x) = x$ and $d(x, g_{1/r}(A)) \geqslant \Sigma_r - \Sigma_1 \geqslant \epsilon_2$. But then $h_t^2(x) = x$ and

$d(x, g_{2/r}(A)) \geqslant \Sigma_r - \Sigma_2 \geqslant \epsilon_3$. Continuing in this fashion we see that $g_t(x) = x$ for all t, so that $g_t \mid M - N_{\Sigma_r}(A) = 1$. Hence, g_t is a Σ_r-push of (M, A) as desired.

Lemma 3.6.3. *Suppose that we are given the following:*

(1) *An infinite sequence of positive numbers* $\{\epsilon_1, \epsilon_2, ...\}$ *such that* $\epsilon_1 + \epsilon_2 + \cdots$ *converges;*

(2) *An infinite sequence of isotopies of* M, $\{h_t^1, h_t^2, ...\}$, *such that* h_t^1 *is an* ϵ_1*-push of* (M, A) *and* h_t^i *is an* ϵ_i*-push of* $(M, g_1^{i-1}(A))$ *for* $i \geqslant 1$, *where* $g_1^i = h_1^i h_1^{i-1} \cdots h_1^1$ $(h_1^0 = 1)$;

(3) $\Sigma_i = \epsilon_1 + \cdots + \epsilon_i$, $\Sigma = \epsilon_1 + \epsilon_2 + \cdots$.

Then,

(1) $g_1^1, g_1^2, ...$ *converges uniformly to a continuous map* $g\colon M \to M$;

(2) g *is onto;*

(3) g *is a* Σ*-map, that is,* $d(x, g(x)) \leqslant \Sigma$ *for all* $x \in M$;

(4) $g \mid M - N_\Sigma(A) = 1$.

(5) *Let*

$$g_t = \begin{cases} h^{i+1}_{(i+2)(i+1)/(2-i^2)}\left(t - \dfrac{i}{i+1}\right) g_1^i, & t \in \left[\dfrac{i}{i+1}, \dfrac{i+1}{i+2}\right], \quad i = 0, 1, ..., \\ g_1(x) = g(x). \end{cases}$$

Then, $G\colon M \times [0, 1] \to M$ *defined by* $G(x, t) = g_t(x)$ *is continuous.*

(6) *If* $t \in [0, 1)$, *then* g_t *is a* Σ*-homeomorphism of* M *which restricts to the identity on* $M - N_\Sigma(A)$.

(7) *If* g *is one-to-one, then* g_t *is a* Σ*-push of* (M, A).

PROOF.

(1) If $x \in M$, then $d(g_n(x), g_{n+p}(x)) \leqslant \epsilon_{n+1} + \epsilon_{n+2} + \cdots + \epsilon_{n+p} \leqslant \Sigma - \Sigma_n$. Then, since M is complete, the sequence of homeomorphisms $g_1^1, g_1^2, ...$ converges uniformly to a continuous map $g\colon M \to M$. (One might hope that g is the last stage of an ϵ-push of (M, A); however, it is easy to see that g may fail to be one-to-one.)

(2) Since $\bar{A}$ is compact and g is continuous, $g(\bar{A})$ is compact and therefore closed in M. Hence, if $x \in M - g(\bar{A})$, then $\delta = d(x, g(\bar{A})) > 0$. Choose n so large that $d(x, g_1^{i-1}(\bar{A})) > \delta/2$ and $\epsilon_i < \delta/2$ for $i \geqslant n$. Then, $h_1^i(x) = x$ for $x \geqslant n$. But then $x = g((g_1^n)^{-1}(x))$, so g is onto.

(3) Certainly g_1^i is a Σ_i-map, hence a Σ-map for $i = 1, 2, ...$. But then g, as the uniform limit of the g_i, is also a Σ-map.

(4) If $d(x, A) \geqslant \Sigma \geqslant \epsilon_1$, then $x = h_1^1(x) = g_1^1(x)$, and $d(x, g_1^1(A)) \geqslant \Sigma - \Sigma_1 \geqslant \epsilon_2$. Then, $x = h_1^2(x) = h_1^2 h_1^1(x) = g_1^2(x)$, and $d(x, g_1^2(A)) \geqslant$

$\Sigma - \Sigma_2 \geqslant \epsilon_3$. Continuing in this way, $g_1{}^i(x) = x$ for $i = 1, 2, \ldots$. Hence, $g(x) = x$.

(5) It is only necessary to check that G is continuous at a point $(x, 1)$. If $\epsilon > 0$ is given, we must find a $\delta > 0$ and a t_0 such that if $d(x, y) < \delta$ and $t > t_0$ then

$$d(G(x, 1), G(y, t)) = d(g(x), g_t(y)) < \epsilon.$$

First choose n so large that $\Sigma - \Sigma_n < \epsilon/3$, and let $t_0 = n/(n + 1)$. Then choose $\delta > 0$ so small that $d(x, y) < \delta$ implies

$$d(g_{n/(n+1)}(x), g_{n/(n+1)}(y)) < \epsilon/3.$$

Finally, if $t > t_0$ and $d(x, y) < \delta$, then

$$\begin{aligned} d(g(x), g_t(y)) &\leqslant d(g(x), g_n(x)) + d(g_n(x), g_n(y)) + d(g_n(y), g_t(y)) \\ &< \epsilon/3 + \epsilon/3 + \epsilon/3 = \epsilon. \end{aligned}$$

(6) This follows in the same way as the proof of Lemma **3.6.2**.

(7) If g is one-to-one, then g is a one-to-one map of a manifold onto itself and hence a homeomorphism. It follows from Properties 1–6 that g_t is a Σ-push of (M, A).

The problem that now remains is to find reasonable conditions which force g to be one-to-one. This is done in the next lemma. (This lemma, in effect, says that if the sequence $\epsilon_1, \epsilon_2, \ldots$ converges fast enough, then g is one-to-one.)

Lemma 3.6.4. *For each integer $i \geqslant 1$, let $\mathscr{C}_i$ be a finite covering of $\bar{A}$ by compact subsets of $\bar{A}$ of diameter less than $1/i$. In addition to the hypotheses of Lemma* **3.6.3**, *suppose that the following conditions hold*:

(1) $2\epsilon_{i+1} < \epsilon_i$,

(2) $4\epsilon_{i+1} < d(g_1{}^i(U), g_1{}^i(V))$ *for any disjoint sets U, V of the covering* $\mathscr{C}_i$,

(3) $N_{\epsilon_{i+1}}(g_1{}^i(\bar{A})) \subset g_1{}^i(N_{1/i}(\bar{A}))$.

Then, g is one-to-one, and hence g_t is a Σ-push of (M, A).

Proof. Suppose x and y are distinct points of $\bar{A}$. Then, for some integer i, there are disjoint sets U and V of the covering $\mathscr{C}_i$ such that $x \in U$ and $y \in V$. Then,

$$\begin{aligned} d(g(x), g(y)) &\geqslant d(g_1{}^i(x), g_1{}^i(y)) - d(g_1{}^i(x), g(x)) - d(g_1{}^i(y), g(y)) \\ &\geqslant d(g_1{}^i(U), g_1{}^i(V)) - (\Sigma - \Sigma_i) - (\Sigma - \Sigma_i) \\ &> 4\epsilon_{i+1} - 2\epsilon_{i+1} - 2\epsilon_{i+1} = 0, \end{aligned}$$

by Conditions (1) and (2). Therefore, $g(x) \neq g(y)$, so $g \mid \bar{A}$ is one-to-one.

Suppose $x \notin \bar{A}$. Then, $x \notin N_{1/i}(\bar{A})$ for sufficiently large i. By Condition (3), $g_1{}^i(x) \notin N_{\epsilon_{i+1}}(g_1{}^i(\bar{A}))$. But this means that

$$g_1{}^i(x) = g_1^{i+1}(x) = g_1^{i+2}(x) = \cdots = g(x).$$

Therefore if $x, y \notin \bar{A}$ and $x \neq y$, then for sufficiently large i,

$$g(x) = g_1{}^i(x) \neq g_1{}^i(y) = g(y),$$

so $g \mid M - \bar{A}$ is also one-to-one.

Suppose again that $x \notin \bar{A}$. Then there is an open neighborhood N of x such that for sufficiently large i, $d(N, \bar{A}) > 1/i$. By Condition (3) above, $g_1{}^i(N)$ does not meet $N_{\epsilon_{i+1}}(g_1{}^i(\bar{A}))$. Hence, $g_1{}^i \mid N = g_1^{i+1} \mid N = \cdots = g \mid N$. Thus, since g is a limit of homeomorphisms $g_1{}^i$, $g(N)$ does not meet $g(\bar{A})$, that is, $g(\bar{A})$ and $g(M - \bar{A})$ are disjoint. Therefore, g is one-to-one.

Proof of Theorem 3.6.3. Let F and F' be dense, solvable subsets of $\mathrm{Hom}_f(X, A; M)$. Then, $F \cup F'$ is certainly dense in $\mathrm{Hom}_f(X, A; M)$, so it remains to show that $F \cup F'$ is solvable. Since F is solvable, there is, for each $\epsilon > 0$, a $\delta(\epsilon)$ satisfying the condition of the definition of solvability. Similarly, since F' is solvable there is, for each $\epsilon' > 0$, a corresponding $\delta'(\epsilon')$. Now let $\epsilon > 0$ be given. Let $\delta(\epsilon/6)$ be chosen for F and $\delta'(\epsilon/6)$ for F'. Finally, let

$$\delta = \min(\delta(\epsilon/6), \delta'(\epsilon/6)).$$

We will show that any two elements of $F \cup F'$ which are closer than δ can be related by an ϵ-push. If the two elements are both from F or both from F', then the existence of such an ϵ-push follows immediately from the individual solvability of F and F' and the choice of δ. So we assume that $f \in F$ and $f' \in F'$ and $d(f, f') < \delta$ (see Fig. **3.6.1**).

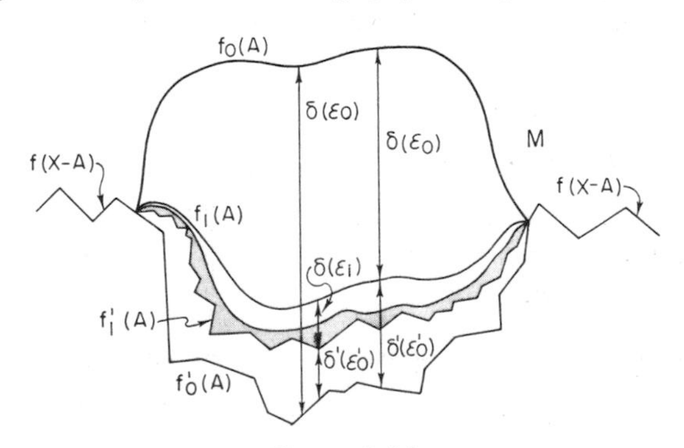

Figure 3.6.1

Since $\bar{A}$ is compact, let $\mathscr{C}_i$, for each integer $i \geqslant 1$, be a finite covering of $\bar{A}$ by compact subsets of $\bar{A}$ of diameter less than $1/i$. We will construct

(a) a sequence of elements of F: $f_0, f_1, f_2, \ldots$;
(a′) a sequence of elements of F': $f_0', f_1', f_2', \ldots$;
(b) a sequence of homeomorphisms of M onto itself: $h_1, h_2, h_3, \ldots$;
(b′) another sequence of homeomorphisms of M onto itself: $h_1', h_2', h_3', \ldots$;
(c) a sequence of positive real numbers: $\epsilon_0, \epsilon_1, \epsilon_2, \ldots$;
(c′) another sequence of positive real numbers: $\epsilon_0', \epsilon_1', \epsilon_2', \ldots$.

At the same time, it will be convenient to consider

(d) a sequence of homeomorphisms of M onto itself: $g_1, g_2, g_3, \ldots$, such that $g_i = h_i h_{i-1} \cdots h_2 h_1$;
(d′) another sequence of homeomorphisms of M onto itself: $g_1', g_2', g_3', \ldots$ such that $g_i' = h_i' h_{i-1}' \cdots h_2' h_1'$.

These sequences will be constructed so as to satisfy the following properties:

(1) $d(f'_{i-1}, f_i) < \delta'(\epsilon_{i-1})$, (1′) $d(f_i, f_i') < \delta(\epsilon_i)$,
(2) $d(f_{i-1}, f_i) < \delta(\epsilon_{i-1})$, (2′) $d(f'_{i-1}, f_i') < \delta'(\epsilon'_{i-1})$,
(3) $h_i f_{i-1} = f_i$, (3′) $h_i' f'_{i-1} = f_i'$,
(4) h_i can be realized by an (ϵ_{i-1})-push of $(M, f_{i-1}(A))$,
(4′) h_i' can be realized by an (ϵ'_{i-1})-push of $(M, f'_{i-1}(A))$,
(5) $2\epsilon_i < \epsilon_{i-1}$, $\epsilon_0 = \epsilon/6$,
(5′) $2\epsilon_i' < \epsilon'_{i-1}$, $\epsilon_0' = \epsilon/6$,
(6) $4\epsilon_i < d(g_i(U), g_i(V))$ for any disjoint sets U, V of the covering $\mathscr{C}_i$,
(6′) $4\epsilon_i' < d(g_i'(U), g_i'(V))$ for any disjoint sets U, V of the covering $\mathscr{C}_i$,
(7) $N_{\epsilon_i}(g_i f(\bar{A})) \subset g_i(N_{1/i}(f(\bar{A})))$,
(7′) $N'_{\epsilon_i}(g_i' f'(\bar{A})) \subset g_i'(N_{1/i}(f'(\bar{A})))$.

To start off, let $f_0 = f$, $f_0' = f'$, $\epsilon_0 = \epsilon/6 = \epsilon_0'$. Condition (1′) is satisfied for $i = 0$ because

$$d(f_0, f_0') = d(f, f') < \delta \leqslant \delta(\epsilon/6) = \delta(\epsilon_0)$$

Conditions (5) and (5′) are also satisfied for $i = 0$. None of the other conditions make sense for $i = 0$.

Step 1. Since F is dense in $\mathrm{Hom}_f(X, A; M)$, choose $f_1 \in F$ such that

$$d(f_0', f_1) < \min(\delta'(\epsilon_0'), \delta(\epsilon_0) - d(f_0, f_0')).$$

Then not only do we get

$$d(f_0', f_1) < \delta'(\epsilon_0'),$$

but also

$$d(f_0, f_1) \leqslant d(f_0, f_0') + d(f_0', f_1)$$
$$< d(f_0, f_0') + \delta(\epsilon_0) - d(f_0, f_0') = \delta(\epsilon_0).$$

This takes care of Conditions (1) and (2) for $i = 1$.

Since $d(f_0, f_1) < \delta(\epsilon_0)$, there is, by solvability of F, a homeomorphism h_1 of M which can be realized by an ϵ_0-push of $(M, f_0(A))$ such that $h_1 f_0 = f_1$. This takes care of Conditions (3) and (4) for $i = 1$. Finally, choose ϵ_1 to satisfy Conditions (5), (6), and (7) for $i = 1$, to complete Step 1.

Step 1′. Referring to Condition (1) for $i = 1$, we have

$$d(f_0', f_1) < \delta\,(\epsilon_0').$$

Hence, by denseness of F' in $\mathrm{Hom}_f(X, A; M)$ we can choose an element $f_1' \in F'$ satisfying

$$d(f_1, f_1') < \min(\delta(\epsilon_1), \delta'(\epsilon_0') - d(f_0', f_1)).$$

Then we get

$$d(f_1, f_1') < \delta(\epsilon_1),$$

and also

$$d(f_0', f_1') \leqslant d(f_0', f_1) + d(f_1, f_1')$$
$$< d(f_0', f_1) + \delta'(\epsilon_0') - d(f_0', f_1) = \delta'(\epsilon_0'),$$

thereby satisfying Conditions (1′) and (2′) for $i = 1$. Since $d(f_0', f_1') < \delta'(\epsilon_0')$, there is, by the solvability of F' a homeomorphism h_1' of M which can be realized by an (ϵ_0')-push of $(M, f_0'(A))$ such that $h_1' f_0' = f_1'$. This takes care of Conditions (3′) and (4′) for $i = 1$. Finally choose ϵ_1' satisfying Conditions (5′), (6′), and (7′) for $i = 1$, completing Step 1′.

Now all the Conditions (1) through (7′) have been satisfied for $i = 1$. The six sequences are then constructed inductively in this manner.

Since h_i can be realized by an (ϵ_{i-1})-push of $(M, f_{i-1}(A))$, we can apply Lemmas **3.6.3** and **3.6.4** to learn that the sequence of homeomorphisms

$$g_1, g_2, g_3, \ldots$$

converges uniformly to a homeomorphism g of M which can be realized by an $(\epsilon/3)$-push of $(M, f(A))$. Similarly, the sequence of homeomorphisms

$$g_1', g_2', g_3', \ldots$$

converges uniformly to a homeomorphism g' of M which can be realized by an $(\epsilon/3)$-push of $(M, f'(A))$.

It follows from Condition (3) that

$$g_i f = f_i ,$$

and from Condition (3′) that

$$g_i' f' = f_i'.$$

Then from Condition (1′),

$$d(g_i f, g_i' f') < \delta(\epsilon_i).$$

Since the ϵ_i converge to zero, so do the $\delta(\epsilon_i)$, and therefore

$$gf = g'f',$$

or equivalently,

$$(g')^{-1} gf = f'.$$

Since g' can be realized by an $(\epsilon/3)$-push of $(M, f'(A))$, it follows from Lemma **3.6.1** that $(g')^{-1}$ can be realized by a $(2\epsilon/3)$-push of $(M, g'f'(A))$. Hence, it follows from Lemma **3.6.2** that $(g')^{-1}g$ can be realized by an ϵ-push of $(M, f(A))$. Since $(g')^{-1} gf = f'$, $F \cup F'$ is solvable as desired.

Notice that a theorem more general than the one below and more general than Homma's original theorem will follow immediately once Theorem **3.6.1** is proved.

Theorem 3.6.4 (Modified Homma Theorem). *Let M^n be an unbounded* PL *n-manifold and let $g: E^n \to M^n$ be an embedding. Let P^k be a (possibly infinite) k-polyhedron and $f: P^k \to E^n$ be a* PL *embedding. Suppose that L is a finite subpolyhedron of P^k such that $gf \mid \mathrm{Cl}(P - L)$ is* PL *(see Fig.* **3.6.2**). *If $2k + 2 \leqslant n$, then for any $\epsilon > 0$, there is an ϵ-push h_t of $(M^n, gf(L))$ such that*

$$h_1 gf \mid P^k : P^k \to M^n$$

is PL *and $h_t \mid gf(P^k - L) = 1$.*

Proof. Let F_* denote the set of PL embeddings of P^k into E^n in $\mathrm{Hom}_f(P^k, L; E^n)$. Among these is, of course, $f: P^k \to E^n$. Because $2k + 2 \leqslant n$, it follows from Corollary **1.6.5** that F_* is dense in $\mathrm{Hom}_f(P^k, L; E^n)$. That F_* is solvable follows from Theorem **3.5.1**. Let

$$F = \{gf_* \mid f_* \in F_*\}.$$

Then, F is dense and solvable in $\mathrm{Hom}_{gf}(P^k, L; g(E^n))$.

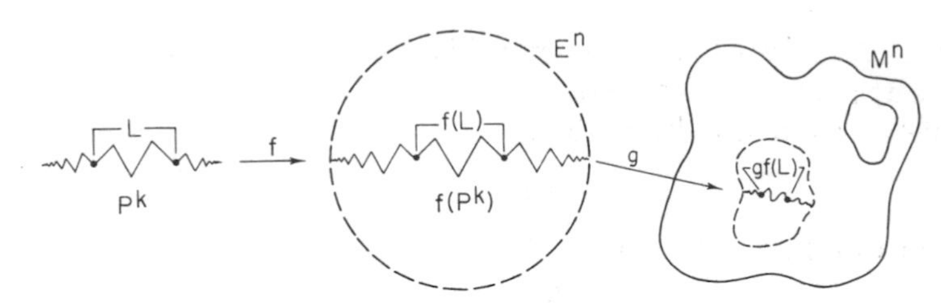

Figure 3.6.2

Since $g(E^n)$ is an open subset of M^n, it inherits a PL structure from M^n. Let F' denote the set of all PL embeddings of $\text{Hom}_{gf}(P^k, L, g(E^n))$. As above F' is also dense and solvable.

By Theorem **3.6.3**, $F \cup F'$ is dense and solvable. Using the denseness of F', first find an embedding $f' \in F'$ within $\delta_{F \cup F'}(\epsilon)$ of the embedding gf. Then, because $F \cup F'$ is solvable, there is a ϵ-push h_t of $(g(E^n), gf(L))$ such that $h_1 gf \mid P^k = f'$ and such that

$$h_t \mid gf(P^k - L) = 1.$$

For sufficiently small ϵ, h_t will extend via the identity to an ϵ-push of $(M^n, gf(L))$. This completes the proof of Theorem **3.6.4**.

Lemma 3.6.5. *Let $F \subset F' \subset \text{Hom}_f(X, A; M)$. Suppose that for each $f' \in F'$ and each $\epsilon > 0$, there is an ϵ-push h_t of $(M, f'(A))$ such that $f = h_1 f'$ is an element of F. If F is solvable, then so is F' solvable.*

Proof. Since F is solvable, there is for each $\epsilon > 0$ a corresponding $\delta(\epsilon)$ satisfying the definition of solvability. To show that F' is solvable, let $\epsilon > 0$ be given. We claim that

$$\delta'(\epsilon) = \delta(\epsilon/4)/2$$

will exhibit the solvability of F'.

Suppose that $f_1', f_2' \in F'$ and $d(f_1', f_2') < \delta'(\epsilon)$. Let

$$\epsilon^* = \min(\delta(\epsilon/4)/4, \epsilon/4).$$

Let h_t^1 and h_t^2 be ϵ^*-pushes of $(M, f_1'(A))$ and $(M, f_2'(A))$ such that $f_1 = h_1^1 f_1'$ and $f_2 = h_1^2 f_2'$ are elements of F, according to the hypothesis of the lemma. Then,

$$\begin{aligned} d(f_1, f_2) &\leqslant d(f_1, f_1') + d(f_1', f_2') + d(f_2', f_2) \\ &< \epsilon^* + \delta'(\epsilon) + \epsilon^* \\ &\leqslant \frac{\delta(\epsilon/4)}{4} + \frac{\delta(\epsilon/4)}{2} + \frac{\delta(\epsilon/4)}{4} = \delta(\epsilon/4). \end{aligned}$$

According to the solvability of F, there is an $(\epsilon/4)$-push h_t of $(M, f_1(A))$ such that $h_1 f_1 = f_2$. Furthermore, since h_t^2 is a $(\epsilon/4)$-push, $(h_t^2)^{-1}$ is an $(\epsilon/2)$-push by Lemma **3.6.1**. Then by Lemma **3.6.2**, $(h_1^2)^{-1} h_1 h_1^1$ can be realized by an ϵ-push h_t' of $(M, f_1'(A))$ such that $h_1' f_1' = f_2'$, and so F' is solvable.

Proof of Theorem 3.6.1

Case 1. (In this case let P^k be finite, let $\tilde{P}^k = P^k$ and let $h = 1$. This will show embeddings of finite polyhedra to be ϵ-tame in the same structure in which they are locally tame.)

Referring back to the definition of locally tame, let us call a subset $A \subset P$ **small** if there is a neighborhood V of A in P^k, a neighborhood U of $f(A)$ in M^n, and a homeomorphism $h_U\colon U \twoheadrightarrow E^n$ such that

(1) $f(V) = U \cap f(P^k)$,
(2) $h_U f\colon V \to E^n$ is PL with respect to the induced PL structure on V as an open subset of P^k.

Take a triangulation K of P^k (which contains a triangulation of R) of mesh so fine that the stars of simplexes are small (see Theorem **1.4.2**). Let $\sigma_0, \sigma_1, ..., \sigma_r$ be the simplexes of K (in order of nondecreasing dimension if you like). Let $H_i = \text{St}(\hat{\sigma}_i, K'')$, where $\hat{\sigma}_i$ is the barycenter of σ_i and K'' is the second barycentric subdivision of K. (If P^k were a PL manifold then the H_i would be the handles constructed in Theorem **1.6.11**.)

The plan is to define, with the help of Theorem **3.6.4**, $k+1$ isotopies of M^n

$$h_t^0, h_t^1, ..., h_t^k$$

with the properties:

(a) h_t^0 is an $(\epsilon/(k+1))$-push of $(M^n, f(H_0 - R))$ such that $h_1^0 f \mid H_0\colon H_0 \to M^n$ is PL;
(b) h_t^i is an $(\epsilon/(k+1))$-push of $(M^n, h_1^{i-1} h_1^{i-2} \cdots h_1^0 f(H_i - R))$ such that $h_1^i h_1^{i-1} \cdots h_1^0 f \mid \bigcup_{j=1}^{i} H_j\colon \bigcup_{j=1}^{i} H_j \to M^n$ is PL for $i = 1, 2, ..., k$.
(c) $h_t^i \mid f(R) = 1$ for all i.

Then, an application of Lemma **3.6.2** with $A = f(P^k)$ will complete the proof.

Step 0. Since H_0 is small, there is a neighborhood V_0 of H_0 in P^k, a neighborhood U_0 of $f(H_0)$ in M^n and a homeomorphism $h_U\colon U_0 \twoheadrightarrow E^n$ satisfying (1) and (2) above. To apply Theorem **3.6.4**, let

$$(M^n, g, P^k, L, f, \epsilon) = (M^n, h_{U_0}^{-1}, \mid H_0 \mid \cup (R \cap V_0), H_0 - R, h_{U_0} f, \epsilon/(k+1)).$$

Theorem **3.6.4** then asserts the existence of an $(\epsilon/(k+1))$-push h_t^0 of $(M^n, h_{U_0}^{-1}h_{U_0}f(H_0 - R)) = (M^n, f(H_0 - R))$ such that

$$h_{U_0}^{-1}h_{U_0}f \mid (H_0) \cup (R \cap V_0) = f \mid \mid H_0 \mid \cup (R \cap V_0)$$

is PL and

$$h_t^0 \mid f((\mid H_0 \mid \cup (R \cap V_0)) - (H_0 - R)) = h_t^0 \mid f(R \cap V_0) = 1.$$

This completes Step 0.

Step i. Suppose that we have constructed isotopies h_t^j, $0 \leqslant j \leqslant i-1$ with the Properties a, b, and c above. Since H_i is small, there is a neighborhood U_i of $f(H_i)$ in M^n and a homeomorphism $h_{U_i}: U_i \twoheadrightarrow E^n$ satisfying Properties (1) and (2) above. To apply Theorem **3.6.4** let

$$(M^n, g, P^k, L, f, \epsilon)$$
$$= \left(M^n, h_1^{i-1} \cdots h_1^1 h_1^0 h_{U_i}^{-1}, \left(\bigcup_{j=1} H_j \cup R\right) \cap V_i, H_i - \left(R \cup \bigcup_{j=0}^{i-1} H_j\right), h_{U_i} f, \epsilon/(k+1)\right).$$

Then, Theorem **3.6.4** asserts that there is an $(\epsilon/(k+1))$-push h_t^i of $(M^n, h_1^{i-1} \cdots h_1^1 h_1^0 f(H_i - R))$ such that

$$h_1^i h_1^{i-1} \cdots h_1^1 h_1^0 f \left| \left(\bigcup_{j=1}^{i} H_j \cup R\right) \cap V_i\right.$$

is PL and

$$h_t^i \left| h_1^{i-1} \cdots h_1^1 h_1^0 f \left(\bigcup_{j=1}^{i-1} H_j \cup R\right) = 1.\right.$$

Thus, $h_1^i h_1^{i-1} \cdots h_1^0 f \mid \bigcup_{j=1}^{i} H_j$ is PL and the inductive step is complete.

Case 2. (In this case allow P^k to be infinite, but again let $\tilde{P}^k = P^k$ and let $h = 1$.) This case follows easily from Case 1 by applying that case an infinite number of times to larger and larger "chunks" of P^k each time staying fixed on the part already fixed up as well as on $f(R)$.

Case 3. (This time we will handle the case that $P^k - R$ is compact and $M^n = E^n$.) In this case we have the following situation: R is a subpolyhedron of the possibly infinite polyhedron P^k and $P^k - R$ is compact; $f: P^k \twoheadrightarrow E^n$, $2k + 2 \leqslant n$, is a locally tame embedding such that $f \mid R$ is PL; $\tilde{P}^k$ is another polyhedron and $h: P^k \twoheadrightarrow \tilde{P}^k$ is a topological homeomorphism such that $h \mid P$ is PL. We must show that fh^{-1} is ϵ-tame keeping $fh^{-1}(R) = f(R)$ fixed.

Let F_1 be the set of all PL embeddings in $\mathrm{Hom}_f(P^k, P^k - R; E^n)$; F_1' the set of all locally tame embeddings in $\mathrm{Hom}_f(P^k, P^k - R; E^n)$; F_2 the set of all embeddings g in $H_f(P^k, P^k - R; E^n)$ such that gh^{-1} is PL; and F_2' the set of all embeddings g in $H_f(P^k, P^k - R; E^n)$ such that gh^{-1} is locally tame. F_1 and F_2 are dense in $\mathrm{Hom}_f(P^k, P^k - R; E^n)$ by general position (see Corollary **1.6.5**) and are solvable in $\mathrm{Hom}_f(P^k, P^k - R; E^n)$ by Theorem **3.5.1**. Thus, by Case 2 and by Lemma **3.6.5**, F_1' and F_2' are also dense and solvable.

Now by Theorem **3.6.3**, $F_1' \cup F_2$ is dense and solvable. Given $\epsilon > 0$, there is a $\delta > 0$ such that any two elements in $F_1' \cup F_2$ which are closer than δ can be related by an ϵ-push. Since F_2 is dense in $\mathrm{Hom}_f(P^k, P^k - R, E^n)$, choose $f_2 \in F_2$ so that $d(f, f_2) < \delta$. Then, there is an ϵ-push h_t of $(E^n, f(P^k - R))$ keeping $f(R)$ fixed such that $h_1 f = f_2$. Thus, $h_1 f h^{-1} = f_2 h^{-1}$ and since $f_2 h^{-1}$ is PL we have shown that h_t is the desired ϵ-push.

Case 4 (The general case). This case follows easily from Case 3 by a technique similar to that used in the proof of Case 1.

Proof of Corollary 3.6.1. If we can show that $f\colon M^k \to M^n$ is locally tame, then the corollary will follow from Theorem **3.6.1**. Since f is locally flat, for each point $x \in M^k$, there is a neighborhood U of $f(x)$ in M^n and a homeomorphism h_U' of $(U, U \cap f(M^k))$ onto either (E^n, E^k) or (E^n, E^k_+). Let $g\colon E^k$ (or E^k_+) $\twoheadrightarrow f^{-1}(U \cap f(M^k))$ be a PL homeomorphism, and consider $h_U' fg$ taking E^k (or E^k_+) onto itself. It is easy to construct a homeomorphism k of E^n onto itself such that $kh_U' fg = 1$. Hence, $kh_U' f = g^{-1}$ is PL and $h_U = kh_U'$ is the desired homeomorphism of U onto E^n which shows f to be locally tame at x.

REMARK 3.6.2. In proving a special case of the corollary, Gluck [1] refers to a certain "product" triangulation which illustrates the fuzziness that can occur if one does not always keep in mind that in order to make sense out of abstract triangulations one must always go back to rectilinear triangulations.

Proof of Theorem 3.6.2. Before the statement of Theorem **3.6.2**, we said that our proof involves a folklore result. That result is that *the set of all* PL *embeddings in* $\mathrm{Hom}_f(P^k, P^k; M^n)$ *is solvable.* (A suggestion for proving this result is to use Theorem **3.5.1** locally. In 1966 the author watched Machusko struggle through a rigorous proof of this result and we hope that the interested reader will carry out a similar struggle.)

Since f and f' are homotopic, there is a sequence of continuous maps

$$f = g_0, g_1, \ldots, g_r = f'$$

of P^k into M^n such that $d(g_i, g_{i+1}) < \delta$, where δ corresponds to some fixed ϵ in the definition of solvability which is assured by our folklore result. Hence by general position there is a sequence of PL embeddings

$$f = f_0, f_1, ..., f_r = f',$$

of P^k into M^n such that $d(f_i, f_{i+1}) < \delta$. Thus, there exist ϵ-pushes h_t^i such that $h_1^i f_{i-1} = f_i$. The theorem then follows from Lemma **3.6.2**.

3.7. ε-TAMING POLYHEDRA IN THE TRIVIAL RANGE WHICH LIE IN HYPERPLANES

In Theorem **2.5.1**, we presented a result of Klee which said that every k-cell D in E^n is flat in E^{n+k}. The principle objective of this section is to generalize that result from cells to polyhedra. This was first done in [Bing and Kister, 1]. In fact, the rough idea of Bing and Kister's proof is to generalize the proof of Klee's result. Let us state the main theorem formally before discussing it further.

Taming Theorem 3.7.1 (Bing and Kister). *Let $h: P^k \to E^n$ be a topological embedding of the k-polyhedron P^k into Euclidean n-space and let $i: E^n \to E^{n+k}$ be the inclusion. Then, $ih: P^k \to E^{n+k}$ is ϵ-tame.*

This theorem lends itself quite readily to generalizations. For instance, one easily sees how to generalize it to embeddings in manifolds (see, for example, Section 7 of [Bing and Kister, 1]). We have already observed that [Bryant, 2] and [Dancis, 1] generalize, to a certain extent, the main theorem of the last section; in addition, they generalize the above theorem. Their effect is to allow $E^n \subset E^{n+k}$ to be replaced by a locally tame polyhedron and not to require all of P^k to be taken into this polyhedron, but to require it to be locally tame where it is not. One might also attempt to improve the dimension restrictions, and results of this nature are contained in [Bryant and Seebeck, 3, Corollary 1.2] and [Rushing, 2, Theorem 6.4].

The proof of Theorem **3.7.1** given here will not be the one of [Bing and Kister, 1], but will use techniques of [Bryant, 1]. This argument has the advantage of being shorter if one is allowed to assume the material of the last two sections. Also, we will have the material in this section available in the next section where we establish the Bryant–Dancis theorem mentioned above.

Suppose that G is an open subset of a polyhedron P^k and $f: P^k \to E^n$,

$2k + 2 \leqslant n$, is an embedding such that $f \mid G$ is PL. Suppose further that $f(P^k - G)$ lies in a polyhedron $X \subset E^n$, $\dim X \leqslant n/2 - 1$. Then, $\mathrm{Hom}_f(P^k, G; E^n)$ denotes the set of all embeddings of P^k into E^n that agree with f on $P^k - G$. Let F be the set of all f' in $\mathrm{Hom}_f(P^k, G; E^n)$ for which there is a polyhedron Z in G such that

(1) $f' \mid P^k - Z = f \mid P^k - Z$, and
(2) $f' \mid Z$ is PL.

Lemma 3.7.1. *The set F defined above is a dense, solvable subset of* $\mathrm{Hom}_f(P^k, G; E^n)$.

Proof. If $\epsilon > 0$ and $g \in \mathrm{Hom}_f(P^k, G; E^n)$ are given, choose a polyhedron Z in G such that $d(f(x), g(x)) < \epsilon$ for all $X \in P^k - Z$. By using standard extension techniques and general position arguments, we can extend $f \mid P^k - Z$ to an embedding f' of P^k into E^n so that $f' \mid Z$ is PL and $d(f', g) < \epsilon$. By definition $f' \in F$; hence F is dense.

To see that F is solvable, let $\epsilon > 0$ be given and suppose that $f_0, f_1 \in F$ with $d(f_0, f_1) < \epsilon$. Notice that f_0 and f_1 agree with f except on a polyhedron Z in G and that both f_0 and f_1 are PL on Z. Since $f(P^k - G)$ $(= f_i(P^k - G)$ for $i = 0, 1)$ is contained in the polyhedron X with $\dim X \leqslant n/2 - 1$, by making an additional general position application in the proof of Lemma **3.5.4**, Theorem **3.5.1** yields an ϵ-push h_t of $(E^n, f_0(Z))$ such that $h_1 f_0 \mid Z = f_1 \mid Z$ and $h_t \mid P \cup f(P^k - Z) = 1$. Thus, F is solvable.

Taming Theorem 3.7.2. *Suppose that L is a subpolyhedron of P^k and f is an embedding of P^k into E^n, $2k + 2 \leqslant n$, such that $f \mid Y$ is locally tame and $f \mid P^k - Y$ is locally tame. Then, f is ϵ-tame.*

Proof. Since Theorem **3.6.1** allows us to assume that $f \mid L$ is PL and $f \mid P^k - L$ is PL, the theorem follows immediately from Lemma **3.7.1** and Theorem **3.6.3**, because the set of PL extensions of $f \mid L$ is dense and solvable in $\mathrm{Hom}_f(P^k, P^k - L; E^n)$.

By using Theorem **3.7.2**, one obtains by induction on k, the following theorem, which was first established for $k = 1$ and $k = 2$ in [Cantrell, 6] and [Edwards, 2], respectively.

Taming Theorem 3.7.3. *An embedding $f\colon P^k \to E^n$, $2k + 2 \leqslant n$, of a k-polyhedron P^k is ϵ-tame if and only if for any triangulation of P^k, $f \mid \sigma$ is tame for each simplex σ.*

Theorem **3.7.1** now follows from the above theorem and Theorem **2.5.1**.

3.8. ε(x)-TAMING EMBEDDINGS WHICH ARE LOCALLY TAME MODULO NICE SUBSETS

The Bryant–Dancis result mentioned in the last two sections is the main theorem of this section. The formulation here is essentially that of [Bryant, 2]. As pointed out in the last section, generalizations are contained in [Bryant and Seebeck, 3, Corollary 1.2] and [Rushing, 2, Theorem 6.4].

Taming Theorem 3.8.1 (Bryant and Dancis). *Suppose that f is an embedding of a (possibly infinite) k-polyhedron P^k into a* PL *n-manifold M^n, $2k+2 \leqslant n$, and that X is a polyhedron in M^n,* $\dim X \leqslant n/2 - 1$, *such that $f \mid (P^k - f^{-1}(X))$ is locally tame. Then, f is $\epsilon(x)$-tame.*

Before giving the proof of Theorem **3.8.1**, we will establish a preliminary lemma.

Lemma 3.8.1. *Suppose Y is a closed subset of P^k and f_0, f_1 are embeddings of P^k into E^n, $2k+2 \leqslant n$ satisfying the conditions*

(1) *$f_i \mid P^k - Y$ is* PL *for* $i = 0, 1$;
(2) $f_0 \mid Y = f_1 \mid Y$;
(3) *$f_0(Y)$ lies in the polyhedron $X \subset E^n$ where* $\dim X \leqslant n/2 - 1$; *and*
(4) $d(f_0, f_1) < \epsilon$.

Then, there is a 2ϵ-push h_t of $(E^n, f_0(P^k - Y))$ such that $h_1 f_0 = f_1$.

PROOF. Let $G = P^k - Y$ so that $\mathrm{Hom}_{f_0}(P^k, G; E^n)$ is the set of all embeddings of P^k into E^n that agree with f_0 on Y. For $i = 0, 1$, let F_i be the subset of $\mathrm{Hom}_{f_0}(P^k, G; E^n)$ associated with f_i as described immediately preceding Lemma **3.7.1**. Then, F_0 and F_1 are dense, solvable subsets of $\mathrm{Hom}_{f_0}(P^k, G; E^n)$ by Lemma **3.7.1**, hence, by Theorem **3.6.3**, $F_0 \cup F_1$ is also dense and solvable.

Let $\delta = \delta(F_0 \cup F_1, \epsilon) > 0$. Since F_0 is dense in $F_0 \cup F_1$, there exists an f_0' in F_0 such that $d(f_0', f_1) < \delta$ and $d(f_0', f_0) < \epsilon$. From the proof of Lemma **3.7.1**, it follows that there exists an ϵ-push h_t^0 of $(E^n, f_0(G))$ such that $h_1^0 f_0 = f_0'$. By the choice of δ, there exists an ϵ-push h_t^1 of $(E^n, f_0'(G))$ such that $h_1^1 f_0' = f_1$. Thus, $h_t = h_t^1 h_t^0$ is a 2ϵ-push of $(E^n, f_0(G))$ with the required properties.

Proof of Theorem 3.8.1

Case 1 ($M^n = E^n$; P^k compact). Let $Y = f^{-1}(X)$ and let $G = P^k - Y$. We may assume that $f \mid G$ is PL by Theorem **3.6.1**. Let Φ

denote the set of all PL homeomorphisms of E^n onto itself. For each $\phi \in \Phi$, the embedding ϕf of P^k into E^n satisfies the conditions that $\phi f \mid G$ is PL and $\phi f(Y) \subset \phi(X)$, a polyhedron in E^n with $\dim \phi(X) \leqslant n/2 - 1$. For each $\phi \in \Phi$, consider $\mathrm{Hom}_{\phi f}(P^k, G; E^n)$, which is the set of all embeddings of P^k into E^n that agree with ϕf on Y. Let $F_{\phi f}$ denote the subset of $\mathrm{Hom}_{\phi f}(P^k, G; E^n)$ associated with ϕf as described immediately preceding Lemma **3.7.1**. Define

$$F = \bigcup_{\phi \in \Phi} F_{\phi f} .$$

(a) *F is dense in* $\mathrm{Hom}(P^k; E^n)$, *the set of all embeddings of P^k into E^n.*

Suppose $\epsilon > 0$ and $g \in \mathrm{Hom}(P^k; E^n)$. Let $\psi: P \to E^n$ be an extension of $gf^{-1}: f(Y) \to E^n$. Then, by general position, there exists a PL, embedding $\bar{\phi}: X \to E^n$ such that $d(\bar{\phi}, \psi) < \epsilon$. Hence, by Theorem **3.5.1**, there is a PL homeomorphism ϕ of E^n onto itself such that $\phi \mid X = \bar{\phi}$. Thus, for each $x \in Y$, $d(\phi f(x), g(x)) < \epsilon$. It is not hard to obtain an extension $f'': P^k \to E^n$ of $\phi f \mid Y$ such that $d(f'', g) < \epsilon$.

Choose $\delta > 0$ so that each map of P^k into E^n within δ of f'' lies within ϵ of g. By applying general position arguments, we can construct an embedding $f': P^k \to E^n$ such that

(1) $d(f', f'') < \delta$,
(2) $f' \mid Y = f'' \mid Y$,
(3) $f' \mid G$ is PL, and
(4) $f'(G) \cap \phi(X) = \emptyset$.

Then $f' \in F$ and $d(f', g) < \epsilon$.

(b) *F is solvable.*

Given $\epsilon > 0$, choose $\delta = \delta(F, \epsilon) = \epsilon/6$. Suppose that $f_0, f_1 \in F$ and $d(f_0, f_1) < \delta$. Then there exist elements $\phi_0, \phi_1 \in \Phi$ for which

$$\phi_1 f \mid Y = f_i \mid Y \qquad (i = 0, 1).$$

Since $d(\phi_0(y), \phi_1(y)) < \delta$ for each $y \in f(Y)$, there exists a polyhedral neighborhood Q of $f(Y)$ in X such that $d(\phi_0(y), \phi_1(y)) < \delta$ for each $y \in Q$ and

$$f_0(Y) \subset \phi_0(Q) \subset N_\delta(f_0(Y)),$$

where $N_\delta(f_0(Y))$ is the δ-neighborhood of $f_0(Y)$ in E^n.

From Theorem **3.5.1**, we obtain a δ-push h_t^0 of $(E^n, \phi_0(Q))$ such that

$$h_1^0: E^n \to E^n \qquad \text{is PL and} \qquad h_1^0 \phi_0 \mid Q = \phi_1 \mid Q.$$

Since $N_\delta(\phi_0(Q)) \subset N_{2\delta}(f_0(Y))$, h_t^0 is a 2δ-push of $(E^n, f_0(Y))$, and hence a 2δ-push of $(E^n, f_0(P^k))$.

Let $f' = h_t^0 f_0$. Then, $d(f', f_1) < \epsilon/2$, $f' \mid G$ is PL, and $f' \mid Y = f_1 \mid Y$. By Lemma **3.8.1**, there exists a $\frac{2}{3}\epsilon$-push h_t^1 of $(E^n, f'(G))$ such that $h_1^1 f_1' = f_1$. Thus, $h_t = h_t^1 h_t^0$ is an ϵ-push of $(E^n, f_0(P^k))$ and $h_1 f_0 = f_1$.

Case 1 now follows from Theorem **3.6.3**, since the set of PL embeddings of P^k into E^n is dense and solvable.

Case 2 (General case). Suppose $x \in P^k$. Let U be the interior of a PL n-ball in M^n that contains $f(x)$, and let g be a PL homeomorphism of U onto E^n. Choose a polyhedral neighborhood Z of x in P^k and a subpolyhedron Q of X such that

(1) $f(Z) \subset U$ and
(2) $X \cap f(Z) \subset Q \subset U$.

Then, by Case 1, $gf \mid Z\colon Z \to E^n$ is tame and so f is locally tame. Case 2 now follows from Theorem **3.6.1**.

3.9. NONLOCALLY FLAT POINTS OF A CODIMENSION ONE SUBMANIFOLD

The proof of the main theorem of this section makes use of all of the preceding sections of this chapter. Before proving this theorem, we will state and discuss it and mention related work done by various mathematicians. The theorem in the form given below appeared in [Kirby, 1] and our proof follows the lines of proof of that paper.

Flattening Theorem 3.9.1 (Kirby). *Let $f\colon M^{n-1} \to N^n$ be an embedding of a topological $(n-1)$-manifold M without boundary into a topological n-manifold N without boundary. Let E^* be the set of points of M at which f is not locally flat. Then, if $n \geqslant 4$, E^* cannot be a nonempty subset of a Cantor set C^* such that C^* is tame in M and $f(C^*)$ is tame in N.*

(A Cantor set in the interior or in the boundary of a topological manifold is said to be **tame** if it lies in a locally flat arc in the interior or boundary, respectively.)

Corollary 3.9.1. *Let $g\colon M^{n-1} \to N^n$ be an embedding of an $(n-1)$-manifold M into an n-manifold N without boundary such that g is locally flat except on a set E^*. If $n \geqslant 4$, then E^* contains no isolated points.* (Thus, E^* must be uncountable because, by using the Baire category theorem, one can show that a closed countable subset of a manifold must contain isolated points.)

PROOF. No isolated points can occur on the boundary of M by Corollary **3.4.3**. Now suppose that $p \in \operatorname{Int} M$ and that p is an isolated point of E^*. Choose M' to be an open $(n-1)$-cell neighborhood of p in M such that $M' \cap E = p$ and $f(M') \subset V \subset N$, where V is an open n-cell. Let α be a locally flat arc in M' through p. $f(\alpha - p)$ is locally flat by transitivity of local flatness and is tame by Theorem **3.6.1**. Hence, since $f(\alpha) \subset f(M') \subset V \approx E^n$, $f(\alpha)$ is flat by Theorem **3.2.1**. Applying Theorem **3.9.1** to $f \mid M' \colon M' \to N$, we see that $p \notin E^*$.

Recall that in Section **3.3** we presented a result (Theorem **3.3.1**) of Cantrell which said that an $(n-1)$-sphere in S^n, $n \geqslant 4$, which is locally flat except at possibly one point is locally flat. For some time after the appearance of Cantrell's result, it was wondered whether a codimension one sphere which is locally flat modulo two points is locally flat. In connection with this, it was shown in [Cantrell and Edwards, 1, Theorem 3.5] that if no embedding of the closed manifold M into the interior of the manifold N fails to be locally flat at precisely one or two points, then every embedding which is locally flat modulo a countable number of points is locally flat.

It was announced in both [Hutchinson, 1] and [Černavskiĭ, 1] that by using engulfing techniques one could show the answer to the two-point question mentioned above to be affirmative. In fact, Černavskiĭ announced a proof of Corollary **3.9.1**, which is stronger. Other proofs of Corollary **3.9.1** evolved later in [Černavskiĭ, 4, 5] and [Kirby, 2] where proofs of the so-called $\beta(n, n-1, n-1, n-2)$-statement, which for convenience will be given below for the second time, appear. (Corollary **3.9.1** follows easily from $\beta(n, n-1, n-1, n-2)$ and Corollary **3.4.2**. We will prove $\beta(n, n-1, n-1, n-2)$ in Section **5.2**.)

Let D_1 and D_2 be cells in S^n such that $D_1 \cap D_2 = \partial D_1 \cap \partial D_2$ is a cell. We say that $D_1 \cup D_2$ is a **flat pair** if there is a homeomorphism h of S^n such that $h(D_i)$ is a simplex and $h(D_1 \cap D_2)$ is a face of $h(D_i)$, $i = 1, 2$.

$\beta(n, m_1, m_2, k)$: *If* D_1 *and* D_2 *are locally flat cells in* S^n *of dimensions* m_1 *and* m_2, *respectively, and if* $D_1 \cap D_2 = \partial D_1 \cap \partial D_2$ *is a* k*-cell which is locally flat in* ∂D_1 *and* ∂D_2, *then* $D_1 \cup D_2$ *is a flat pair.*

REMARK 3.9.1. Recall that the β-statement was given Section **2.7** and it was mentioned that this statement was first investigated by Doyle [2, 3] in the three dimensional case and by Cantrell [7] in high dimensions and later by Lacher [2], Cantrell and Lacher [1, 2], and the author [5] as well as by Černavskiĭ and Kirby as mentioned above. In Section **2.7**, counterexamples were given for certain β-statements.

In Theorem 5.3 of [Lacher, 3], an interesting equivalence to Corollary **3.9.1** is given which is related to the work of this section.

Before beginning the proof of Theorem **3.9.1**, we establish a lemma. In this lemma, we regard Σ^k as being E^k compactified by adding the point ∞, $rB^k = \{x \in E^k \mid |x| \leqslant r\}$ and $B^k = 1B^k$.

Lemma 3.9.1. *Let $f\colon B^{n-1} \to \Sigma^n$ be an embedding which is locally flat except at $0 \in B^{n-1}$. If $n \geqslant 4$, then f extends to an embedding of Σ^{n-1} into Σ^n which is locally flat except at 0 and ∞.*

PROOF. Let X_n represent the X_n-axis in E^n and let $2\tilde{B}^n$ be the decomposition space $2B^n/(2B^n \cap X_n)$. We would like to say that since $f \mid B^{n-1} - 0$ is locally flat, it follows that f extends to an embedding of $2\tilde{B}^n$. This would follow from Part (c) of Theorem **1.7.6** if we knew that $f(B^{n-1} - 0)$ were two-sided. For $n \geqslant 5$, this result is a corollary of [Rushing, 6]. Since this fact is quite easy to believe, we will not include a proof here. (People who have read the preceding part of this book are well-prepared to read [Rushing, 6].)

It is sufficient to find an embedding $g\colon \Sigma^{n-1} \to \Sigma^n$, locally flat except at 0 and ∞, which agrees with f on some neighborhood of 0, say rB^{n-1}, $r > 0$. This follows since there are homeomorphisms $h_1, h_2\colon \Sigma^n \to \Sigma^n$, h_1 taking $g(B^{n-1})$ to $g(rB^{n-1}) = f(rB^{n-1})$ and h_2 taking $f(rB^{n-1})$ to $f(B^{n-1})$, which means that $f(B^{n-1}) \subset h_2 h_1 g(\Sigma^{n-1})$ and f may be extended as required.

We now will show that we may assume that $f(\partial B^{n-1}) = f(S^{n-2}) = S^{n-2}$. There is an arc $A \subset f(B^{n-1})$ (the image of a radius) from $f(0)$ to a point $b \in f(\partial B^{n-1})$, which is locally flat mod $f(0)$, hence flat by Corollary **3.4.1**. Then, there exists a map $s\colon \Sigma^n \to \Sigma^n$ which is the identity on $f(\partial B^{n-1})$, a homeomorphism off A, and shrinks A to b. Then $sf(B^{n-1})$ is an $(n-1)$-cell with boundary $f(\partial B^{n-1})$ which is locally flat except at the point b in its boundary. By Corollary **3.4.2**, this cell is flat. Therefore $f(\partial B^{n-1})$ is unknotted in Σ^n, so a space homeomorphism takes it onto S^{n-2}. Furthermore, we can assume $f(0) = 0$.

Let $j\colon \Sigma^{n-1} \to 2\tilde{B}^n - \partial B^{n-1}$ be a map satisfying

(a) $j = 1$ in a neighborhood of 0,
(b) $j(\infty) = 0$,
(c) $j \mid (\Sigma^{n-1} - \{0, \infty\})$ is an embedding, and
(d) if ρ is a compactified ray in Σ^{n-1} beginning at 0, then $j(\rho)$ lies

in the plane P determined by X_n and ρ. Furthermore, ρ winds once around ∂B^{n-1} as in Fig. **3.9.1**. In particular, we require that $j(\rho)$

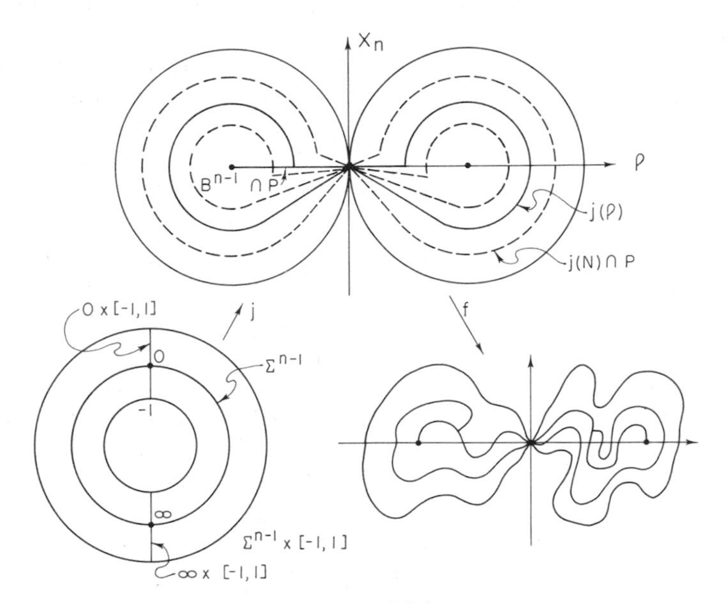

Figure 3.9.1

"represents" a generator of

$$H_1(2\tilde{B}^n - \partial B^{n-1}) \approx Z.$$

To insure local flatness later on, we assume that j is extended to $N = \Sigma^{n-1} \times [-1, 1]/\{0 \times [-1, 1], \infty \times [-1, 1]\}$, where we identify x and $[x \times 0]$ for each $x \in \Sigma^{n-1}$, such that $j \mid N - \{[0], [\infty]\}$ is an embedding into $2\tilde{B}^n - \partial B^{n-1}$. Now we wish to "unwind" $fj(N)$, rotating around S^{n-2}, so that $fj(0)$ and $fj(\infty)$ no longer coincide; then Σ^{n-1} will be flatly embedded except at 0 and ∞ and will contain a neighborhood of $f(0)$ in $f(B^{n-1})$.

There is a natural homeomorphism of Int $B^{n-1} \times S^1$ onto $\Sigma^n - S^{n-2}$ taking $0 \times S^1$ onto the compactified X_n-axis and Int $B^{n-1} \times 0$ onto Int B^{n-1} (here S^1 is $[0, 2\pi]$ with $0 = 2\pi$). Identifying $\Sigma^n - S^{n-2}$ with Int $B^{n-1} \times S^1$ in this way, let p_1: $\Sigma^n - S^{n-2} \to$ Int B^{n-1} and p_2: $\Sigma^n - S^{n-2} \to S^1$ be the projections. Let q: $E^1 \to S^1$ be the universal covering, where $q(x) = x \bmod 2\pi$. (For discussions of the elementary properties of covering spaces used here, see Section 3.3 of [Singer and Thorpe, 1] and Section 16 of Chapter III of [Hu, 1].) Since N is simply connected, the map $p_2 fj\colon N \to S^1$ lifts to a unique map $\lambda\colon N \to E^1$ satisfying $q\lambda = p_2 fj$ and $\lambda(0) = 0$ (see Theorem 16.2 of Chapter III of [Hu, 1]). In effect, λ assigns to each point in N a "winding number" around S^{n-2}.

We now show that $\lambda(\infty) \neq 0$. Let ρ be a compactified ray in Σ^{n-1} as above, so ρ is an arc with end-points 0 and ∞. Recalling that $f(\partial B^{n-1}) = S^{n-2}$, $fj(\rho)$ represents a generator of $H_1(f(2\tilde{B}^n) - S^{n-2})$, since $j(\rho)$ represents a generator of $H_1(2\tilde{B}^n - S^{n-2})$. By excision on the pair $(\Sigma^n, f(2\tilde{B}^n))$, we see that $H_i(\Sigma^n - S^{n-2}, f(2\tilde{B}^n) - S^{n-2}) = 0$, $(i = 1, 2)$, and hence the inclusion $f(2\tilde{B}^n) - S^{n-2} \subset \Sigma^n - S^{n-2}$ induces an isomorphism of first homology groups. Thus, $fj(\rho)$ also represents a generator of $H_1(\Sigma^n - S^{n-2})$. Finally, $p_2\colon \Sigma^n - S^{n-2} \to S^1$ is a homotopy equivalence, so $p_2 fj(\rho)$ represents a generator of $H_1(S^1)$. Thus, $\lambda(\infty) = \pm 2\pi$.

To unwind the map fj, let $\alpha\colon E^1 \to (-\pi, \pi)$ be a homeomorphism which is the identity on $(-\pi/2, \pi/2)$ and define $G\colon N \to \Sigma^n - S^{n-2} = \operatorname{Int} B^{n-1} \times S^1$ by $G(z) = (p_1 fj(z), q\alpha\lambda(z))$. The verification that G is an embedding is routine, noting that $\lambda(0) \neq \lambda(\infty)$. If $z \in B^{n-1}$ is sufficiently close to 0, then $p_1 fj(z) = p_1 f(z)$ and $q\alpha\lambda(z) = q\lambda(z) = p_2 fj(z) = p_2 f(z)$. Thus, $g = G \mid \Sigma^{n-1}$ agrees with f near 0 and is locally flat off $0, \infty \in \Sigma^{n-1}$, completing the proof.

Proof of Theorem 3.9.1. We will break the proof into two parts. In Part 1, we will show that it suffices to prove Theorem **3.9.1** in the special case that $M^{n-1} = S^{n-1}$ and $N^n = S^n$ (or E^n) and then in Part 2, we will establish that special case.

Part 1. Since C^* is tame in M, it lies on a locally flat arc J^* in M. Then, it follows from Theorem **3.4.1** that J^* lies in the interior of an $(n-1)$-ball B^* in M for which there is a homeomorphism $k\colon (B^*, J^*) \twoheadrightarrow (B_2^{n-1}, B^1)$. Also, since $f(C^*)$ is tame, it lies on a locally flat arc J' which lies in a set U in N for which there is a homeomorphism $h_U\colon (U, J') \twoheadrightarrow (E^n, B^1)$.

There exists a closed neighborhood N of C^* in J^* such that $f(N) \subset U$. $h_U f k^{-1} \mid k^{-1}(N - C^*)$ is locally flat by transitivity of local flatness, hence locally tame. Then, by applying Theorem **3.8.1**, we conclude that $h_U f k^{-1} \mid k^{-1}(N)$ is tame. Thus, $f(J^*)$ is locally flat everywhere. We may assume that $f(B^*)$ is contained in a set V in N for which there is a homeomorphism $h_V\colon (V, f(J^*), fk^{-1}(0)) \twoheadrightarrow (E^n, B^1, 0)$.

Let s denote a map of E^n which shrinks B^1 to 0, which is a homeomorphism from $E^n - B^1$ to $E^n - 0$ and which is the identity on a neighborhood of $h_V f(\partial B^*)$. Let g be a map of B_2^{n-1} which shrinks B^1

to 0, which is a homeomorphism from $B_2^{n-1} - B^1$ to $B_2^{n-1} - 0$, and which is the identity on a neighborhood of ∂B_2^{n-1}. Then, $h_V f k^{-1} g^{-1}: B_2^{n-1} \to E^n \subset \Sigma^n$ is an embedding which is locally flat except at $0 \in B_2^{n-1}$ (see Fig. **3.9.2**). We apply Lemma **3.9.1** to extend $h_V f k^{-1} g^{-1}$ to an embedding

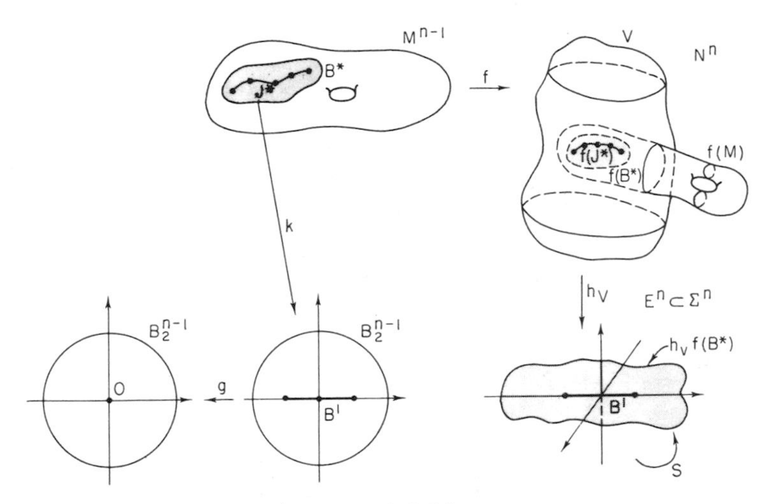

Figure 3.9.2

$F': \Sigma^{n-1} \to \Sigma^n$ which is locally flat except at 0 and ∞. Define $F: \Sigma^{n-1} \to \Sigma^n$ by piecing together $h_V f k \mid B_2^{n-1}$ and $s^{-1}F' \mid \mathrm{Cl}(\Sigma^{n-1} - B_2^{n-1})$. F is locally flat except at $k(C^*) \cup \infty$. But $k(C^*) \cup \infty$ is a subset of a tame Cantor set in Σ^{n-1} whose image is tame in Σ^n. Hence, if the theorem held for $M^{n-1} = S^{n-1}$, we would have that F is locally flat from which it would follow that f is locally flat.

Part 2 (Notation). Let E be a subset of the standard Cantor set C in $[-1, 1]$ gotten by deleting middle thirds, and suppose $-1 \in E$ and $1 \in E$. Let $D = \{d_1, d_2, \ldots\}$ be a countable dense subset in the complement of the Cantor set C in $[-1, 1]$. Let $[-1, 1]$ be identified with the subset

$$J = \{(x_1, \ldots, x_n) \in S^{n-1} \subset E^n \mid x_{n-1}^2 + x_n{}^2 = 1,$$
$$x_{n-1} \geqslant 0, x_1 = x_2 = \cdots = x_{n-2} = 0\}$$

by the identification $t \leftrightarrow (0, 0, \ldots, (1 - t^2)^{1/2}, t)$. Then, E and C can be thought of as subsets of J. Let A be an annulus pinched at E, that is

$$A = S^{n-1} \times [-1, 1]/[(x, t) = (x, 0) \quad \text{if} \quad x \in E, \quad -1 \leqslant t \leqslant 1].$$

Let $S_t^{n-2} = \{(x_1, ..., x_n) \in S^{n-1} \mid x_n = t\}$ and $\Sigma_t = S_t^{n-2} \times \frac{1}{2}$ in A. Also, let $A_{1/2} = \{(x, s) \in A, \mid 0 \leqslant s \leqslant \frac{1}{2}\}$. (See Fig. **3.9.3**.) (Strictly speaking, we regard A as a nice subset of E^n as indicated in Fig. **3.9.3** rather than the decomposition space given.)

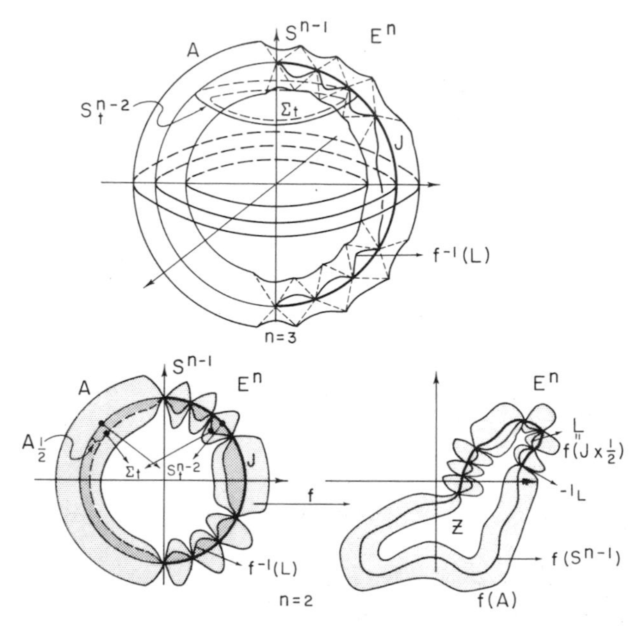

Figure 3.9.3

Preliminaries to Proof. By the Jordan–Brouwer separation theorem, $f(S^{n-1})$ separates E^n into a bounded open set Z "inside" $f(S^{n-1})$ and an unbounded set "outside". Since C^* is tame in S^{n-1}, there is a homeomorphism of S^{n-1} taking C^* onto C, so we assume $C^* = C$ and $E^* = E$. Since $f \mid (S^{n-1} - E)$ is locally flat and since $f(S^{n-1} - E)$ is two-sided in E^n by the Jordan–Brouwer separation theorem (or, more simply, [Rushing, 6]), it follows from Theorem **1.7.6**, Part (c), that f may be considered to embed all of the pinched annulus A into E^n with $f(S^{n-1} \times 1)$ in $\bar{Z}$.

Since $f \mid J - E$ is locally flat and $f(E)$ is tame, we can conclude from Theorem **3.8.1** that $f(J)$ is locally flat in E^n by the method used in Part 1. By the same reasoning, $L = f(J \times \frac{1}{2})$ is locally flat. Denote the points of E^n lying on L by t_L, $t \in [-1, 1]$.

Idea of Proof. The idea of the proof is to find a map $K: E^n \twoheadrightarrow E^n$ such that K is a homeomorphism outside $f(S^{n-1} \times \frac{1}{2})$, is the identity outside $f(S^{n-1})$, and collapses $f(S^{n-1} \times \frac{1}{2})$ to L, taking $f(\Sigma_t)$ to t_L

(see Step D below). This "fills up" $\bar{Z}$ which is then shown to be an n-cell (see Step E below). One then notices that the same thing could be done on the other side of $f(S^{n-1})$ in S^n. Thus, by Exercise **1.8.4**, f is flat (hence locally flat) as desired.

Steps of Proof.

A. Z is homeomorphic to E^n.

B. "How to pull the cloth together." Given $\epsilon > 0$, there exists a homeomorphism $G: E^n \to E^n$ for which

(1) $G =$ identity outside $f(S^{n-1})$,

(2) $Gf(S^{n-1} \times \frac{1}{2}) \subset N_\epsilon(L)$.

C. "How to take a stitch." For any $t \in (-1, 1) - E$ and hence for any $t \in D$, there exists a map $H: E^n \twoheadrightarrow E^n$ which satisfies

(1) $Hf(\Sigma_t) = t_L$,

(2) $H =$ identity outside $f(S^{n-1})$,

(3) H is a homeomorphism outside $f(S^{n-1} \times \frac{1}{2})$ and on a neighborhood of $f(S^{n-1} \times \frac{1}{2} - \Sigma_t)$.

[It is implied by (3) that Hf is a locally flat embedding of $(S^{n-1} \times \frac{1}{2}) - E - \Sigma_t$].

D. "How to sew." There exists a continuous map $K: E^n \to E^n$ which satisfies

(1) $K =$ identity outside $f(S^{n-1})$,

(2) K is a homeomorphism outside $f(S^{n-1} \times \frac{1}{2})$,

(3) $Kf(\Sigma_t) = t_L$ for all $t \in [-1, 1]$.

E. $f \mid S^{n-1}$ extends to an embedding f' of B^n into E^n.

Proof of A. Since $f(J)$ is flat, we may shrink it to a point p by a map $g: E^n \to E^n$, where g is a homeomorphism off $f(J)$. Then $gf(S^{n-1})$ is an $(n-1)$-sphere which is locally flat except at p. By Theorem **3.3.1**, $gf(S^{n-1})$ bounds an n-cell. In particular, the interior $g(Z)$ is homeomorphic to E^n and so Z is also homeomorphic to E^n.

Proof of B. Let s be the map which shrinks L to a point p and is a homeomorphism elsewhere. Then, $sf(S^{n-1} \times \frac{1}{2})$ is an $(n-1)$-sphere with only one nonlocally flat point p. Thus, it bounds an n-cell by Theorem **3.3.1**. Hence, each point $x \in sf(S^{n-1} \times \frac{1}{2})$ is joined by a ray r_x to p. Under s^{-1} (not defined at p), these rays are taken to rays called $s^{-1}r_x$, which may not converge to any point of L, but do get arbitrarily close to L. By using these rays, we may slide $f(S^{n-1} \times \frac{1}{2})$ into $N_\epsilon(L)$, fixing L. The result of this slide is called G.

Proof of C. The set $S_t^{n-2} \times (0, \frac{1}{2}]$ in A may be identified with $B^{n-1} - 0$ (see Fig. **3.9.4**). Then, $f(B^{n-1} - 0) = f(S_t^{n-2} \times (0, \frac{1}{2}])$ lies

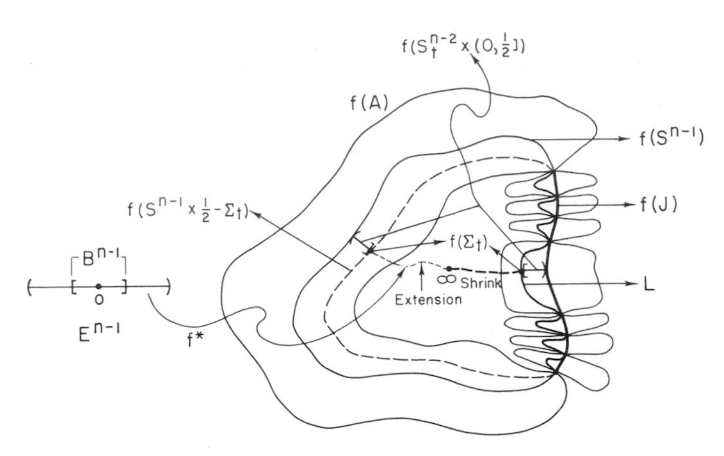

Figure 3.9.4

in Z as a closed, locally flat submanifold. By adding ∞ to $Z = E^n$, we obtain Σ^n. Letting $f(0) = \infty$, f embeds B^{n-1} and is locally flat on $B^{n-1} - 0$. By Lemma **3.9.1**, f extends to an embedding $f^*\colon \Sigma^{n-1} - \Sigma^n$, locally flat on $\Sigma^{n-1} - \{0, \infty\}$. We have $f^*(\Sigma^{n-1} - 0) \subset Z$. We can ensure that

$$f^*(\Sigma^{n-1} - 0) \cap f(A_{1/2}) = f^*(B^{n-1} - 0) = f(S_t^{n-2} \times (0, \tfrac{1}{2}])$$

by pushing $f^*(\Sigma^{n-1} - B^{n-1})$ off $f(A_{1/2})$ using the collaring of $f(S^{n-1})$.

Now let α be the shortest arc in Σ^{n-1} joining $(f^*)^{-1}(t_L)$ to ∞. Then, $f^*(\alpha)$ is locally flat mod $f(\infty)$, hence flat by Corollary **3.4.2** (or by combining Theorem **3.2.1** and the corollary to Theorem **3.6.1**). Let $s\colon Z \to Z$ be a map which shrinks $f^*(\alpha)$ to t_L, is the identity on $f(A_{1/2})$, and is a homeomorphism off $f^*(\alpha)$. Then, $f(S^{n-2} \times \tfrac{1}{2}) = f(\Sigma_t)$ bounds $D = sf^*(\Sigma^{n-1} - \operatorname{Int} B^{n-1})$ which is an $(n-1)$-cell which is locally flat except at t_L, and whose interior misses $f(A_{1/2})$. By applying [Rushing, 6] or by using the Jordan–Brouwer separation theorem and Part (c) of Theorem **1.7.6**, one can obtain a bicollar which is pinched at t_L of a slightly larger $(n-1)$-cell. It is now easy to construct H by shrinking D to t_L with a map which is the identity outside the pinched bicollar.

Proof of D. A rough idea of the proof is first to pull the cloth close together, then take a stitch, then pull the cloth closer together, then take another stitch, etc.

More precisely, the idea of the proof is to deform $f(A_{1/2})$ onto $\bar{Z}$ by collapsing $f(S^{n-1} \times \tfrac{1}{2})$ onto L, taking $f(\Sigma_t)$ to t_L for all t. First, using B, we move $f(S^{n-1} \times \frac{1}{2})$ close to L. Then, using C, we pinch $f(\Sigma_t)$ to t_L for $t = d_1$. This splits the embedding into two parts, the part between -1 and d_1, and the part between d_1 and 1. We proceed in the same way with each part. Thus, each Σ_t is pulled closer and closer to t_L and one by one each $f(\Sigma_t)$ for $t \in D$ is taken to t_L. The limit of this process will be K.

Recall that in B, we were able to move $f(S^{n-1} \times \frac{1}{2})$ toward L, but in a nonconvergent manner; that is, $f(\Sigma_t)$ would not go to t_L, but would oscillate up and down L. We now prevent this by splitting the embedding into smaller and smaller parts and thereby restrict the motion of $f(\Sigma_t)$ to its appropriate part. This forces $f(\Sigma_t)$ to converge to t_L.

For a given integer $i > 0$, let $e_1, ..., e_{i-1}$ be a reordering of $d_1, ..., d_{i-1}$ so that $e_j < e_{j+1}$ for $j = 1, ..., i-2$. Then, $[-1, 1]$ is subdivided into intervals $I_1 = [-1, e_1]$, $I_2 = [e_1, e_2], ..., I_{i-1} = [e_{i-2}, e_{i-1}]$ and $I_i = [e_{i-1}, 1]$. Let $I_{jL} = \{t_L \in L \mid t \in I_j\}$. Let $\alpha^i: [-1, 1] \to [0, \frac{1}{2}]$ be a continuous map for which $\alpha^i(-1) = \alpha^i(1) = \alpha^i(e_j) = \frac{1}{2}$, for $j = 1, ..., i-1$, and $\alpha^i(t) \neq \frac{1}{2}$ otherwise. Also, let $\alpha^i(t) \geqslant \alpha^{i-1}(t)$ for all i and $t \in [-1, 1]$. Let $\alpha_j^i = \alpha^i \mid I_j$. Let

$$A(\alpha_j^i) = \{(x, s) \in A \mid x_n \in I_j, \alpha_j^i(x_n) \leqslant s \leqslant \tfrac{1}{2}\},$$

where x_n is the nth coordinate of $x \in S^{n-1}$. Then each $A(\alpha_j^i)$ looks like a cylinder $S^{n-2} \times I$ fattened except at the top and bottom and at points of C [except that $A(\alpha_0^i)$ has its bottom pinched and $A(\alpha_i^i)$ its top pinched to a point], see Fig. **3.9.5**. Furthermore, if the top and

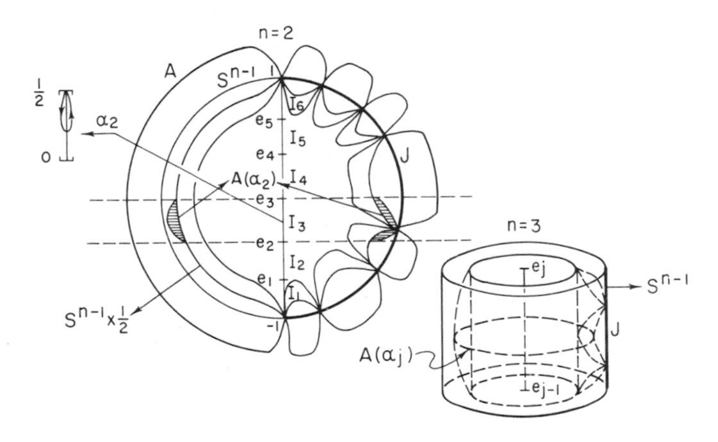

Figure 3.9.5

bottom are pinched to points, then the resulting $A^*(\alpha_j{}^i)$, is homeomorphic to A', where A' is constructed as A was, but using a different subset E of the standard Cantor set. Finally, let $A(\alpha^i)$ be the union of the $A(\alpha_j{}^i)$, $j = 1, ..., i$ and let $A^*(\alpha^i)$ be the union of the $A^*(\alpha_j{}^i)$, $j = 1, ..., i$.

We will construct inductively a sequence of continuous maps $\{K_i\}$, K_i: $E^n \to E^n$, see Fig. **3.9.6**, satisfying

(1_i) K_i = identity on $f(S^{n-1})$ for $i \geqslant 1$,
(2_i) $K_i = K_{i-1}$ on $f(A_{1/2} - A(\alpha^i))$,
(3_i) $K_i f(\Sigma_t) = t_L$ for $t = d_1, ..., d_i$,
(4_i) $K_i f(\Sigma_t) \subset N_{1/i}(I_{jL})$ if $t \in I_j$, $j = 1, ..., i$,
(5_i) K_i is a homeomorphism on a neighborhood of $f(A_{1/2} - \bigcup_{j=1}^{i} \Sigma_{d_j})$.

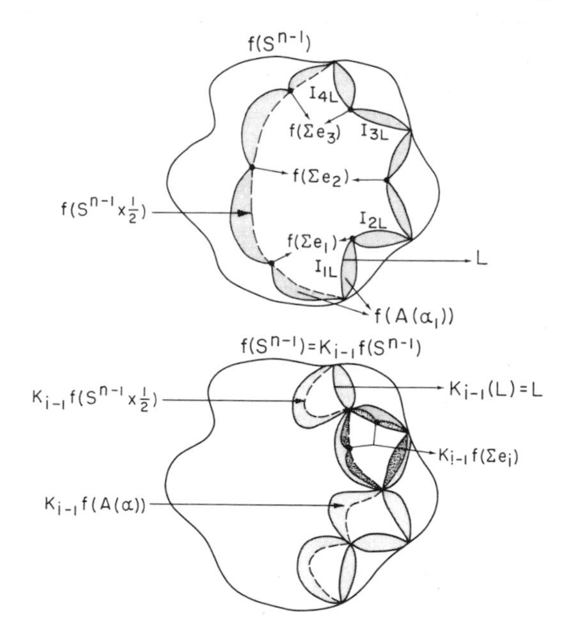

Figure 3.9.6

Start the induction with K_0 = identity.

Assume by induction that K_{i-1} has been constructed. Then, f embeds $A(\alpha^i)$ and $K_{i-1}f$ embeds $A^*(\alpha^i)$. We can apply B to each $A^*(\alpha_j{}^i)$ with $\epsilon = 1/i$ and $f = K_{i-1}f$ to obtain G_i, where $G_i K_{i-1}$ will satisfy (1_i), (2_i), (4_i), and (5_i).

Next, we apply C to just the $A^*(\alpha_j{}^i)$ which contains d_i, letting the f of C be $G_i K_{i-1} f$ and $t = d_i$. We obtain a map H_i, and $K_i = H_i G_i K_{i-1}$ will clearly satisfy (1_i)–(5_i).

Let $K = \lim K_i$. If we choose the $\alpha_j{}^i$ so that $\lim_{i\to\infty} \alpha^i(t) = \frac{1}{2}$ for all t, then K exists and is a homeomorphism outside $f(S^{n-1} \times \frac{1}{2})$ because in a neighborhood of any such point, $K_i = K_{i+1}$ for i large enough [see (2_i)]. By (3_i), $Kf(\Sigma_t) = t_L$ for $t \in D$. Furthermore, K is continuous on $f(S^{n-1} \times \frac{1}{2}) = \bigcup_t \Sigma_t$ by (4_i). Since D is dense, $Kf(\Sigma_t) = t_L$ for all $t \in [-1, 1]$.

Proof of E. Let

$$F\colon B^n - (J \times \tfrac{1}{2}) \to A_{1/2} - (S^{n-1} \times \tfrac{1}{2})$$

be the obvious homeomorphism which expands radially away from $J \times \frac{1}{2}$. (Recall that we are regarding $A_{1/2}$ as a nice subset of E^n as indicated in Fig. **3.9.3**.) Define f' by

$$f(x) = \begin{cases} KfF(x) & \text{if } \; x \in B^n - (J \times \frac{1}{2}), \\ f(x) & \text{if } \; x \in J \times \frac{1}{2}. \end{cases}$$

Since F = identity on S^{n-1}, f' is an extension of f. It is easy to see that f' is well defined by using (3) of D. Also, it is easily verified that f' is an embedding and this completes the proof.

Exercise 3.9.1. Show that a closed, locally flat embedding f of $S^{k-1} \times E^1$ into E^n, $n \geqslant 5$, unknots; that is, show that there is a homeomorphism h of E^n onto itself such that hf = identity.

Exercise 3.9.2. Can the requirement that C^* be tame in M be eliminated from Theorem **3.9.1**? (The author does not know the answer to this question.)

CHAPTER **4**

Engulfing and Applications

4.1. INTRODUCTION

The concept of engulfing has been one of the most useful discoveries to topological embeddings (as well as to piecewise linear topology) during the past decade. It is impossible to state a reasonable engulfing theorem that applies to all situations; thus, it is important to understand the various engulfing methods. (A similar statement was made concerning general position in Chapter **1**.) In this chapter, we will first consider Stallings' engulfing and then use this type of engulfing to prove the (weak) generalized Poincaré theorem, to prove the *hauptvermutung* for open cells, and to flatten topological sphere pairs and cell pairs. Next, we will discuss Zeeman's engulfing and its relationship to Stallings' engulfing. Zeeman's engulfing will be employed to give a proof of Irwin's embedding theorem. We will use both Stallings' and Zeeman's engulfing (although this is not necessary by results of Section **4.6**) to show that McMillan's cellularity criterion implies cellular. As an application of this result on the cellularity criterion, we will show that locally nice codimension-one spheres in S^n are weakly flat. Radial engulfing, which is a modification formulated by Connell and Bing of Stallings' engulfing, is developed next. Radial engulfing is applied in establishing the PL approximation of stable homeomorphisms of E^n. We also develop topological engulfing and use it to prove a topological H-cobordism theorem and to prove the topological Poincaré theorem. Finally, we develop infinite engulfing. (An important application of infinite engulfing will be given in Section **5.2**.)

4.2. STALLINGS' ENGULFING

Stallings' engulfing theorem [Stallings, 3] says that an open subset of a piecewise linear manifold can expand to engulf, like a piecewise linear amoeba, any given subpolyhedron, provided that certain dimension, connectivity, and finiteness conditions are met. Let M^n be a connected PL n-manifold without boundary, U an open set in M^n, K a (possibly infinite) complex in M^n of dimension at most $n - 3$, and L a (possibly infinite) subcomplex of K in U such that $\mathrm{Cl}(|K| - |L|)$ is the polyhedron of a finite subcomplex R of K. The idea of the proof is to let U act as an amoeba and send out feelers to engulf the vertices of R one at a time, all the while keeping L covered. Once a simplex is engulfed, it is added to L so that none of the previously considered simplexes are uncovered. After the vertices of R are covered, the engulfing is extended to the 1-simplexes of R, one at a time. In this case it is not so much like sending out feelers, but more like sliding the new U sideways along a singular disk bounded by the 1-simplex to be engulfed and an arc in the extended U joining the ends of the simplex. This process is extended to all simplexes of R.

As more and more of R is engulfed, care must be taken not to uncover any of L or any of the essential part of R already covered. Fig. **4.2.1** illustrates what must be avoided.

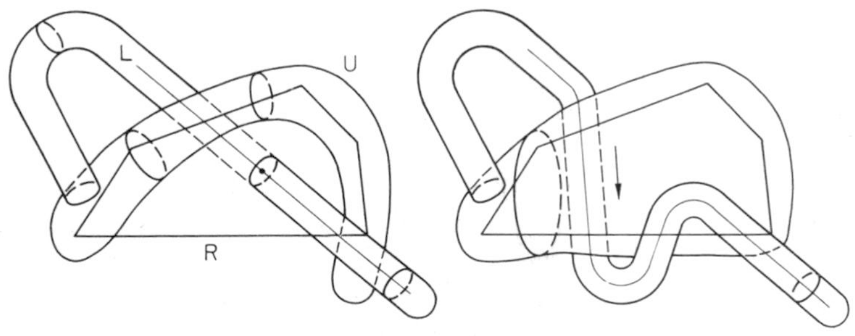

Figure 4.2.1

As simplexes of higher and higher dimensions are engulfed, the task becomes increasingly difficult since shadows leading out to singularities must be identified and engulfed to prevent uncovering something that has already been considered and added to L. In fact, if R is $(n - 3)$ dimensional, when U is finally required to reach out and engulf an $(n - 3)$-dimensional simplex of R, the amoeba may be required to regurgitate some of the $(n - 2)$-dimensional path it has previously eaten so as to be able to hold L and eat its way forward.†

† The preceding discussion follows [Bing, 11].

A pair (M, U) is **p-connected** if $\pi_i(M, U) = 0$ for all $i \leqslant p$. It follows that if the pair (M, U) is p-connected, and Δ is an i-simplex, $i \leqslant p$, then any map f taking $(\Delta \times 0, \text{Bd}\, \Delta \times 0)$ into (M, U) can be extended to a map $\bar{f}$ taking $(\Delta \times [0, 1], (\text{Bd}\, \Delta \times [0, 1]) \cup (\Delta \times 1))$ into (M, U).

Stallings' Engulfing Theorem 4.2.1. *Let M^n be a* PL *n-manifold without boundary, U an open set in M^n, K a (possibly infinite) complex in M^n of dimension at most $n - 3$ such that $|K|$ is closed in M^n, and L a (possibly infinite) subcomplex of K in U such that* $\text{Cl}(|K| - |L|)$ *is the polyhedron of a finite r-subcomplex R of K* (see Fig. **4.2.2**). *Let (M^n, U) be r-connected. Then, there is a compact set $E \subset M^n$ and an ambient isotopy e_t of M^n such that $|K| \subset e_1(U)$ and*

$$e_t \mid (M^n - E) \cup |L| = 1 \mid (M^n - E) \cup |L|.$$

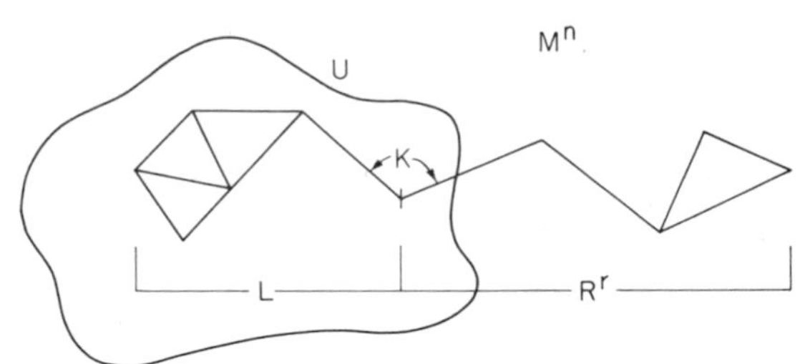

Figure 4.2.2

PROOF. The proof will be by induction on r. Certainly the theorem is true if $r = -1$. In order not to start the induction with a completely trivial case, let us prove the theorem for $r = 0$. In this case R consists of a bunch of vertices. Clearly, it will suffice to establish the case R is one vertex v since then we could engulf any number of vertices one at a time each time adding the previously engulfed ones to L. Since (M, U) is 0-connected there is a map g: $[0, 1] \to M$ such that $g(0) = v$ and $g(1) \in U$. By simplicial approximation and general position, we may assume that g is PL and that $g([0, 1]) \cap L = \emptyset$. Let $\tilde{M}$ be an open subset of M containing $g([0, 1])$. Then, since $g([0, 1]) \searrow g(1)$ the desired conclusion follows from Exercise **1.6.12** by substituting $(\tilde{M}, g([0, 1]), g(1), U \cap \tilde{M})$ for (M, P, Q, U) and extending the resulting isotopy to M by the identity.

We now inductively assume that the theorem is true for $r = 1, 2, ..., i - 1$ and show it true for $r = i \leqslant n - 3$. It will suffice to establish the case when R is one i-simplex, Δ, (that is, $|K| = |L| \cup \Delta$) since by induction we could engulf the $(i - 1)$-skeleton of R and then engulf any number of i-simplexes one at a time, each time adding to L the part of R previously engulfed.

Identify each point x of Δ with the pair $(x, 0)$, see Fig. **4.2.3**. Then,

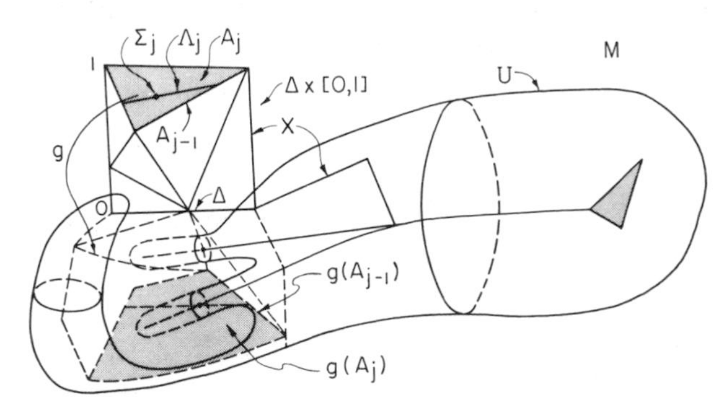

Figure 4.2.3

since the inclusion takes $(\Delta \times 0, \text{Bd } \Delta \times 0)$ into (M, U), it can be extended to a map g taking $(\Delta \times [0, 1], (\text{Bd } \Delta \times [0, 1]) \cup (\Delta \times 1))$ into (M, U) by i-connectivity of (M, U). Extend g over $|L|$ by the identity. We may assume that g is PL by the relative simplicial approximation theorem (Theorem **1.6.11**). By Part 1 of Theorem **1.6.10**, we can assume that there are triangulations (K', L'), L' subdividing L, of $(|K| \cup (\Delta \times [0, 1]), |L|)$ and M' of M such that $g: K' \to M'$ is a nondegenerate simplicial map. By Theorem **1.6.3**, there is a triangulation (X, Y) of $(|K| \cup (\Delta \times [0, 1], |L| \cup (\text{Bd } \Delta \times [0, 1]) \cup (\Delta \times 1))$, where X subdivides K', such that $X \searrow Y$. Clearly, $g: X \to M'$ is also a linear homeomorphism on each simplex of X. Thus, by Part 2 of Theorem **1.6.10**, we can assume that $g: X \to M'$ embeds each simplex of X and that $\dim S(g \mid \sigma \cup \tau) \leqslant \dim \sigma + \dim \tau - n$ for each two simplexes $\sigma, \tau \in X$.

Let $A_1, ..., A_s$ denote the simplexes of the triangulation X such that $(|L| \cup (\text{Bd } \Delta \times [0, 1]) \cup (\Delta \times 1) \cup A_1 \cup \cdots \cup A_{i-1}) \cap A_j = v_j * \text{Bd } B_j$, where $A_j = v_j * B_j$. Let D_j denote the part of the i-skeleton of X contained in $|L| \cup (\text{Bd } \Delta \times [0, 1]) \cup (\Delta \times 1) \cup A_1 \cup \cdots \cup A_j$, if $i = n - 3$, otherwise let D_j be all of

$$|L| \cup (\text{Bd } \Delta \times [0, 1]) \cup (\Delta \times 1) \cup A_1 \cup \cdots \cup A_j .$$

The idea now is to engulf $g(D_s)$, which contains $\Delta \cup |L| = |K|$ by a sequence of moves which follow backwards the image under g of the sequence of elementary simplicial collapses obtained above. In the case $i < n - 3$, we will in fact engulf all of $g(\Delta \times [0, 1]) \cup |L|$. In the case

$i = n - 3$ the regurgitation discussed in the preliminary remarks of this section takes place.

The proof that we can engulf $g(D_s)$ is by induction on j. Certainly, we already have $g(D_0) = g(|L| \cup (\text{Bd } \Delta \times I) \cup (\Delta \times 1))$ engulfed, so suppose that we have $g(D_{j-1})$ engulfed. Let

$$\Sigma_j = \bigcup_{B \in D_{j-1}} \{S(g \mid A_j \cup B)\}.$$

Then,

$$\dim \Sigma_j = \max_{B \in D_{j-1}} [\dim(S(g \mid A_j \cup B)] \leqslant \max_{B \in D_{j-1}} [\dim A_j + \dim B - n]$$

$$\leqslant (i + 1) + (n - 3) - n = i - 2.$$

Hence, by Lemma **1.6.3**, there is a polyhedron Λ such that $A_j \searrow (v_j * \text{Bd } B_j) \cup \Lambda \searrow v_j * \text{Bd } B_j$, $\Sigma \subset \Lambda \cup v_j * \text{Bd } B_j$ and $\dim \Lambda \leqslant i - 1$. Since $g \mid A_j$ is a PL embedding, we have that

$$g(A_j) \searrow g(v_j * \text{Bd } B_j) \cup g(\Lambda) \searrow g(v_j * \text{Bd } B_j),$$

$g(\Sigma) \subset g(\Lambda) \cup g(v_j * \text{Bd } B_j)$ and $\dim g(\Lambda) \leqslant i - 1$. Thus, by the $(i - 1)$-inductive hypothesis, we can engulf $g(\Lambda) \cup g(D_{j-1})$ staying fixed on $g(D_{j-1})$. Now, by Exercise **1.6.12**, we can engulf $g(D_j)$ staying fixed on $g(\Lambda) \cup g(D_{j-1})$. This completes the induction on j and so we can engulf $g(D_s)$, consequently Δ. Hence, the induction on $i = r$ is also completed and we are through.

Figure **4.2.4** illustrates how the incorrect procedure depicted in Fig. **4.2.1** is avoided if one first engulfs Λ.

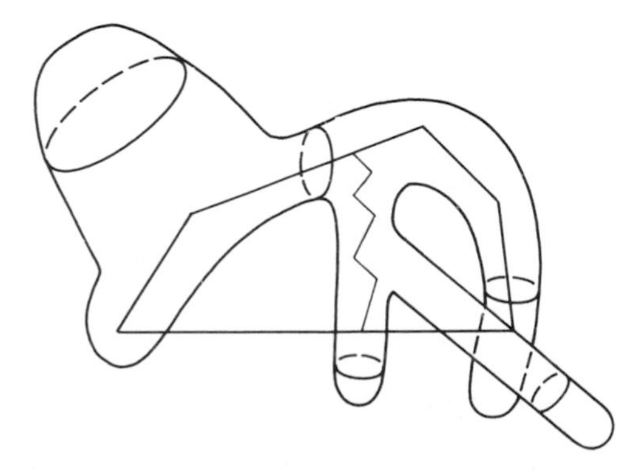

Figure 4.2.4

EXERCISE 4.2.1. Give an example to show that it is necessary that $|K|$ be closed in M^n in Theorem **4.2.1**. (A simple such example due to Bing was communicated to the author by Fred Crary.)

4.3. THE GENERALIZED POINCARÉ THEOREM

The first person to conceive a proof of the Poincaré conjecture in high dimensions was Smale [1]. Smale not only proved that a connected, closed PL manifold which is a homotopy sphere is topologically a sphere in high dimensions, but is in fact a PL sphere. The topological version is sometimes called the "weak Poincaré theorem" and is the version that we shall prove in this section. Our proof will be based on the generalized Schoenflies theorem and Stallings' engulfing theorem. For other proofs refer to Corollary **4.6.1** and to Corollary **4.13.2**.

Lemma 4.3.1. *Let M be a closed* PL *n-manifold such that every subpolyhedron of dimension at most $n/2$ is contained in the interior of a* PL *n-ball in M. Then, M is topologically homeomorphic to S^n.*

Proof. Let T be a triangulation of M, let p be the greatest integer less than or equal to $n/2$, denoted $[n/2]$, and let $q = n - p - 1 \leqslant n/2$. Let T_p be the p-skeleton of T and let T^q be the dual q-skeleton (which is defined to be the maximal subcomplex of the first barycentric subdivision of T which does not intersect T_p). By hypothesis T_p and T^q lie in the interior of PL n-balls B_* and B^*, respectively, in M.

Let N_* and N^* be the simplicial neighborhoods of T_p and T^q in the second barycentric subdivision of T. Then, $M = N_* \cup N^*$. Let N_0 be a regular neighborhood of T_p in Int B_*. By the regular neighborhood theorem there is a PL homeomorphism h_* of M such that $h_*(N_0) = N_*$. Then, $h_*(B_*)$ is a PL n-ball which contains N_* in its interior. Similarly, we can construct a PL n-ball $h^*(B^*)$ which contains N^* in its interior. Then, $M = h_*(\text{Int } B_*) \cup h^*(\text{Int } B^*)$. The lemma now follows from an application of Theorem **1.8.4**. (Recall that the proof of Theorem **1.8.4** used the generalized Schoenflies theorem.)

Weak Generalized Poincaré Theorem 4.3.1. *If M^n, $n \geqslant 5$, is a connected, closed* PL*-manifold which has the homotopy type of S^n (actually we only need $\pi_i(M) = 0$ for $i \leqslant [n/2]$), then M is (topologically) homeomorphic to S^n.*

Proof. By Lemma **4.3.1**, it will suffice to show that every k-polyhedron P^k in M, $k \leqslant n/2$, is contained in the interior of an n-ball in M. Let B^n be a n-ball in M. Then we will engulf P with Int B by using Stallings' engulfing theorem. Certainly $k \leqslant [n/2] \leqslant n - 3$ since $n \geqslant 5$. Thus, we will be through it we can show that the pair $(M, \text{Int } B)$

is k-connected. In order to do this consider the following homotopy sequence of the pair $(M, \text{Int } B)$, (see [Hu, 1 p. 109]):

$$\pi_k(\text{Int } B) \to \pi_k(M) \to \pi_k(M, \text{Int } B) \to \pi_{k-1}(\text{Int } B) \to \cdots$$
$$\to \pi_1(\text{Int } B) \to \pi_1(M) \to \pi_1(M, \text{Int } B) \to \pi_0(\text{Int } B) \to \pi_0(M).$$

By hypothesis and the fact that Int B is contractible, the above sequence becomes:

$$0 \to 0 \to \pi_k(M, \text{Int } B) \to 0 \to \cdots \to 0 \to 0 \to \pi_1(M, \text{Int } B) \to 0 \to 0.$$

Hence, the pair $(M, \text{Int } B)$ is k-connected as desired.

4.4. THE HAUPTVERMUTUNG FOR OPEN CELLS

The following is the main result of this section and is a slight strengthening of Theorem 4 of [Stallings, 3].

Theorem 4.4.1. *If* M^n *is a* k*-connected* PL n*-manifold without boundary which is* $(n - k - 2)$*-connected at infinity where* $[n/2] \leqslant k \leqslant n - 3$ *and* $n \geqslant 5$, *then* $M^n \overset{\text{PL}}{\approx} E^n$.

Corollary 4.4.1 (*Hauptvermutung* for open cells). *Any two open topological* n*-cells*, $n \geqslant 5$, *which are* PL *manifolds are* PL *homeomorphic.*

The corollary is usually stated by saying that E^n, $n \geqslant 5$, has a unique PL structure, however this is misleading since it is quite easy to show that any two triangulations of E^n are PL equivalent. Of course, the statement takes into account abstract triangulations, and when one translates it into terms of (rectilinear) triangulations he comes up with the hauptvermutung for open cells. The hauptvermutung for open n-cells, $n \leqslant 3$, is also known (see [Moise, 1]), but is still open for $n = 4$.

A manifold M is said to be k**-connected at infinity**, if for every compact $C \subset M$, there is a compact D, where $C \subset D \subset M$, such that $M - D$ is k-connected; that is $\pi_i(M - D) = 0$ for all $i \leqslant k$.

Our first lemma is a special case of the regular-neighborhood annulus theorem proved in [Hudson and Zeeman, 1]. We will actually only use the corollary to Lemma **4.4.1**. Stallings in [3] observed that the corollary also follows easily from results of either [Newman, 1] or [Gugenheim, 1].

Lemma 4.4.1 (Hudson and Zeeman). *Let N and N_1 be regular neighborhoods of the polyhedron P which lies in the interior of the* PL *manifold M. If $N_1 \subset \operatorname{Int} N$, then* $\mathrm{Cl}(N - N_1) \overset{\mathrm{PL}}{\approx} \mathrm{Bd}\, N \times I$.

PROOF. Let (K, L) be a triangulation of (M, P) such that L is full in K. Let $f\colon K \to [0, 1]$ be the unique simplicial map which takes all of the vertices of L to 0 and all of the other vertices of K to 1. Let K' be the derived subdivision of K obtained by starring each simplex $\sigma \in K$ at some point of $f^{-1}(\frac{1}{2}) \cap \operatorname{Int} \sigma$ if $f(\sigma) = [0, 1]$ and arbitrarily otherwise. Then, $f^{-1}([0, \frac{1}{2}]) = |N(P, K')|$ and $|N(P, K')|$ is a regular neighborhood of P in M by Part 1 of the regular neighborhood theorem (Theorem **1.6.4**). Let K'' be the derived subdivision of K' obtained by starring each simplex $\sigma \in K'$ at some point of $f^{-1}(\frac{1}{4}) \cap \operatorname{Int} \sigma$ if $f(\sigma) = [0, \frac{1}{2}]$ and arbitrarily otherwise. As above $f^{-1}([0, \frac{1}{4}]) = |N(P, K'')|$ is a regular neighborhood of P in M.

Let

$$|N(P, K')| = f^{-1}([0, \tfrac{1}{2}]) = N_2,$$

and

$$|N(P, K'')| = f^{-1}([0, \tfrac{1}{4}]) = N_3.$$

Then, $\mathrm{Cl}(N_2 - N_3)$ is PL homeomorphic to $\mathrm{Bd}\, N_3 \times I$ as follows. Let $\sigma_1, \ldots, \sigma_s$ be a listing in order of nondecreasing dimension of the simplexes of K meeting P but not contained in P. The PL homeomorphism is constructed inductively on $\sigma_i \cap \mathrm{Cl}(N_2 - N_3)$, which is a "skew" prism with walls $\mathrm{Bd}\, \sigma_i \cap \mathrm{Cl}(N_2 - N_3)$, top $\sigma_i \cap \mathrm{Bd}\, N_2$ and bottom $\sigma_i \cap \mathrm{Bd}\, N_3$. By induction the PL homeomorphism has already been defined on the walls, so extend it to the top, bottom and interior of the respective PL cells.

Part 2 of the regular neighborhood theorem (Theorem **1.6.4**) gives a PL homeomorphism h of N_2 onto N which leaves P fixed. Then, N_1 and $h(N_3)$ are both regular neighborhoods of P in $\operatorname{Int} N$ and so by Part 3 of the regular neighborhood theorem there is a PL homeomorphism h' of N onto itself which leaves $\mathrm{Bd}\, N \cup P$ fixed such that $h'h(N_3) = N_1$. Hence,

$$\mathrm{Cl}(N - N_1) = h'h(\mathrm{Cl}(N_2 - N_3)) \overset{\mathrm{PL}}{\approx} \mathrm{Cl}(N_2 - N_3) \overset{\mathrm{PL}}{\approx} \mathrm{Bd}\, N_3 \times I \overset{\mathrm{PL}}{\approx} \mathrm{Bd}\, N \times I.$$

Corollary 4.4.2. *Let $B_1 \subset \operatorname{Int} B_2 \subset B_2 \subset \operatorname{Int} B_3 \subset \cdots$ be a sequence of* PL *n-cells. Then,* $\bigcup_{i=1}^{\infty} B_i \overset{\mathrm{PL}}{\approx} E^n$.

For the topological version of Corollary **4.4.2**, see Exercise **3.4.3**.

Let us now digress to establish two more corollaries of Lemma **4.4.1** which will not be used in this section, but which will be used later.

Corollary 4.4.3. *If M is a bounded compact* PL *n-manifold and Q is a component of ∂M such that $M \searrow Q$, then $M \overset{\text{PL}}{\approx} Q \times I$.*

Proof. (The author saw this proof presented by C. H. Edwards.) By Theorem **1.6.8**, we may suppose that $M \subset E^k$ for some k. Consider the PL n-manifold $M \cup (Q \times I) \subset E^{k+1}$. Define $X = Q \times \frac{1}{3}$, $N_1 = Q \times [0, \frac{2}{3}]$ and $N = M \cup (Q \times I)$. Then, clearly

$$N \searrow Q \times I \searrow N_1 \searrow X,$$

so that Lemma **4.4.1** implies that

$$N - \operatorname{Int} N_1 \overset{\text{PL}}{\approx} \partial N \times I \approx \partial N_1 \times I.$$

But, $N - \operatorname{Int} N_1$ is the disjoint union of M and $Q \times [\frac{2}{3}, 1]$, and $\partial N_1 \times I$ consists of two copies of $Q \times I$. Hence, $M \overset{\text{PL}}{\approx} Q \times I$.

Analogously to the topological definition in Section **1.7**, we say that ∂M is **PL-collared** in M if there exists a PL homeomorphism h of $\partial M \times I$ into M such that $h(x, 0) = x$ for $x \in \partial M$. We call $h(\partial M \times I)$ a **PL-collar** of ∂M in M. The next corollary follows directly from the preceding corollary.

Corollary 4.4.4. *If M is a bounded compact* PL *manifold and N is a regular neighborhood of ∂M in M, then N is a* PL*-collar of ∂M in M.*

Remark 4.4.1. For a proof of Corollary **4.4.4** by different techniques, see Theorem **1.7.7** and Remark **1.7.3**.

Lemma 4.4.2. *Let M^n be a k-connected* PL *n-manifold without boundary and let P^k be a k-polyhedron in M where $k \leqslant n - 3$. Then, P is contained in the interior of a* PL *n-cell in M.*

Lemma **4.4.2** follows from an application of Stallings' engulfing theorem (Theorem **4.2.1**) as in the proof of Theorem **4.3.1**.

Proposition 4.4.1. *Let M^n be a k-connected* PL *n-manifold which is $(n - k - 2)$-connected at infinity where $[n/2] \leqslant k \leqslant n - 3$. Then, for every compact $C \subset M$ there is a compact D, where $C \subset D \subset M$, such that $(M, M - D)$ is $(n - k - 1)$-connected.*

Proof. Since M is $(n - k - 2)$-connected at infinity, there is a compact D such that $C \subset D \subset M$ and $M - D$ is $(n - k - 2)$-connected. Consider the following exact sequence of the pair $(M, M - D)$:

$$\pi_{n-k-1}(M) \to \pi_{n-k-1}(M, M - D) \to \pi_{n-k-2}(M - D)$$
$$\to \pi_{n-k-2}(M) \to \cdots \to \pi_0(M - D) \to \pi_0(M) \to \pi_0(M, M - D).$$

Notice that $\pi_i(M) = 0$ for $i \leqslant n - k - 1$ since $k \geqslant n - k - 1$ follows from the hypothesis that $[n/2] \leqslant k$. Thus, the above sequence becomes

$$0 \to \pi_{n-k-1}(M, M - D) \to 0 \to 0 \to \cdots \to 0 \to 0 \to \pi_0(M, M - D).$$

It follows that $(M, M - D)$ is $(n - k - 1)$-connected as desired.

Lemma 4.4.3. *Let M^n be a k-connected* PL *n-manifold without boundary which is $(n - k - 2)$-connected at infinity where $[n/2] \leqslant k \leqslant n - 3$ and $n \geqslant 5$. Let T be a combinatorial triangulation of M and let T_{n-k-1} denote its $(n - k - 1)$-skeleton. Then, for every compact set $C \subset M$ there is a compact set $E_1 \subset M$ and a* PL *homeomorphism h_1: $M \to M$ such that*

$$C \subset E_1 \subset M \quad \textit{and} \quad T_{n-k-1} \subset h_1(M - C) \quad \textit{and}$$
$$h_1 \mid (M - E_1) = 1 \mid (M - E_1).$$

PROOF. By Proposition **4.4.1** there is a compact D with $C \subset D \subset M$ and $(M, M - D)$ $(n - k - 1)$-connected. Thus, Lemma **4.4.3** follows immediately from Stallings' engulfing theorem (Theorem **4.2.1**) by substitution of $M - D$ for U and T_{n-k-1} for K.

Proof of Theorem 4.4.1. We first observe that it will suffice to show that if C is a compact subset of M, then $C \subset \operatorname{Int} F \subset M$, where F is a PL n-cell. Let T be a combinatorial triangulation of M and let $\{v_i\}_{i=1}^{\infty}$ enumerate the vertices of T. Let $B_1 = \operatorname{St}(v_1, T)$. Inductively, suppose that a PL n-ball $B_{j-1} \subset M$ is defined and let B_j be a PL n-ball which contains the compact set $B_{j-1} \cup \bigcup_{i=1}^{j} \operatorname{St}(v_i, T)$. Clearly, $B_1 \subset \operatorname{Int} B_2 \subset \operatorname{Int} B_3 \subset \cdots$ is a sequence of PL n-balls such that $\bigcup_{i=1}^{\infty} B_i = M$. It then follows from Corollary **4.4.2** that $M^n \overset{\text{PL}}{\approx} E^n$.

We now will show that if C is a compact subset of M, then $C \subset F \subset M$ where F is a PL n-cell. Let T_{n-k-1} denote the $(n - k - 1)$-skeleton of T and apply Lemma **4.4.3**. Let

$$K = T_{n-k-1} \cup \{\sigma \in T \mid \sigma \subset M - E_1\}.$$

Then, $K \subset h_1(M - C)$.

Let T^k denote the finite complex which is dual to K. Since $k \leqslant n - 3$, it follows from Lemma **4.4.2** that T^k is contained in the interior of a PL n-cell $A \subset M$. Since $K \subset h_1(M - C)$ and $T^k \subset \operatorname{Int} A$, it is easy to construct a PL homeomorphism h_2: $M \twoheadrightarrow M$ such that $h_2(\operatorname{Int} A) \supset h_1(C)$. Thus, $h_1^{-1}h_2(\operatorname{Int} A) \supset C$ and so $F = h_1^{-1}h_2(A)$ is the desired PL n-cell containing C.

4.5. FLATTENING TOPOLOGICAL SPHERE PAIRS AND CELL PAIRS

Theorem **1.8.2** (the generalized Schoenflies theorem), said that a locally flat $(n-1)$-sphere in S^n is flat. In fact, Theorem **3.3.1** said that if the $(n-1)$-sphere is locally flat except possibly at one point and $n \geqslant 4$, then it is flat. It is natural to wonder whether an analogous statement is true for spheres of codimensions other than one. In this section, we shall establish the following result which was first proved in [Stallings, 2].

Flattening Theorem 4.5.1 (Stallings). (a) *A k-sphere in S^n which is locally flat except possibly at one point is flat if $k \leqslant n-3$.* (b) *A locally flat $(n-2)$-sphere in S^n, $n \geqslant 5$ is flat if its complement has the homotopy type of S^1.*

We shall also establish a related result (Theorem **4.5.2**) concerning the unknotting of cell pairs which was proved in [Glaser and Price, 1]. If the k-cell D^k is properly contained in the n-cell D^n, that is, boundary contained in boundary and interior in interior, then (D^n, D^k) is said to be a **cell pair**. A cell pair (D^n, D^k) is said to be **flat** if it is homeomorphic to (I^n, I^k).

Unknotting Cell Pairs Theorem 4.5.2 (Glaser and Price). (a) *A cell pair (D^n, D^k), $n \geqslant 5$, $k \neq n-2$, which is locally flat except possibly at one point of* $\text{Int}(D^k)$ *is flat.* (b) *A locally flat cell pair (D^n, D^{n-2}), $n \geqslant 6$, is flat if $D^n - D^{n-2}$ and* $\text{Bd}\, D^n - \text{Bd}\, D^{n-2}$ *each have the homotopy type of S^1.* (c) *A cell pair (D^n, D^{n-1}), $n \geqslant 4$, which is locally flat except possibly at one point of* $\text{Int}(D^{n-1})$ *is flat.*

The idea of the proof of Theorem **4.5.1** is similar to the proof given in Section **4.4**, although more delicate. In particular, a point is removed from the k-sphere in S^n so that a closed k-string X^k in E^n results. The objective now is to express (E^n, X^k) as the monotone union of cell pairs so that by the proof of Theorem **3.4.4**, X^k is flat in E^n and its one-point compactification (the original k-sphere) is flat in S^n. To show that (E^n, X^k) is the monotone union of cell pairs, it suffices to show that if C is a compact set in E^n, then there is an open set U such that $C \subset U \subset E^n$ and $(U, U \cap X^k)$ is homeomorphic to (E^n, E^k). In order to obtain such a U, a flattening neighborhood W in E^n of an arbitrary point of X^k is taken and by a "horizontal" engulfing process is stretched over $C \cap X^k$ and then by a "vertical" engulfing process is stretched over the rest of C.

The horizontal engulfing becomes trivial by applying Theorem **3.4.4**. The meat of the proof is the vertical engulfing. It will be necessary to use some algebraic topology to eastablish the required connectivity. Here again, however, our work is made simpler than Stallings' original proof by applying Theorem **3.4.4**. (These uses of Theorem **3.4.4** evolved from the observation of Price that Theorem **3.4.4** could be used to obtain a simple proof of Corollary 3.5 of [Stallings, 2].)

Lemma 4.5.1 (Horizontal engulfing). *Let X be a closed, locally flat string in E^n, let B be a compact set in E^n, and let W be a flattening neighborhood in E^n of an arbitrary point x of X. Then, there is a homeomorphism h of (E^n, X) onto itself such that $B \cap X \subset h(W)$.*

PROOF. By Theorem **3.4.4**, there is an open set U in E^n containing X such that there is a homeomorphism g: $(U, X) \twoheadrightarrow (E^n, E^k)$. Let W' be a flattening n-cell neighborhood of x contained in $W \cap U$. It is easy to get a homeomorphism f: $(E^n, E^k) \twoheadrightarrow (E^n, E^k)$ which is the identity outside of a compact set such that $g(B \cap X) \subset f(g(W'))$. Then, $h = g^{-1}fg$:$(U, X) \twoheadrightarrow (U, X)$ can be extended to E^n by the identity and this is the desired homeomorphism.

We next state Stallings' engulfing theorem in the form which will be needed to do the vertical engulfing.

Lemma 4.5.2 (Modified Stallings' engulfing). *Let M^n be a* PL *n-manifold without boundary, let $C \subset U \subset M$, where C is closed in M and U is open in M, let P be a polyhedron in M, such that the dimension of $P - C$ is p, where $p \leqslant n - 3$. Let $(M - C, U - C)$ be p-connected and let $P - U$ be compact. Then, there is a compact $E \subset M - C$, and there is a* PL *homeomorphism h: $M \twoheadrightarrow M$ such that $P \subset h(U)$ and $h\,|(M - E) = 1\,|(M - E)$* (see Fig. **4.5.1**).

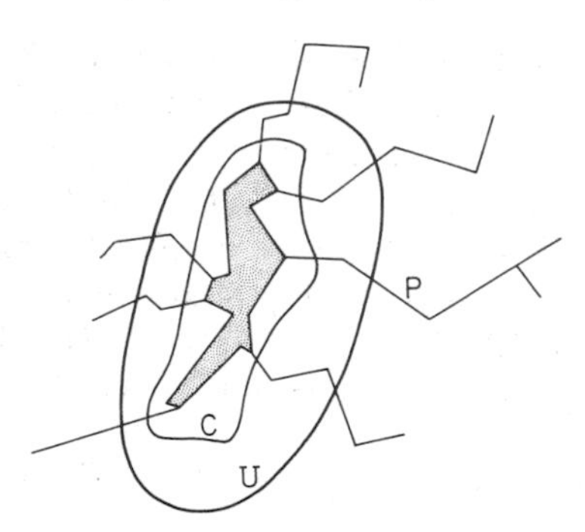

Figure 4.5.1

Proof. Apply Theorem **4.2.1** to $M_1 = M - C$, $U_1 = U - C$ and K_1 which is some triangulation of $P - C$. That theorem says there exists a compact $E \subset M_1$ and a PL homeomorphism $h_1: M_1 \twoheadrightarrow M_1$ (which can be realized be an ambient isotopy) such that $P_1 \subset h_1(U_1)$ and $h_1 |(M_1 - E) = 1 |(M_1 - E)$. Let $h: M \twoheadrightarrow M$ be the extension of h_1 by the identity over $M - E$. Then, h clearly satisfies the lemma.

A closed string $X \subset E^n$ is said to be **unraveled at infinity** if for each compact $C \subset E^n$ there is a compact D, where $C \subset D \subset E^n$, such that $\pi_i(E^n - X, E^n - (X \cup D)) = 0$ for $i = 0, 1, 2$.

The next two lemmas contain the connectivity necessary to do the vertical engulfing.

Lemma 4.5.3 (Connectivity in codimensions greater than two). *Let $X^k \subset E^n$, $k \leqslant n - 3$, be a closed, locally flat k-string. Let $V \subset E^n$ be such that $(V, V \cap X) \approx (E^n, E^k)$. Then, $\pi_i(E^n - X, V - X) = 0$ for all i and X is unraveled at infinity.*

Lemma 4.5.4 (Connectivity in codimension two). *Let $X^{n-2} \subset E^n$ be a closed, locally flat $(n - 2)$-string such that $E^n - X$ has the homotopy type of S^1. Let $V \subset E^n$ be such that $(V, V \cap X) \approx (E^n, E^{n-2})$. Then, $\pi_i(E^n - X, V - X) = 0$ for all i. If, furthermore, $X \cup \infty$ is locally flat in the one-point compactification of the pair (E^n, X), then X is unraveled at infinity.*

Before proving the above two lemmas, a couple of preliminary propositions will be established.

Proposition 4.5.1. *Let $X^k \subset E^n$, $k \leqslant n - 3$, be a closed, locally flat k-string and for $r \geqslant 0$ define $B(r) = \{x \in E^n \mid \| x \| \leqslant r\}$. Then, $\pi_1(E^n - (X \cup B(r))) = 0$ for all $r \geqslant 0$.*

Proof. Let l be a loop in $E^n - (X \cup B(r))$. Consider l to be a map of Bd Δ into $E^n - (X \cup B(r))$ where Δ is a 2-simplex. Since $E^n - B(r)$ is simply connected, l can be extended to a map $\bar{l}$ of Δ into $E^n - B(r)$. By Theorem **3.4.4**, there is an open set U in E^n containing X and a homeomorphism h: $(U, U \cap X) \twoheadrightarrow (E^n, E^k)$. Certainly we can assume that $U \cap l(\text{Bd } \Delta) = \emptyset$. Then $\bar{l}^{-1}(U)$ is an open subset of Int Δ and as such is a PL manifold. Also $E^n - h(B(r))$ is a PL manifold which contains $E^k - h(B(r))$ as a PL submanifold. We have the map

$$h\bar{l} \mid \bar{l}^{-1}(U): \bar{l}^{-1}(U) \to E^n - h(B(r)).$$

By the techniques of Section **1.6** (Parts **C** and **D**), $h\bar{l} \mid \bar{l}^{-1}(U)$ may be replaced by a map $f\colon \bar{l}^{-1}(U) \to E^n - B(r)$ which is PL and in general position with respect to $E^k - B(r)$ and is such that $\bar{f}\colon \Delta \to E^n$ defined by $\bar{f} \mid \Delta - \bar{l}^{-1}(U) = \bar{l} \mid \Delta - \bar{l}^{-1}(U)$ and $\bar{f} \mid \bar{l}^{-1}(U) = h^{-1}f$ is the desired extension of l which takes Δ onto $E^n - (X \cup B(r))$.

Proposition 4.5.2. *Let $\Sigma \subset S^n$ be a k-sphere and let B be a closed subset of S^n such that $\Sigma - B \neq \emptyset$. Then, the maps*

$$H_i(S^n - (\Sigma \cup B)) \to H_i(S^n - \Sigma)$$

are onto.

Proof. By the Alexander duality theorem (see p. 177 of [Greenberg, 1] or p. 296 of [Spanier, 1]), it will suffice to show that $H^j(\Sigma \cup B) \to H^j(\Sigma)$ is onto. This is trivial for $j \neq k$, since Σ is homeomorphic to S^k, and hence acyclic in dimensions other than k. Since $\Sigma - B \neq \emptyset$, there is a map $f\colon \Sigma \twoheadrightarrow \Sigma$ of degree one (see p. 304 of [Eilenberg and Steenrod, 1]) taking $\Sigma \cap B$ into a single point x_0 which can be obtained by smashing the complement of the interior of a locally flat k-cell in $\Sigma - B$ to x_0 and stretching the interior of the k-cell over $\Sigma - x_0$. Extend f to a map $g\colon \Sigma \cup B \to \Sigma$ by defining $g \mid \Sigma = f$, $g(B) = x_0$, and let i be the inclusion of Σ into $\Sigma \cup B$. Then, we have the following commutative diagram

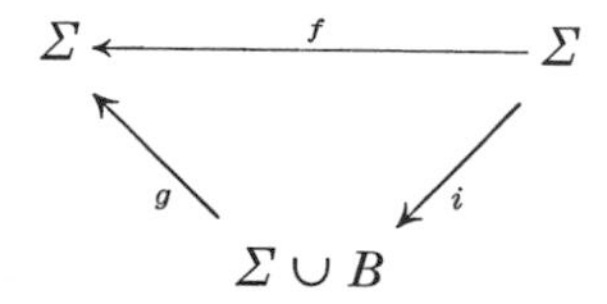

By taking induced maps we obtain the diagram,

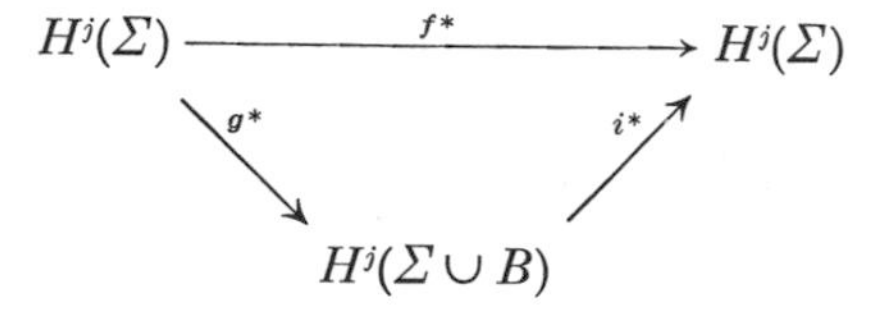

Since f^* is an isomorphism, it follows that i^* is onto as desired.

Proof of Lemma 4.5.3. It follows from Proposition 4.5.2, upon one-point compactifying (E^n, X), that the maps $H_i(V - X) \to H_i(E^n - X)$ are onto for each i. Since for each i, these groups are cyclic and isomorphic, it follows that $H_i(V - X) \to H_i(E^n - X)$ is an isomorphism for all i, and hence from the exact homology sequence of a pair that

$H_i(E^n - X, V - X) = 0$ for all i. Clearly, $\pi_1(V - X) = 0$, and by Proposition **4.5.2**, $\pi_1(E^n - X) = 0$. It now follows from the relative Hurewicz theorem (Theorem 4, p. 397 of [Spanier, 1]) that $\pi_i(E^n - X, V - X) = 0$ for all i, and this is the desired first conclusion of Lemma **4.5.3**. (The relative Hurewicz theorem may be deduced easily from Theorem 3 of [Whitehead, 2] and this Whitehead theorem must be used in the corresponding part of the proof of Lemma **4.5.4**.)

To show that X is unraveled at infinity, it is enough to show that for any r, $\pi_i(E^n - X, E^n - (X \cup B(r)) = 0$, $i = 0, 1, 2$. (Recall that $B(r) = \{x \in E^n \mid \| x \| \leqslant r\}$.) First, we would like to see that

$$\pi_2(E^n - (X \cup B(r))) \to \pi_2(E^n - X)$$

is onto. Well, by Proposition **4.5.1**, we know that

$$\pi_1(E^n - (X \cup B(r))) = 0$$

and $\pi_1(E^n - X) = 0$. Utilizing the Hurewicz isomorphism for 1-connected spaces (see Theorem 5, p. 398 of [Spanier, 1]), it suffices to see that $H_2(E^n - (X \cup B(r))) \to H_2(E^n - X)$ is onto, and this follows from Proposition **4.5.2** on taking the one-point compactification. Now consider the following part of the exact sequence of the pair $(E^n - X, E^n - (X \cup B(r)))$ (see p. 115 of [Hu, 1]):

$$\cdots \longrightarrow \pi_2(E^n - (X \cup B(r))) \overset{i_*}{\twoheadrightarrow} \pi_2(E^n - X) \overset{j_*}{\longrightarrow} \pi_2(E^n - x, E^n - (X \cup B(r)))$$
$$\overset{\partial}{\longrightarrow} \pi_1(E^n - (X \cup B(r))) \longrightarrow \cdots.$$

Since i_* is onto, the kernel of j_* is all of $\pi_2(E^n - X)$ and so the image of j_* is the zero of $\pi_2(E^n - X, E^n - (X \cup B(r)))$. However, $\pi_1(E^n - (X \cup B(r)))$ is zero and so the kernel of ∂ is all of $\pi_2(E^n - X, E^n - (X \cup B(r)))$. Thus, by exactness,

$$\pi_2(E^n - X, E^n - (X \cup B(r))) = 0$$

as desired.

Proof of Lemma 4.5.4. The first assertion of Lemma **4.5.4** comes from an argument like the first part of the above proof by utilizing the fact that the Hurewicz homeomorphism $\pi_i(S^1) \to H_i(S^1)$ is an isomorphism for all i. (See Proposition 2, p. 394 of [Spanier, 1].)

To prove the second part, compactify (E^n, X) and let ∞ denote the added point. There are arbitrarily small neighborhoods V of ∞ such that $(V, (X \cup \infty) \cap V)$ is homeomorphic to (E^n, E^{n-2}). By the first part of this lemma (on removing a point of $X - V$ so as to get a string again), $\pi_i(E^n - X, V - X) = 0$ for all i. This more than proves X unraveled at infinity.

Lemma 4.5.5 (Main lemma). *Let $X^k \subset E^n$, $n - k \geqslant 2$, $n \geqslant 5$ be a closed, locally flat k-string. If $k = n - 2$, assume further that $E^n - X$ has the homotopy type of S^1 and that X is unraveled at infinity. Then, given any compact set $C \subset E^n$, there is a set $V \subset E^n$ such that $C \subset V$ and $(V, V \cap X) \approx (E^n, E^k)$.*

Proof. By Lemma **4.5.3**, or by assumption if $k = n - 2$, X is unraveled at infinity. Hence, there is a compact $B \subset E^n$ such that $C \subset B$ and $\pi_i(E^n - X, E^n - (X \cup B)) = 0$, for $i = 0, 1, 2$. Let W' be a flattening neighborhood of an arbitrary point x of X. Then, by Lemma **4.5.1**, there is a homeomorphism of (E^n, X) onto itself that takes W' onto a neighborhood W such that $B \cap X \subset W$ and $(W, W \cap X) \approx (E^n, E^k)$. Let $\alpha\colon (W, W \cap X) \twoheadrightarrow (E^n, E^k)$ be a homeomorphism and define $W(r)$ to be $\alpha^{-1}(B(r))$, where $B(r)$ is the closed ball of radius r with center at the origin in E^n. Since $B \cap X$ is compact, there is an r_0 such that $B \cap X \subset \operatorname{Int} W(r_0)$, see Fig. **4.5.2**.

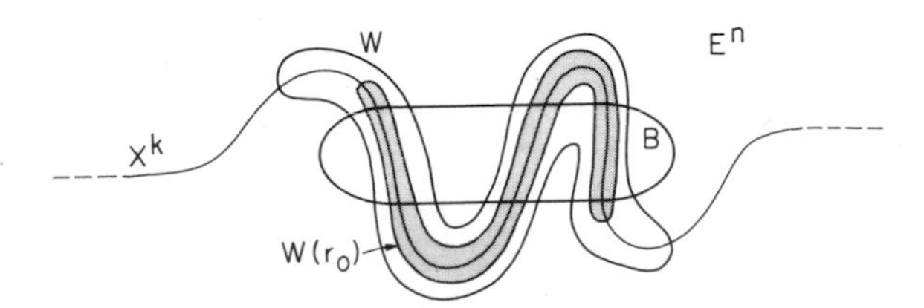

Figure 4.5.2

Since $E^n - X$ is an open subset of E^n, it is a PL manifold. Let T denote a combinatorial triangulation of $E^n - X$ with the property that if Δ is any simplex of T, then the diameter of Δ is less than the distance from Δ to X. Let K be the 2-skeleton of $N(E^n - (X \cup \operatorname{Int} W(r_0)), T)$.

We would now like to get a PL homeomorphism h_1 of E^n onto itself that is the identity outside of some compact subset E_1 of $E^n - X$ and that throws B off of K, that is, $h_1 |(E^n - E_1) = 1 |(E^n - E_1)$ and $K \subset h_1(E^n - B)$. To obtain h_1 apply Theorem **4.2.1** substituting $(E^n - X, E^n - (X \cup B), K, E_1, h_1)$ for (M^n, U, K, E, e_1) and extend the resulting h_1 to E^n by the identity on X. Note that in order to apply Theorem **4.2.1** in this way, we need to know that

$$\pi_i(E^n - X, E^n - (X \cup B)) = 0 \qquad \text{for} \quad i = 0, 1, 2$$

and this follows by definition of B.

Define K_1 to be the union of K and those simplexes Δ of T such that $\Delta \subset E^n - (E_1 \cup B)$. (Then, $K_1 \subset h_1(E^n - (X \cup B))$.) Let L be the subcomplex of the barycentric subdivision of T dual to K_1. It is easy to see that $|L| - \operatorname{Int} W(r_0)$ is compact and that $\dim(L - W(r_0)) \leqslant n - 3$.

We now would like to engulf L with W and in doing so stay the identity outside some compact subset E_2 of $E^n - X$, that is, it will be shown that there is a PL homeomorphism $h_2\colon E^n \twoheadrightarrow E^n$ such that

$$h_2 \mid (E^n - E_2) = 1 \mid (E^n - E_2)$$

and $L \subset h_2(W)$. To obtain h_2 apply Lemma **4.5.2** substituting

$$(E^n - X, W - X, W(r_0) - X, L, E_2, h_2)$$

for (M^n, U, C, P, E, h) and extend the resulting h_2 to E^n by the identity on X. In order to apply Lemma **4.5.2** in this situation, we need to know that

$$\pi_1(E^n - (X \cup W(r_0)), W - (X \cup W(r_0))) = 0 \qquad \text{for} \quad i \leqslant n - 3.$$

This follows from Lemmas **4.5.3** and **4.5.4** since it is easy to see that $(E^n - (X \cup W(r_0)), W - (X \cup W(r_0)))$ is a deformation retract of $(E^n - X, W - X)$ and that $\pi_1(E^n - X, W - X) = 0$ for all i by those lemmas.

By using h_1 and h_2, we proceed to engulf C with W as in the proof of Theorem **4.4.1** of the last section.

Flattening Theorem 4.5.3. *Let $X^k \subset E^n$, $k \leqslant n - 2$, $n \geqslant 5$, be a closed, locally flat k-string. If $k = n - 2$, let X be unraveled at infinity and let $E^n - X$ have the homotopy type of S^1. Then, X is flat.*

Proof. Lemma **4.5.5** is the analog of Lemma **3.4.1**. The proof of Theorem **4.5.3** is now essentially the same as that of Theorem **3.4.2** and the analogs of the preliminary Lemmas **3.4.2** and **3.4.3** are proved essentially the same way as those lemmas.

Proof of Theorem 4.5.1. Theorem **4.5.1** follows directly from Theorem **3.6.1** and Theorem **3.5.1** for the case $n = 4$. For $n \geqslant 5$ Theorem **4.5.1** is in fact a corollary of Theorem **4.5.3**. Let Σ be the k-sphere in S^n which is locally flat except possibly at some point ∞. We can assume that the point ∞ is an element of $\Sigma \cap S^k$. Remove ∞, apply Theorem **4.5.3**, and extend the flattening homeomorphism resulting from Theorem **4.5.3** by the identity on ∞ to obtain the desired flattening homeomorphism for Σ.

Proof of Theorem 4.5.2 (Parts a and b). Let (D^n, D^k), $n \geqslant 5$, $k \leqslant n - 2$, be a cell pair which is locally flat except possibly at one point $x \in \text{Int}(D^k)$. If $k = n - 2$, we suppose that $n \geqslant 6$ and that $D^n - D^k$ and $\text{Bd}\, D^n - \text{Bd}\, D^k$ each have the homotopy type of S^1. We may assume that $D^n = I^n$ and by Theorem **4.5.1** that $\text{Bd}\, D^k = \text{Bd}\, I^k$. Also, we may easily assume that x is origin, 0, in E^n. It follows from Exercise **1.7.8**, that the pair $(\text{Bd}\, I^n, \text{Bd}\, I^k)$ is collared in (I^n, I^k). By making use of this collar, we see that it can be assumed that there is a neighborhood U of $\text{Bd}\, D^k$ in D^k which is contained in I^k. Hence, there is an n-cube $I_*{}^n$ contained in $\text{Int}\, I^n$ which is concentric with I^n and is such that $(I^n - \text{Int}\, I_*{}^n) \cap D^k \subset U$ (see Fig. **4.5.3**).

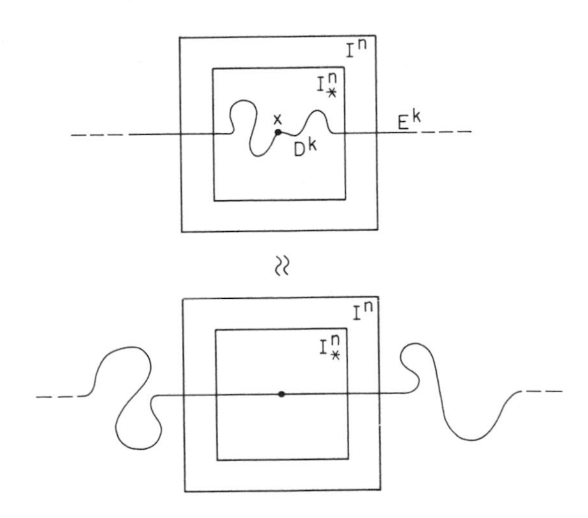

Figure 4.5.3

After one-point compactifying E^n with ∞ and removing 0, we can consider the resulting space to be E^n where ∞ corresponds to 0, $(E^k - 0) \cup \infty$ corresponds to E^k, $(E^n - \text{Int}\, I^n) \cup \infty$ corresponds to $I_*{}^n$, $(E^n - \text{Int}\, I_*{}^n) \cup \infty$ corresponds to I^n, and $\text{Int}\, I_*{}^n - 0$ corresponds to $E^n - I^n$. Let Y denote the closed string which corresponds to $((E^k - I^k) \cup D^k \cup \infty) - 0$. It is easy to see that Y can be unknotted by a homeomorphism which is the identity on $I_*{}^n$ by the same techniques used to prove Theorem **4.5.3**. Therefore, by compactifying with 0 and removing ∞, we see that we have unknotted the ball pair (I^n, D^k) staying the identity on $\text{Bd}\, I^n$ as desired.

Exercise 4.5.1. Use Theorem **3.3.1** to establish Part c of Theorem **4.5.2**.

4.6. ZEEMAN ENGULFING

Zeeman's work on engulfing [Zeeman, 3] and Chapter 7 of [Zeeman, 1], preceded that of Stallings which was presented in Section **4.2**. Zeeman engulfing has a number of important applications. In the next section we will discuss very nice piecewise linear embedding and unknotting theorems which follow from Zeeman's engulfing theorem. We shall see, in this section, a simple proof, due to Stallings and based on Zeeman engulfing, of the weak Poincaré conjecture (see [Zeeman, 3]). Also, McMillan's proof that the cellularity criterion implies cellularity, which will be discussed in Section **4.8**, involves a result of [McMillan and Zeeman, 1] that uses Zeeman engulfing.

At the end of this section we show that Zeeman engulfing follows easily from Stallings' engulfing theorem. Nevertheless, the bulk of this section will be devoted to developing Zeeman engulfing by Zeeman's original techniques. Our reasons for including Zeeman's techniques are two-fold. First, since these techniques are adaptable to a number of situations, an understanding of them, in addition to Stallings' techniques, is quite valuable as a research tool. Secondly, they provide simple proofs to several interesting results (such as the weak Poincaré theorem for $n \geqslant 7$). For a discussion of Zeeman and Stallings' engulfing, see [Hirsch and Zeeman, 1].

Only an intuitive exposition of "piping" will be presented here. Piping is used by Zeeman to obtain the smallest codimension (codimension three) in his engulfing. We will present all of the techniques necessary to do the engulfing theorem in codimensions four and greater. It is very interesting to note that piping is not necessary if one obtains Zeeman engulfing via Stallings' engulfing theorem as we do at the end of this section.

A subspace X of a manifold M is **inessential** in M if the inclusion map $X \subset M$ is homotopic to a constant. A manifold M is k-**connected** if $\pi_i(M) = 0$ for $i \leqslant k$.

EXERCISE 4.6.1. Show that any k-dimensional subpolyhedron of a k-connected manifold M is inessential in M.

Engulfing Theorem 4.6.1 (Zeeman). *Let M^m be a k-connected* PL *manifold, and X^x be a compact polyhedron in the interior of M such that $x \leqslant m - 3$ and $k \geqslant 2x - m + 2$. Then, X is inessential in M if and only if it is contained in an m-ball in the interior of M.*

Before proving the above engulfing theorem we will prove one lemma.

Lemma 4.6.1. *If the polyhedron X^x is inessential in* Int M^m, *then there exist polyhedra Y^y and Z^z in* Int M *such that* $X \subset Y \searrow Z$, $y \leqslant x + 1$, *and* $z \leqslant 2x - m + 2$.

PROOF

Case 1 (Proof of the weaker result $z \leqslant 2x - m + 3$). Let C be the cone on X. Since X is inessential, we can extend the inclusion $X \subset \text{Int } M$ to a continuous map $f\colon C \to \text{Int } M$. By the relative simplicial approximation theorem (Theorem **1.6.11**) we can make f piecewise linear keeping $f \mid X$ fixed. By Corollary **1.6.5**, we can homotop f into general position keeping $f \mid X$ fixed. Therefore, the singular set $S(f)$ of f will be of dimension $\leqslant 2(x + 1) - m$.

Let D be the subcone of C through $S(f)$; that is to say D is the union of all rays of C that meet $S(f)$ in some point other than the vertex of the cone. Then, $\dim D \leqslant 2x - m + 3$. Since a cone collapses to any subcone (Exercise **1.6.8**, Part b), we have $C \searrow D$, and since $D \supset S(f)$ we have $Y \searrow Z$ where $Y = f(C)$ and $Z = f(D)$. Since $f(X) = X$, $X \subset Y \searrow Z$ and the proof of the weaker result is complete.

Case 2 (Proof of the stronger result $z \leqslant 2x - m + 2$). This improvement of one dimension over Case 1 represents an improvement of the Poincaré conjecture from $n \geqslant 7$ down to $n \geqslant 5$. This case involves piping and as mentioned earlier in this section our discussion here will be an intuitive sketch. For a rigorous development of piping see Chapter 7 of [Zeeman, 1].

As in the proof of Case 1, let C be the cone on X and obtain a map $f\colon C \to \text{Int } M$ such that $f \mid X$ is the inclusion and dimension $S(f) \leqslant 2x - m + 2$. Again let D be the subcone of C through $S(f)$. By a little juggling, we can arrange that the top-dimensional simplexes of $S(f)$ lie inside the top-dimensional simplexes of C and that they get identified in pairs under f. For each pair (not through the vertex of C) we pipe away the middle of one of the pair over the edge of the cone. This alters $f\colon C \to M$ globally, leaving X fixed. (The previous alterations by simplicial approximation and general position were only local.) The pictures of Fig. **4.6.1** are drawn for $m = 5$, $x = 2$, but, of course, they are very inadequate dimensionwise, and they do not depict the fact we have to pipe C not merely D, nor the fact that the path along which the pipe must run can be very kinky. The pictures do, however, illustrate the purpose of the pipe. The effect has been to alter f by piercing holes in each of the pair of top-dimensional simplexes of $S(f)$, thereby enabling us to collapse away the stuff in D of dimension more than $2x - m + 2$ without touching the singular set of the altered f.

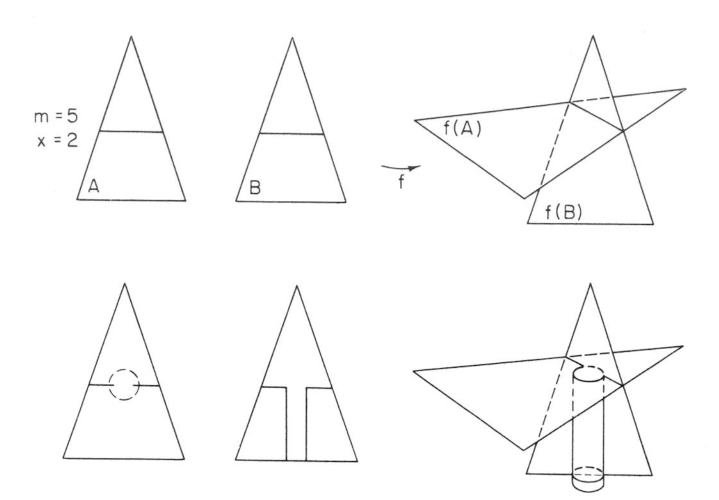

Figure 4.6.1

More precisely, let E^{2x-m+2} denote the union of $S(f')$, where f' is the altered f, together with the $(2x - m + 2)$-dimensional subcone through the $(2x - m + 1)$-skeleton of $S(f)$. Then, $D \searrow E$. (It does not matter that we didn't pierce holes in those pairs of top simplexes of $S(f)$ through the vertex of C, because they were not obstructing anything anyway.) We can now define $Z^{2x-m+2} = f'(E)$ and echo the collapses under f',

$$Y = f'(C) \searrow f'(D) \searrow f'(E) = Z.$$

The proof of Lemma **4.6.1** is complete.

Proof of Engulfing Theorem 4.6.1. We have X^x inessential in Int M, and have to show that X is contained in an m-ball in Int M. The proof is by induction on x, starting trivially with $x = -1$. Assume the result true for dimensions less than x.

By Lemma **4.6.1**, we can choose Y and Z in Int M such that $X \subset Y \searrow Z$. where

$$z \leqslant 2x - m + 2 \leqslant k.$$

Therefore, by Exercise **4.6.1**, Z is inessential in M. But $z < x$ by the hypothesis $x \leqslant m - 3$. (This is one of the places where codimension $\geqslant 3$ is crucial.) Therefore, Z is contained in an m-ball in Int M by induction. By Exercise **1.6.12**, Y (hence X) is also contained in an m-ball as desired.

As a corollary of Theorem **4.6.1**, we get another proof of the weak generalized Poincaré theorem which we have already proved as Theorem

4.3.4. This proof via Theorem **4.6.1** is probably the most simple proof known of the weak generalized Poincaré theorem.

Corollary 4.6.1 (Stallings). *If M^n, $n \geqslant 5$, is a connected, closed* PL-*manifold which has the homotopy type of S^n (actually we only need $\pi_i(M) = 0$ for $i \leqslant [n/2]$), then M is (topologically) homeomorphic to S^n.*

PROOF. The corollary follows directly from Theorem **4.6.1** and Lemma **4.3.1**.

Engulfing Theorem 4.6.2 (Zeeman). *Let M^m be a k-connected* PL *manifold. Let C be a collapsible subpolyhedron and X^x, $x \leqslant m - 3$, a subpolyhedron, both in the interior of M. If $k \geqslant x$, then there is a polyhedron Λ in the interior of M such that $X \subset C \cup \Lambda \searrow C$ and* $\dim \Lambda \leqslant x + 1$.

Before proving Theorem **4.6.2**, we prove a necessary lemma.

Lemma 4.6.2. *If $P \subset C \subset \operatorname{Int} B \subset B$ are polyhedra with B a* PL *m-manifold and if $C \searrow P$ and $B \searrow P$, then $B \searrow C$.*

PROOF. Let N be a regular neighborhood of C in Int B. Then, $N \searrow C \searrow P$ and so N is a regular neighborhood of P. By Lemma **4.4.1**, $\operatorname{Cl}(B - N) \overset{\text{PL}}{\approx} \operatorname{Bd} B \times I$. Therefore, $B \searrow N \searrow C$ and $B \searrow C$ as desired.

Proof of Theorem 4.6.2. By Lemma **1.6.4**, there is a subpolyhedron X_0 of $C \cup X$ such that $C \cup X \searrow X_0$ and $\dim X_0 \leqslant \dim X = x$. Now X_0 is contained in an m-ball B in Int M by Theorem **4.6.1**. Hence, by Exercise **1.6.12**, $C \cup X$ is also contained in an m-ball in Int M. We may assume that $C \cup X \subset \operatorname{Int} B$ by taking a regular neighborhood if necessary. By Lemma **4.6.2**, $B \searrow C$. The existence of Λ now follows directly from Lemma **1.6.3**, and the proof is complete.

Theorem **4.6.2** will suffice for most applications of Zeeman type engulfing. The proof of Theorem **4.6.2** presented here was quite simple except for the piping necessary to do the codimension 3 case. We are now going to prove an even stronger Zeeman type engulfing theorem by the same technique used to prove Theorem **4.6.2**, except that we will use Stallings' engulfing in place of Theorem **4.6.1**. Thus, piping will be avoided since it was not used in the proof of Stallings' engulfing theorem. (Theorem **4.6.2** could be slightly improved with a little more work. For instance, one could consider the case $C \cap \partial M \neq \emptyset$ and $X \cap \partial M \neq \emptyset$.)

Engulfing Theorem 4.6.3 (Zeeman). *Let C be a subpolyhedron of the interior of the* PL *manifold M^m such that (M, C) is k-connected and such that $C \searrow P^p$, where $p \leqslant n - 3$. Let X^x, $x \leqslant m - 3$, be an arbitrary polyhedron in* Int M. *If $k \geqslant x$, then there is a polyhedron Λ in the interior of M such that $X \subset C \cup \Lambda \searrow C$ and* $\dim \Lambda \leqslant x + 1$.

Proof (via Stallings engulfing). By Lemma **1.6.4** there is a subpolyhedron X_0 of $C \cup X$ such that $P \subset X_0$, $C \cup X \searrow X_0$ and $\dim \mathrm{Cl}(X_0 - P) \leqslant x$. Let N be a regular neighborhood of P in Int M and let $U = \mathrm{Int}\, N$. It follows from Lemma **1.6.2** that the inclusion of P into C is a homotopy equivalence and from Lemma **4.4.1** that the inclusion of P into U is a homotopy equivalence. Hence, (M, U) is k-connected since (M, C) is k-connected by hypothesis. By Stallings' engulfing theorem (Theorem **4.2.1**), there is a PL homeomorphism f of M onto itself such that $f \mid P = 1$ and $X_0 \subset f(U)$. By Exercise **1.6.12**, there is a PL homeomorphism f_* of M onto itself such that $f_* \mid X_0 = 1$ and $C \cup X \subset f_* f(U)$. Since $f_* f \mid P = 1$, $f_* f(N)$ is a regular neighborhood of P. Thus, $f_* f(N) \searrow P$ and $C \searrow P$ and so by Lemma **4.6.2** $f_* f(N) \searrow C$. The existence of Λ now follows directly from Lemma **1.6.3**, and the proof is complete.

4.7. THE PENROSE, WHITEHEAD, ZEEMAN EMBEDDING THEOREM AND IRWIN'S EMBEDDING THEOREM

Let us begin by stating the main result of this section, Irwin's embedding theorem.

Codimension Three Embedding Theorem 4.7.1 (Irwin). *Let M^m and Q^q be* PL *manifolds and let $f\colon M \to Q$ be a (topological) map such that $f \mid \partial M$ is a* PL *embedding of ∂M in ∂Q. Then, f is homotopic to a proper embedding keeping ∂M fixed, provided*

$$m \leqslant q - 3,$$

$$M \text{ is } (2m - q)\text{-connected, and}$$

$$Q \text{ is } (2m - q + 1)\text{-connected.}$$

Corollary 4.7.1. *Any element of $\pi_m(Q)$, where Q is $(2m - q + 1)$-connected and $m \leqslant q - 3$, may be represented by an embedding of an m-sphere in Q.*

Corollary 4.7.2. *Any closed r-connected* PL *manifold* M, $r \leqslant m - 3$, *may be embedded in* E^{2m-r}.

Proofs of forms of Theorem **4.7.1** appear in [Irwin, 1], [Zeeman, 5] and Chapter 5 of [Zeeman, 1]. (Hudson has generalized Theorem **4.7.1** by putting a connectivity requirement on f rather than Q. This work of Hudson was announced in [Hudson, 2] and the complete proof is given in [Hudson, 3].) Theorem **4.7.1** may be regarded as a generalization of an embedding theorem of Penrose, Whitehead, and Zeeman [1] which emerges as Corollary **4.7.2**. Actually, as can be ascertained from [Zeeman, 5], the techniques of [Penrose *et al.*, 1] more than suffice to prove the following embedding theorem.

Metastable Range Embedding Theorem 4.7.2 (Penrose *et al.*) *Same as Theorem* **4.7.1** *except require that* $m \leqslant \frac{2}{3}q - 1$, *that is,* m *is in the metastable range with respect to* q.

Before describing the organization of this section let us state a lemma.

Lemma 4.7.1. *Let* M^m *and* Q^q *be* PL *manifolds and let* $f\colon M \to Q$ *be a proper map (that is,* $f^{-1}(\partial Q) = \partial M$) *such that* $S(f) \subset \operatorname{Int} M$ *and such that* f *is in general position (that is,* f *is nondegenerate and* dim $S(f) \leqslant 2m - q$). *Then, there exist collapsible subpolyhedra* C *and* D *of* Int M *and* Int Q, *respectively, such that* $S(f) \subset C = f^{-1}(D)$, *provided*

$$m \leqslant q - 3,$$

$$M \text{ is } (2m - q)\text{-connected, and}$$

$$Q \text{ is } (2m - q + 1)\text{-connected.}$$

The organization of the remainder of this section is as follows: (a) proof of Theorem **4.7.1** (for closed manifolds) modulo Lemma **4.7.1**; (b) proof of Lemma **4.7.1**, via [Penrose, *et al.* 1], for $m \leqslant \frac{2}{3}q - 1$ (thus giving a proof of Theorem **4.7.2** for closed manifolds); (c) proof of Lemma **4.7.1** via Irwin's technique (thus giving a proof of Theorem **4.7.2** for closed manifolds); (d) statement and proof of Lemma **4.7.2**; (e) proof of Theorem **4.7.1** for bounded case via Lemma **4.7.2**.

Proof of Theorem 4.7.1 (for M a Closed Manifold) Modulo Lemma 4.7.1. We are given a continuous map $f\colon M \to Q$ which we have to homotop to a piecewise linear embedding in the interior, and we are given that

$$m \leqslant q - 3,$$

$$M \text{ is } (2m - q)\text{-connected, and}$$

$$Q \text{ is } (2m - q + 1)\text{-connected.}$$

By Theorem **1.7.4**, ∂Q is collared in Q. By shrinking this collar to half its length (the inner half) homotop Q into $\operatorname{Int} Q$. This homotopy carries $f(M)$ into $\operatorname{Int} Q$. Now use the simplicial approximation theorem (Theorem **1.6.11**) to homotop the new f to a PL map. Finally, apply Corollary **1.6.5** to homotop the resulting f to a map (renamed f) into general position, $\dim S(f) \leqslant 2m - q$.

It now follows from Lemma **4.7.1** that there are collapsible subpolyhedra C and D of $\operatorname{Int} M$ and $\operatorname{Int} Q$, respectively, such that

$$S(f) \subset C = f^{-1}(D)$$

provided

$$m \leqslant q - 3,$$

$$M \quad \text{is} \quad (2m - q)\text{-connected, and}$$

$$Q \quad \text{is} \quad (2m - q + 1)\text{-connected.}$$

Choose a compact submanifold Q_* of Q containing $f(M) \cup D$ in its interior. Triangulate M and Q_* so that f is simplicial and C and D are subcomplexes. If we pass to the second barycentric subdivisions, then f remains simplicial because f is nondegenerate since it is in general position. Let B^m and B^q denote the second derived neighborhoods of C and D in M and Q_*, respectively. These are balls by Corollary **1.6.4,** because C and D are collapsible. Lemma **4.7.1** implies that $S(f) \subset B^m = f^{-1}(B^q)$. In fact that lemma implies more. It implies that f maps $\operatorname{Int} B^m$ into $\operatorname{Int} B^q$, that f embeds ∂B^m in ∂B^q, and that f embeds $M - B^m$ in $Q - B^q$.

We have localized the singularities of f inside balls where it is easy to straighten them out. More precisely, let $h: B^q \twoheadrightarrow I^q$ be a PL homeomorphism and consider $hf(\partial B^m) \subset \partial I^q$. Let C^m represent the m-ball in I^q which is the cone over $hf(\partial B^m)$ from the origin. Then, $hf \mid \partial B^m$ extends to a PL homeomorphism g_* of B^m onto C^m. Let $g: B^m \to B^q$ be defined by $h^{-1}g_*$. Then, $g \mid \partial B^m = h^{-1}g_* \mid \partial B^m = h^{-1}hf \mid \partial B^m = f \mid \partial B^m$, and we can extend g to an embedding $g: M \to Q$ by making g equal to f outside B^m. Notice that g_* was homotopic to $hf \mid B^m$ keeping $hf(\partial B^m)$ fixed simply by taking each point $hf(x)$ linearly to the point $g_*(x)$ inside of I^q. Hence, $g \mid B^m$ is homotopic to $f \mid B^m$ by a homotopy which takes place inside B^q and which is fixed on $f(\partial B^m)$. Consequently, f is homotopic to g, and this completes the proof of Theorem **4.7.1** modulo Lemma **4.7.1** in the case M is closed.

Proof of Lemma 4.7.1 for $m \leqslant \frac{2}{3}q - 1$ (Penrose *et al.*). Since f is in general position, $\dim S(f) \leqslant 2m - q$. Let $\mathscr{C}(S(f))$ denote the cone on

$S(f)$. Since M is $(2m - q)$-connected, we can extend the inclusion $S(f) \subset M$ to a PL map $g: \mathscr{C}(S(f)) \to \operatorname{Int} M$ in general position by Theorem **1.6.11** and Corollary **1.6.5**. Hence,

$$\begin{aligned} \dim S(g) &\leqslant 2(2m - q + 1) - q = 4m - 3q + 2 \\ &\leqslant 4(\tfrac{2}{3}q - 1) - 3q + 2 = -\tfrac{1}{3}q - 2 < -1, \end{aligned}$$

and g embeds $\mathscr{C}(S(f))$. Consider $\mathscr{C}(f(g(\mathscr{C}(S(f)))))$. Since Q is $(2m-q+1)$-connected the inclusion $f(g(\mathscr{C}(S(f)))) \subset Q$ extends to a PL map $g_*: \mathscr{C}(f(g(\mathscr{C}(S(f))))) \to Q$ in general position with respect to $f(M)$ by Theorem **1.6.11** and Corollary **1.6.5**. Hence,

$$\begin{aligned} \dim S(g_*) &\leqslant m + (2m - q + 1) - q = 3m - 2q + 1 \\ &\leqslant 3(\tfrac{2}{3}q - 1) - 2q + 1 = -2, \end{aligned}$$

and so g_* embeds $\mathscr{C}(f(g(\mathscr{C}(S(f)))))$ and $g_*(\mathscr{C}(f(g(\mathscr{C}(S(f)))))) \cap f(M) = f(g(\mathscr{C}(S(f))))$. Since cones are collapsible the polyhedra

$$C = g(\mathscr{C}(S(f))) \qquad \text{and} \qquad D = g_*(\mathscr{C}(f(g(\mathscr{C}(S(f))))))$$

satisfy the conditions of Lemma **4.7.1** and the proof is complete.

Proof of Lemma 4.7.1 for $m \leqslant q - 3$ (Irwin). The main idea of the proof is to use engulfing Theorem **4.6.2** several times in an inductive process.

Since M is $(2m - q)$-connected, we can start by engulfing $S(f)$ in a collapsible subpolyhedron C_1^{2m-q+1}, $S(f) \subset C_1 \subset \operatorname{Int} M$ (see Fig. **4.7.1**). Of course, when C_1 is mapped by F into Q it may no longer remain collapsible, because bits of $S(f)$ may get glued together. Nevertheless, since Q is $(2m - q + 1)$-connected, we can engulf $f(C_1)$ in a collapsible subpolyhedron D_1^{2m-q+2}, $f(C_1) \subset D_1 \subset \operatorname{Int} Q$. We are not finished yet because although $f^{-1}(D_1)$ contains C_1, it may contain other stuff as well. The idea is to move D_1 so as to minimize the dimension of this other stuff and then engulf it.

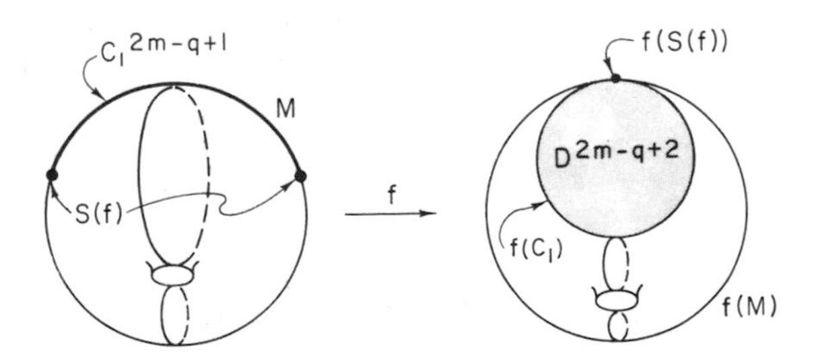

Figure 4.7.1

More precisely, we shall define an induction on i, where the ith-inductive statement is as follows:

i^{th}-INDUCTIVE STATEMENT. There exist three collapsible subpolyhedra C_i, E_i, and D_i such that $C_i \subset \operatorname{Int} M$ and $E_i \subset D_i \subset \operatorname{Int} Q$, such that

(1) $S(f) \subset C_i$,

(2) $f^{-1}(E_i) \subset C_i \subset f^{-1}(D_i)$, and

(3) $\dim(D_i - E_i) \leqslant (2m - q) - i + 3$.

The induction begins at $i = 1$, by constructing C_1 and D_1 as above, and choosing E_1 to be a point of $f(C_1)$. The induction ends at $i = (2m - q) + 4$, because then $D_i = E_i$ and so we have $S(f) \subset C_i = f^{-1}(D_i)$ as required.

There remains to prove the inductive step, and so assume the ith-inductive statement is true, where $1 \leqslant i < (2m - q) + 4$. Then, $f(C_i) \cup E_i \subset D_i$ by (2). Let $f = D_i - (f(C_i) \cup E_i)$. Then,

$$\dim F \leqslant (2m - q) - i + 3$$

by (3). By the technique of proof of Theorem **1.6.10**, we can ambient isotop D_i in $\operatorname{Int} Q$ keeping $f(C_i) \cup E_i$ fixed until F is in general position with respect to $f(M)$, that is,

$$\begin{aligned} \dim(F \cap f(M)) &\leqslant ((2m - q) - i + 3) + m - q \\ &\leqslant (2m - i + 3) + (q - 3) - q \\ &\leqslant ((2m - q) - i + 3) + (q - 3) - q = (2m - q) - i. \end{aligned}$$

Therefore, $\dim f^{-1}(F) \leqslant (2m - q) - i$, because f is nondegenerate, being in general position. Let E_{i+1} denote the new position of D_i after the isotopy. Then, E_{i+1} is collapsible because it is PL homeomorphic to D_i. Since M is $(2m - q)$-connected, we use Theorem **4.6.2** to engulf $\operatorname{Cl}(f^{-1}(F))$ in a subpolyhedron C_{i+1} of $\operatorname{Int} M$ such that

$$f^{-1}(F) \subset C_{i+1} \searrow C_i \qquad \text{and} \qquad \dim(C_{i+1} - C_i) \leqslant (2m - q) - i + 1.$$

Then, C_{i+1} is collapsible because $C_{i+1} \searrow C_i \searrow 0$; $S(f) \subset C_{i+1}$, because $S(f) \subset C_i \subset C_{i+1}$; and $f^{-1}(E_{i+1}) \subset C_{i+1}$ because

$$f^{-1}(E_{i+1}) = C_i \cup f^{-1}(E_i) \cup f^{-1}(F) = C_i \cup f^{-1}(F) \subset C_{i+1}.$$

Since Q is $((2m - q) + 1)$-connected we can now use Theorem **4.6.2** to engulf $f(C_{i+1} - C_i)$ in a subpolyhedron D_{i+1} in Int Q such that

$$f(C_{i+1} - C_i) \subset D_{i+1} \searrow E_{i+1},$$

and

$$\dim(D_{i+1} - E_{i+1}) \leqslant (2m - q) - i + 2 = (2m - q) - (i + 1) + 3.$$

Then, D_{i+1} is collapsible because $D_{i+1} \searrow E_{i+1} \searrow 0$.

We have constructed the three spaces, and verified all the conditions of the $(i + 1)^{\text{st}}$-inductive statement except that $C_{i+1} \subset f^{-1}(D_{i+1})$. This follows because $f(C_{i+1}) \subset E_{i+1}$, since the isotopy kept $f(C_i)$ fixed, and so

$$f(C_{i+1}) = f(C_i) \cup f(C_{i+1} - C_i) \subset E_{i+1} \cup D_{i+1} = D_{i+1}.$$

This completes the proof of the inductive step, and hence the proof of Lemma **4.7.1**.

Before proving Theorem **4.7.1** for the bounded case we will prove the following preliminary lemma.

Lemma 4.7.2. *Let M^m and Q^q be* (PL) *manifolds and let $f: M \to Q$ be a* (PL) *map such that $f \mid \partial M$* (PL) *embeds ∂M in ∂Q. Then, f is homotopic to a proper* (PL) *map $g: M \to Q$ keeping ∂M fixed such that $S(g) \subset$* Int M. (See Fig. **4.7.2**.)

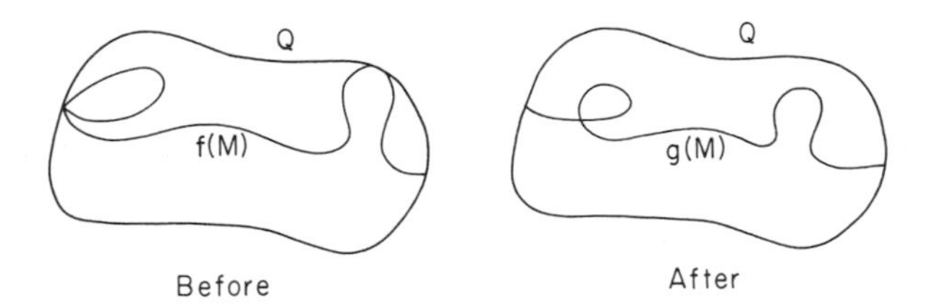

Figure 4.7.2

Proof. By Theorem **1.7.4** (or Corollary **4.4.4**), we can choose a collar of ∂M in M, that is, there is an embedding

$$c_M: \partial M \times I \to M$$

such that $c_M(x, 0) = x$ for all $x \in \partial M$. Let $M_0 = \text{Cl}(M - c_M(\partial M \times I))$ and let $h_t: M \to M$ be a homotopy that starts with the identity and finishes with a map h_1 that shrinks the collar onto ∂M and maps M_0

homeomorphically onto M. Such a homotopy can be easily defined by stretching the inner half of a collar twice as long. In particular,

$$h_1 c_M(x, u) = x$$

for all $x \in \partial M$ and $u \in I$. Then, h_t keeps ∂M fixed and therefore f is homotopic to fh_1 keeping ∂M fixed.

Again apply Theorem **1.7.4** to get a collar of ∂Q in Q, that is, there is an embedding

$$c_Q\colon \partial Q \times I \to Q$$

such that $c_Q(y, 0) = y$ for all $y \in \partial Q$. (In the PL case we would apply Corollary **4.4.4** to a regular neighborhood Q_1 of $f(M)$ and do the remainder of the proof in Q_1 rather than Q.) Let $Q_0 = \mathrm{Cl}(Q - c_Q(\partial Q \times I))$. Let $k_t\colon Q \to Q$ be a homotopy that shrinks $c_Q(\partial Q \times I)$ onto ∂Q_0 keeping Q_0 fixed. We now use k_t to construct a homotopy $g_t\colon M \to Q$ that moves M_0 into Q_0 and stretches the collar c_M out again compatibly with c_Q. More precisely, for $0 \leqslant t \leqslant 1$, define

$$g_t c_M(x, u) = \begin{cases} c_Q(f(x), u), & 0 \leqslant u \leqslant t \\ c_Q(f(x), t), & t \leqslant u \leqslant 1 \end{cases} \quad x \in \partial M,$$

$$g_t(x) = \{k_t f h_1(x)\} \; x \in M_0 .$$

Notice that g_t keeps ∂M fixed. It is easy to check that $g = g_1$ is the desired map.

Proof of Theorem 4.7.1 in the Case that *M* Has Boundary. After one applies Lemma **4.7.2**, the theorem follows by the same arguments as in the unbounded case to eliminate the singularities of g by working entirely in the interior of Q.

EXERCISE 4.7.1. Observe that the techniques of this section suffice to prove the following fact which is Lemma 5.10 of [Rushing, 2]: Let $f\colon M^m \to Q^q$ be a proper map (that is, boundary to boundary, interior to interior) of the compact, PL m-manifold M into the topological q-manifold Q (whose interior has a PL structure) such that $f \mid V$ is an embedding and $f \mid (V - \partial M)$ is PL, where V is a closed PL collar of ∂M in M. Let $U \subset \mathrm{Int}\, Q$ be an open set such that either

(1) $M - f^{-1}(U) \subset \mathrm{Int}\, V \cup \partial M$ and is a collar of ∂M, or

(2) $M - f^{-1}(U) = \partial M$.

Then, f is homotopic to a proper embedding f' which is PL on Int M, where the track of the homotopy lies inside $U \cup f(M)$, provided,

(1) $q - m \geqslant 3$,

(2) M is $(2m - q)$-connected, and

(3) U is $(2m - q + 1)$-connected.

Furthermore, if Q is a PL manifold and if f is PL on V, then f' is PL on M.

EXERCISE 4.7.2. Let M and Q be PL manifolds of dimensions m and q, respectively. Let $I = [0, 1]$. A **concordance** of M in Q is a PL embedding $h: M \times I \rightarrow Q \times I$ such that $h(M \times 0) \subset Q \times 0$ and $h(M \times 1) \subset Q \times 1$. An **isotopy** of M in Q is a concordance which is level-preserving, that is, $h(M \times t) \subset Q \times t$ for all $t \in I$. $h_t: M \rightarrow Q$ is used to denote the embedding defined by $h(x, t) = (h_t(x), t)$ for all $x \in M$. An embedding $f: M \rightarrow Q$ is **proper** if $f^{-1}(\partial Q) = \partial M$. A concordance (or isotopy) h of M in Q is **proper** if $h^{-1}(Q \times 0) = M \times 0$, $h^{-1}(Q \times 1) = M \times 1$, h_0 is a proper embedding and $h^{-1}(\partial Q \times I) = (h_0^{-1}(\partial Q)) \times I$.

The following theorem was announced in [Hudson, 4] and proofs appear in [Hudson, 1, 5].

Concordance Theorem (Hudson). *Let $h: M \times I \rightarrow Q \times I$ be a proper concordance of the compact* PL *manifold M^m into the* PL *manifold Q^q, $q - m \geqslant 3$, which is fixed on ∂M.*

(a) *Then, there exists an ambient isotopy H of $Q \times I$, fixed on $\partial(Q \times I)$, such that $H_1 h$ is level preserving.*

(b) *Then, there exists an ambient isotopy H of $(Q \times I)$, fixed on $(Q \times 0) \cup (\partial Q \times I)$, such that $H_1 h = h_0 \times 1$.*

The exercise is to use Theorem **4.7.1** and the concordance theorem to prove the following unknotting theorem. (A different proof of this unknotting theorem appears in [Zeeman, 1]. A stronger unknotting theorem is proved in [Hudson, 2].)

Unknotting Theorem (Zeeman). *Let $f, g: M^m \rightarrow Q^q$ be two proper embeddings such that $f \mid \partial M = g \mid \partial M$. If f and g are homotopic keeping ∂M fixed, then they are ambient isotopic keeping ∂M fixed, provided*

$$m \leqslant q - 3,$$

M is $(2m - q + 1)$-connected, and

Q is $(2m - q + 2)$-connected.

4.8. THE CELLULARITY CRITERION IN HIGH DIMENSIONS

Recall that in Section **1.8**, we defined a set X in an n-dimensional manifold M to be **cellular** in M if $X = \bigcap_{i=1}^{\infty} D_i^n$, where each D_i^n is an n-cell such that $\text{Int}\, D_i^n \supset D_{i+1}^n$. If M is a PL manifold and each D_i can be chosen as a PL n-cell, we say that X is **cellular with respect to PL-cells**. The notion of cellularity is a handy weapon in the study of manifolds. One reason for this usefulness is that cellular subsets behave essentially like points (see Corollary **1.8.2**).

The main result (Theorem **4.8.1**) of this section gives a necessary and sufficient condition (the cellularity criterion of McMillan [1]) for a compact set in the interior of a PL manifold of dimension five or more to be cellular with respect to PL cells. Although McMillan obtains a modified version of the main result for 3-manifolds, we will not present that version here since in this book we are mainly concerned with high-dimensional phenomena. Although the main theorem is stated for X a compact absolute retract (abbreviated, CAR), the only property of CAR's relevant to the proof is that for each open set U of the manifold containing X, there is another open set V such that $X \subset V \subset U$, and the inclusion $V \to U$ is null homotopic. (We show that a CAR embedded in a manifold always has this property in the proof of Lemma **4.8.1**.) In the next section we will present one of the many interesting applications of Theorem **4.8.1**, which shows that a 1-ULC complementary domain of a topological $(n-1)$-sphere in the n-sphere is topologically an open cell. Other results related to Theorem **4.8.1** will be mentioned at the end of this section.

Let us now state the important property that we shall deal with throughout this section.

The Cellularity Criterion. Let X be a compact set in the interior of an n-manifold M^n, $n \geqslant 3$. Then X satisfies the **cellularity criterion** in M if for each open set U containing X, there exists an open set V such that $X \subset V \subset U$ and each loop in $V - X$ is null-homotopic in $U - X$.

The proof of the main theorem, which we are about to state, depends heavily on engulfing and a trick of John Stallings which we have already used three times in previous sections of this chapter. (Our proof of Theorem **4.8.1** is based primarily on [Curtis and McMillan, 1], [McMillan and Zeeman, 1] and [McMillan, 1].)

Theorem 4.8.1 (McMillan). *A necessary and sufficient condition* *that a* CAR X *in the interior of a* PL *manifold* M^n, $n \geqslant 5$, *be cellular with respect to* PL *cells is that it satisfy the cellularity criterion in* M.

Corollary 4.8.1. *Let* X *be a* CAR *in the interior of a* PL *manifold* M^n, $n \geqslant 5$, *then,* X *is cellular if and only if it is cellular with respect to* PL-*cells.*

PROOF. By the first paragraph of the proof of Theorem **4.8.1**, if X is cellular, then it satisfies the cellularity criterion. Hence, by Theorem **4.8.1**, X is cellular with respect to PL cells.

Before beginning the proof of Theorem **4.8.1**, let us establish a few preliminary lemmas.

Lemma 4.8.1. *Let* A *be a* CAR *in the interior of the* PL *n-manifold* M^n. *Then, there exists a sequence* $\{H_i\}$ *of compact* PL *n-manifolds, with nonempty boundaries, such that* $H_{i+1} \subset \operatorname{Int} H_i$, $A = \cap_i H_i$ *and each inclusion* $H_{i+1} \to H_i$ *is homotopically trivial.*

PROOF. Let us first see the following.

(a) *If* U *is an arbitrary open neighborhood of* A. *Then, there exists a finite* PL *n-manifold* H, *with nonempty boundary, such that*

$$A \subset \operatorname{Int} H \subset H \subset U.$$

Such an H may be obtained as a small regular neighborhood of the closed simplicial neighborhood of A in a sufficiently fine subdivision of M.

After we give the next statement (b), one will immediately see that Lemma **4.8.1** follows from (a) and (b) and so all that will remain is to prove (b).

(b) *Let* $A \subset \operatorname{Int} H$ *as in* (a). *Then, there exists a neighborhood* V *of* A *such that* $V \subset \operatorname{Int} H$ *and the inclusion* $i\colon V \to H$ *is null-homotopic.*

Since H is an absolute neighborhood retract, there exists (by the reasoning in the proof of Theorem **1.6.11**) an $\epsilon > 0$ with the property that if f and g are maps of a space K into H such that $d(f(k), g(k)) < \epsilon$ for each $k \in K$, then f and g are homotopic in H. Let r be a retraction of H onto A, and choose V to be an open set such that $A \subset V \subset \operatorname{Int} H$ and $d(x, r(x)) < \epsilon$ for each x in V. Since A is contractible, V is the required neighborhood of A.

Lemma 4.8.2. *Suppose* $M_1, M_2, \ldots, M_{k+2}$ *is a sequence of compact* PL *n-manifolds such that each* M_i *is a subspace of* M_{i+1} *and each inclusion*

$M_i \to M_{i+1}$ is homotopically trivial. If X^x is a subpolyhedron of M_1 such that $x \leqslant k$ and $n - k \geqslant 3$, then X lies in a PL *n-ball in M_{k+2}.*

REMARK 4.8.1. Our proof of Lemma **4.8.2** will be based on Zeeman engulfing (Lemma **4.6.1**) developed in Section **4.6**; however, the lemma can be proved just as well using Stallings' engulfing techniques developed in Section **4.2**. The setup of Stallings' engulfing which we shall present later to do radial engulfing is particularly adaptable here. Of course, the proof via Stallings' engulfing would not involve piping.

Proof of Lemma 4.8.2. The proof will be by induction on dim X. Trivially start the induction with dim $X = -1$. Assume the lemma true for all X of dimension $< k$, $n - k \geqslant 3$. Given $X^k \subset M_1$, then X is null-homotopic in Int M_2. Therefore, by Lemma **4.6.11** there exist polyhedra Y^y and Z^z in Int M_2 such that $X \subset Y \searrow Z$, $y \leqslant x + 1$ and $z < 2x - n + 2$. But,

$$z < 2x - n + 2 \leqslant 2k - n + 2 \leqslant 2(n-3) - n + 2 = n - 4 < k.$$

Hence, by induction Z is contained in a PL n-ball in Int M_{k+2}. Since $Y \searrow Z$, it follows from Exercise **1.6.12** that Y is contained in a PL n-ball in Int M_{k+2} and hence also X.

Lemma 4.8.3. *Let A be a* CAR *in the interior of the* PL *n-manifold M^n. Then, there exists a sequence $\{H_i\}$ of compact* PL *n-manifolds with nonempty boundaries, such that $H_{i+1} \subset$* Int *H_i, $A = \cap_i H_i$, each inclusion $H_{i+1} \to H_i$ is homotopically trivial, and if Y is a subcomplex of H_{i+1} and* dim *$Y \leqslant n - 3$, then Y lies in a* PL *n-ball in H_i. (Furthermore,* Bd *H_{i+1} may be assumed to be connected which shows that X cannot separate H_i.)*

PROOF. All except the last sentence of Lemma **4.8.3** follows by combining Lemma **4.8.1** and Lemma **4.8.2**.

EXERCISE 4.8.1. Establish the last sentence of Lemma **4.8.3**.

EXERCISE 4.8.2. Prove the following fact (**homotopy extension theorem for maps of polyhedra**). *Let K be a complex, L a subcomplex, $f_0\colon |K| \to X$ a map of $|K|$ into a topological space X and $g_t\colon |L| \to X$ a homotopy such that $g_0 = f_0 \mid |L|$. Then, there is a homotopy $f_t\colon |K| \to X$ of the map f_0 such that $g_t = f_t \mid |L|$.* (Although this fact is easy to prove, one can find a complete proof on p. 33 of [Hilton and Wylie, 1].)

Proof of Theorem 4.8.1. Let us first show that the cellularity criterion is necessary for topological cellularity, in fact, for $n \geqslant 3$.

Let U be an open set such that $X \subset U$. By cellularity there is an open n-cell V such that $X \subset V \subset U$. Then, by Corollary **1.8.2**, $V - X$ is simply connected and necessity follows.

Now, assuming the conditions of the theorem, let $X \subset W$, where W is an open set. The proof will be complete after we show the existence of a PL n-ball F such that $X \subset \text{Int}\, F \subset F \subset W$. By Lemma **4.8.3**, we may assume that M has been subdivided so as to contain subcomplexes M_0, M_1, M_2, M_3 with the properties:

$$X \subset \text{Int}\, M_{i+1} \subset M_{i+1} \subset \text{Int}\, M_i \subset M_i \subset W \qquad (i = 0, 1, 2),$$

where M_i is a compact PL n-manifold with nonempty, connected boundary; each inclusion $M_{i+1} \to M_i$ is homotopic to a constant; and if Y is a finite polyhedron in M_{i+1} with $\dim Y \leqslant n - 3$, then there is a PL n-ball in Int M_i whose interior contains Y. Also, the cellularity criterion hypothesis on X allows us to assume that M_3 is so close to X that each loop in Int $M_3 - X$ is null-homotopic in Int $M_2 - X$.

Let P^2 be the 2-skeleton of M_1. We apply Stallings' engulfing theorem (Theorem **4.2.1**) to the PL manifold Int M_1 to obtain a PL homeomorphism $h: M_1 \to M_1$ such that h is the identity in a neighborhood of Bd M_1 and $P \subset h(M_1 - X)$. We have only to varify that the pair $\pi = (\text{Int}\, M_1, \text{Int}\, M_1 - X)$ is 2-connected. Since M_3 is arcwise connected, π is 0-connected.

Now consider any path $f: (\Delta^1, \text{Bd}\, \Delta^1) \to (\text{Int}\, M_1, \text{Int}\, M_1 - X)$. If $f(\Delta^1) \subset \text{Int}\, M_3$, then since $M_3 - X$ is arcwise connected (by the statement in parenthesis in Lemma **4.8.3**) and the inclusion $M_3 \to M_2$ is homotopically trivial, f is homotopic (rel Bd Δ^1) in M_2 to a path in $M_3 - X$. In the general case, cover $f^{-1}(X)$ with the interiors of a finite number of disjoint arcs $\sigma_1, \ldots, \sigma_k$ so close to $f^{-1}(X)$ that each $f(\sigma_i) \subset \text{Int}\, M_3$. By the special case above, each path $f \mid \sigma_i$ is homotopic (rel Bd σ_i) in M_2 to a path in $M_3 - X$. By piecing these homotopies together, we see that f is homotopic (rel $\Delta^1 - \cup_i \text{Int}\, \sigma_i$) to a path in Int $M_1 - X$. Hence, π is 1-connected.

Now consider any map $f: (\Delta^2, \text{Bd}\, \Delta^2) \to (\text{Int}\, M_1, \text{Int}\, M_1 - X)$. If $f(\text{Bd}\, \Delta^2) \subset M_3 - X$ and $f(\Delta^2) \subset \text{Int}\, M_2$, then by our special hypothesis there is a map $g: \Delta^2 \to (\text{Int}\, M_2 - X)$ with $g \mid \text{Bd}\, \Delta^2 = f \mid \text{Bd}\, \Delta^2$, and f and g are homotopic (rel Bd Δ^2) in Int M_1. In the general case, take a simplicial neighborhood K of $f^{-1}(X)$ in a fine triangulation of Δ^2 and let N be the regular neighborhood of K in that triangulation. We may assume that $f(N) \subset \text{Int}\, M_3$. Let A = 1-skeleton of N and $B = \Delta^2 - \text{Int}\, N$.

By using the results of the 0-dimensional and 1-dimensional cases, one can obtain a homotopy $\varphi_t: A \to \text{Int}\, M_2$ such that $\varphi_0 = f \mid A$, $\varphi(A) \subset (\text{Int}\, M_3 - X)$, and $\varphi_t \mid A \cap B = f \mid A \cap B$. (This is possible since $f(A \cap B) \subset (\text{Int}\, M_3 - X)$). By the homotopy extension theorem (Exercise **4.8.2**), we may suppose that φ_t is defined on N. That is, there is $\Phi_t: N \to \text{Int}\, M_2$ such that $\Phi_0 = f \mid N$, $\Phi_1(A) \subset (\text{Int}\, M_3 - X)$ and $\Phi_t \mid A \cap B = f \mid A \cap B$. We now extend Φ_t to all of Δ^2 by defining $\Phi_t \mid B = f \mid B$ for each t. Let ρ be a 2-simplex of N. Then, $\Phi_1 \mid \rho$ takes $(\rho, \text{Bd}\, \rho)$ into $(\text{Int}\, M_2, \text{Int}\, M_3 - X)$ and so by applying to each ρ the special case already considered, we find that Φ_1 is homotopic (rel $A \cup B$) in Int M_1 to $G: \Delta^2 \to (\text{Int}\, M_1 - X)$. Thus, f is homotopic to G (rel B and hence rel Bd Δ^2) and π is 2-connected. Hence, the required homeomorphism h exists. Extend h to all of M_0 by $h \mid M_0 - M_1 = \text{identity}$.

The proof may now be completed essentially as the proof of Theorem **4.4.1**. Briefly, form the complex Q by adding to P^2 all the closed simplexes of $\text{Cl}(M_0 - M_1)$ and note that $Q \subset h(M_0 - X)$. Let R be the complex in a first derived subdivision of M_0 dual to Q. Then, $\dim R \leqslant n - 3$ and $R \subset \text{Int}\, M_1$, so that there is a PL n-ball F^* with

$$R \subset \text{Int}\, F^* \subset F^* \subset \text{Int}\, M_0 \subset W.$$

There is a PL homeomorphism $h^*: M_0 \to M_0$ such that

$$h(M_0 - X) \cup h^*(\text{Int}\, F^*) = M_0 .$$

We then take $F = h^{-1}h^*(F^*)$.

EXERCISE 4.8.3. Use the technique of proof of Theorem **4.8.1** to show that for $n \geqslant 3$ if $\alpha_n \subset E^n$ is an arc for which $\pi_1(E^3 - \alpha_n) \neq 0$, then α_n does not satisfy the cellularity criterion, or in other terminology, $E^n - \alpha_n$ is not projectively 1-connected. (Such arcs α_n are explicit in [Blankenship, 1]. For related arcs see Example **2.6.4**.)

A closed embedding $f: P^k \to \text{Int}\, Q^n$, $n - k \geqslant 3$, of a k-polyhedron P into an n-manifold Q is **locally nice** if $Q - f(P)$ is 1-LC at $f(x)$ for each $x \in P$. Hempel and McMillan [1] proved the following important theorem related to Theorem **4.8.1**.

Theorem 4.8.2 (Hempel and McMillan). *Let X be a* CAR *in the* PL *k-manifold M^k and let $f: M^k \to N^n$, $n - k \geqslant 3$, be a locally nice embedding of M^k into the* PL *n-manifold N^n. Then, $f(X)$ satisfies the cellularity criterion. Hence, if $n \geqslant 5$, X is cellular in N^n.*

We shall mention and prove results related to the above theorem in the next section. Let us mention here that [Hempel and McMillan, 1] continues work of [Eilenberg and Wilder, 1], [Harrold, 1, 2]. In conclusion, we point out that Lacher [4] has studied cellularity criteria for maps.

4.9. LOCALLY NICE CODIMENSION ONE SPHERES IN $S^{n \geqslant 5}$ ARE WEAKLY FLAT

Let us first consider more closely the concept of local connectivity given in Section **2.6**. Let W be an open connected subset of the PL manifold M^n. If $x \in (\text{Cl } W - W) \cap \text{Int } M$, then W is **locally 1-connected at x (1-LC at x)** if for each open set U in M containing x there is an open set V in M such that $x \in V \subset U$ and each loop in $V \cap W$ is null-homotopic in $U \cap W$. We say that W is **uniformly locally 1-connected (1-ULC)** in M if for each $\epsilon > 0$, there is a $\delta > 0$ such that each loop in W of diameter $< \delta$ is null-homotopic in W on a set of diameter $< \epsilon$. Analogous definitions for **0-LC** and **0-ULC** are obtained by replacing "each loop" by "each pair of points" and "is null-homotopic" by "can be joined by an arc" in the above.

Exercise 4.9.1. Show that if Cl W is a compact subset of Int M, then W is 1-LC at each point of Cl $W - W$ if and only if W is 1-ULC.

Recall that in the last section we defined a closed embedding $f: P^k \to \text{Int } Q^n$, $n - k \geqslant 3$, of a k-polyhedron P into an n-manifold Q to be **locally nice** if $Q - f(P)$ is 1-LC at $f(x)$ for each $x \in P$. A subpolyhedron P^k of Q^n is **locally nice** if the inclusion $P \to Q$ is locally nice.

In this section we will give a sufficient condition for the complementary domains of a topologically embedded $(n - 1)$-sphere in S^n to be open n-cells. In general, if $\Sigma^k \subset S^n$ is a topologically embedded k-sphere, one may ask for conditions which guarantee that the complement $S^n - \Sigma^k$ is homeomorphic to $S^n - S^k$, that is, when $S^n - \Sigma^k \approx S^{n-k-1} \times E^{k+1}$. When this is the case we follow Rosen [1] and say that Σ^k is **weakly flat.**

Our main theorem, which was first proved in [McMillan, 1], follows.

Theorem 4.9.1 (McMillan). *Let W be a component of $S^n - \Sigma^{n-1}$, where Σ^{n-1} is a topologically embedded $(n - 1)$-sphere and $n \geqslant 5$. If W is* 1-ULC, *then it is an open n-cell.*

Corollary 4.9.1. *Each locally nice codimension one sphere Σ^{n-1} in S^n, $n \geqslant 5$, is weakly flat.*

Before beginning the proof of Theorem **4.9.1**, let us mention some related results. It is shown in [Bing, 6] that a complementary domain W of a topological 2-sphere in S^3 is 1-ULC if and only if Cl W is a closed 2-cell. Thus, codimension one spheres in S^3 whose complementary domains are 1-ULC are flat. (This result is still unknown in higher dimensions and is a good research problem.† Seebeck [3] has shown that codimension one spheres in S^n whose complements are 1-ULC are flat if they can be approximated by locally flat embeddings. Also, partial results on the codimension one approximation problem have been obtained in [Price and Seebeck, 1, 2].) Properties of 1-ULC complementary domains of $(n-1)$-spheres in S^n were studied in [Eilenberg and Wilder, 1] and were shown to be contractible. (For $n \geqslant 5$, we will show here that they are in fact homeomorphic to E^n.) Conditions are given in [Rosen, 1], [Hempel and McMillan, 1] and [Duvall, 1] which imply that k-spheres in S^n are weakly flat. In particular, Duvall shows that a k-sphere in S^n, $n \geqslant 5$, $2 \leqslant k \leqslant n-3$, is weakly flat if and only if it satisfies the cellularity criterion. Daverman [1] obtains the analog of this result for 1-spheres.

Exercise 4.9.2. Let Z be a complementary domain of an arbitrary $(n-1)$-sphere in S^n. Show that Cl Z is a CAR. (To do this you may want to refer to [Bing, 7].)

Proof of Theorem 4.9.1. We are given W a component of $S^n - \Sigma^{n-1}$. Let Z be the other component of $S^n - \Sigma^{n-1}$. Then, by Exercise **4.9.2**, Cl Z is a CAR, and so if we can show that Cl Z satisfies the cellularity criterion, it will follow from Theorem **4.8.1** that Cl Z is cellular. Hence, it will follow from Corollary **1.8.2** that $S^n - \text{Cl } Z = W$ is an open n-cell.

Let an open set U containing Cl Z be given. Choose $\epsilon > 0$ so that $N_\epsilon(\text{Cl } Z, S^n) \subset U$, where $N_\epsilon(\text{Cl } Z, S^n) = \{x \in S^n \mid \text{dist}(x, \text{Cl } Z) < \epsilon\}$. Let $\delta > 0$ be the number corresponding to $\epsilon/4$ and promised by the fact that W is 1-ULC. We take $\delta < \epsilon/4$.

Since Σ^{n-1} is a simply connected compact absolute neighborhood retract (CANR) in Cl W, there is an open set V^* is S^n such that

$$\Sigma^{n-1} \subset V^* \cap \text{Cl } W \subset N_{\epsilon/4}(\Sigma^{n-1}, \text{Cl } W),$$

and each loop in $V^* \cap \text{Cl } W$ is null-homotopic in $N_{\epsilon/4}(\Sigma^{n-1}, \text{Cl } W)$. We assert that the open set required by Theorem **4.8.1** is $V = Z \cup V^*$.

Let a loop $f\colon \text{Bd } \Delta^2 \to V - \text{Cl } Z \subset V^* \cap W$ be given. Then, f extends to

$$F\colon \Delta^2 \to N_{\epsilon/4}(\Sigma^{n-1}, \text{Cl } W).$$

† This problem has since been solved by Daverman. See the Appendix.

Let δ_1 be so that $0 < \delta_1 < \delta$ and each pair of points of W within δ_1 of each other can be joined by an arc in W of diameter $< \delta/2$. (Such a δ_1 exists since W is 0-ULC (see [Wilder, 2], p. 66, Theorem 5.35).) Let $\delta_2 > 0$ be such that if X is a subset of Δ^2 with diameter $< \delta_2$ then $F(X)$ has diameter $< \delta_1/2$.

Consider a triangulation T of Δ^2 in which each 1-simplex has diameter $< \delta_2$. We will now complete the proof by defining a map $G: \Delta^2 \to N_\epsilon(\Sigma^{n-1}, W) \subset U - \text{Cl } Z$, such that $G \mid \text{Bd } \Delta^2 = f$.

Let T^i be the i-skeleton of T. If $v \in T^0$ and $F(v) \in \Sigma^{n-1}$, let $G(v)$ be a point of W within $\delta_1/4$ ($< \epsilon/16$) of $F(v)$. Otherwise, let $G(v) = F(v)$. Note that $G(T^0) \subset N_{\epsilon/4}(\Sigma^{n-1}, W)$, $G = f$ on the 0-simplexes of Bd Δ^2, and G(boundary of 1-simplex) has diameter $< \delta_1/2 + \delta_1/4 + \delta_1/4 = \delta_1$.

Now G extends to T^1 in the obvious fashion, mapping a 1-simplex of T into an arc in W of diameter $< \delta/2$ ($< \epsilon/8$) joining the images of its end-points. This can be done so that $G = f$ on Bd Δ^2 (since $\delta_1/2 < \delta/2$). Note that $G(T^1) \subset N_{\epsilon/2}(\Sigma^{n-1}, W)$ and G(boundary of 2-simplex) has diameter $< \delta$.

Finally, use this last fact to extend G to T^2 so that G(2-simplex) $\subset W$ and has diameter $< \epsilon/4$. Clearly, $G = f$ on Bd Δ^2 and $G(\Delta^2) \subset N_\epsilon(\Sigma^{n-1}, W)$, which completes the proof.

4.10. RADIAL ENGULFING

The first radial engulfing theorem appeared as Lemma 1 of [Connell, 1]. That radial engulfing theorem was for codimensions greater than three and Connell used it to show that stable homeomorphisms of E^n, $n \geqslant 7$, can be approximated by PL homeomorphisms. A little later, Bing [10] announced that he could do Connell's radial engulfing theorem in codimensions greater than two, which gave the PL approximation of stable homeomorphisms of E^n for $n \geqslant 5$. (The next section concerns such approximations.) An excellent presentation of Bing's work on radial engulfing is given in [Bing, 11]. The fundamental results of this section are based on the part of that paper involving radial engulfing in codimensions greater than two. A somewhat refined (though easily proved) radial engulfing theorem for codimensions greater than three is also given in [Bing, 11]. Wright [1] later proved a form of that theorem for codimension three. We shall discuss these more refined engulfing theorems at the conclusion of this section.

Radial engulfing has found a number of important applications since Connell first used it to obtain his approximation theorem. For instance, see [Connelly, 1], [Edwards and Glaser, 1], [Wright, 2], and [Seebeck, 3].

Before stating our main engulfing theorem, let us make a definition. Let M^n be a connected PL n-manifold, let U be an open set in M and let $\{A_\alpha\}$ be a collection of sets in M^n. We define what is meant by the statement that **r-polyhedra in M^n can be pulled in U along $\{A_\alpha\}$.** This means that if P^k is any k-polyhedron whatsoever in M^n and Q is a subpolyhedron of P^k that lies in U and is such that $\dim(P - Q) \leqslant r$, then there is a homotopy H: $P^k \times [0, 1] \to M$ such that $H_0 = 1$, $H_1(P^k) \subset U$, $H_t = 1$ on Q and for each point $x \in P^k$, $H(x \times [0, 1])$ lies in an element of $\{A_\alpha\}$. For example, if $M^n = E^n$, $U = g(U')$ (where g is a fixed homeomorphism of E^n onto itself and U' is an open ball with center at the origin), and $\{A_\alpha\}$ is $\{g(A_\alpha')\}$ (where $\{A_\alpha'\}$ is the collection of all open rays leading from the origin), then for each $i \leqslant n$, i-polyhedra can be pulled into U along $\{A_\alpha\}$.

Radial Engulfing Theorem 4.10.1 (Bing). *Let M^n be a connected PL n-manifold without boundary, U an open set in M^n, P an $(n - 3)$-dimensional polyhedron (not necessarily compact), $Q \subset P$ a subpolyhedron in U such that $R = \mathrm{Cl}(P - Q)$ is compact and of dimension r. Suppose that f is a map of M^n onto a finite-dimensional metric space Y such that for each open covering $\{G_\alpha\}$ of Y, r-polyhedra ($r \leqslant n - 3$) in M^n can be pulled into U along $\{f^{-1}(G_\alpha)\}$. Then, for each open covering $\{G_\beta'\}$ of Y there is an engulfing isotopy H: $M^n \times [0, 1] \to M^n$ such that*

(1) $H_0 = 1$,

(2) $H_t \mid Q = 1$,

(3) $P \subset H_1(U)$,

(4) *for each $x \in M^n$, there is a $G_\gamma' \in \{G_\beta'\}$ such that $H(x \times [0, 1])$ lies in $f^{-1}(G_\gamma')$,*

(5) *H_t is the identity outside a compact subset of M.*

In order to state a couple of corollaries to Theorem **4.10.1** which will be used in the next section, let us formulate a little notation. Let $O_r^n = \{x \in E^n \mid \|x\| < r\}$ and let ${}_rO^n$ denote the complement of O_r^n in E^n. If 0 is the origin in E^n and $x \neq 0 \neq y$, then $\theta\{x, y\}$ will represent the angle in radians between the two line intervals, one joining 0 to x and the other joining 0 to y. Thus, $0 \leqslant \theta\{x, y\} \leqslant \pi$.

Corollary 4.10.1 (Connell). *Suppose that a, b, and ϵ are numbers such that $0 < a < b$ and $\epsilon > 0$. Let g: $E^n \twoheadrightarrow E^n$ be a homeomorphism and suppose that P is a polyhedron such that $P \subset g(O_b^n)$ and $\dim P \leqslant n - 3$. Then, there is a PL homeomorphism h: $E^n \twoheadrightarrow E^n$ such that $h \mid g(0^n_{(a-\epsilon)}) = 1$, $h \mid g({}_bO^n) = 1$, $h(g(O_a^n)) \supset P$ and $\theta\{g^{-1}(h(x)), g^{-1}(x)\} < \epsilon$ for all $x \in E^n$.*

Corollary 4.10.2 (Connell). *Suppose that a, b, and ϵ are numbers such that $0 < a < b$ and $\epsilon > 0$. Let $g: E^n \twoheadrightarrow E^n$ be a homeomorphism and suppose that P is a polyhedron such that $P \subset g(\mathrm{Cl}_a O^n))$ and $\dim P \leqslant n - 3$. Then, there is a PL homeomorphism $h : E^n \twoheadrightarrow E^n$ such that*

$$h \mid g({}_{(b+\epsilon)}O^n) = 1, \qquad h \mid g(O_a{}^n) = 1, \qquad h(g(\mathrm{Cl}({}_b O^n))) \supset P,$$

and $\theta\{ g^{-1}(h(x)), g^{-1}(x)\} < \epsilon$ or all $x \in E^n$.

Let us establish a couple of preliminary lemmas before beginning the proof of Theorem **4.10.1**. The following point-set exercise will be used in the proof of Lemma **4.10.1**.

EXERCISE 4.10.1. Suppose that $g: P \times [0, 1] \to M$ is a homeomorphism where P is a polyhedron and M is a manifold. Show that there is a positive number δ such that any connected set which lies in the δ-neighborhood of the union of two $g(x \times [0, 1])$'s actually lies in the ϵ-neighborhood of one of them.

Lemma **4.10.1** will play a similar role in the proof of the radial engulfing theorem **4.10.1** to that played by Exercise **1.6.12** in the proofs of Stallings' engulfing theorem **4.2.1** and Zeeman's engulfing theorems in Section **4.6**. Lemma **4.10.1** is more refined than Exercise **1.6.12** in that its conclusion contains a restriction on the paths of points.

Engulfing Lemma 4.10.1. *Suppose that*

P is a polyhedron,

T is a triangulation of $P \times [0, 1]$,

$g: P \times [0, 1] \to M^n$ is a map into the PL *n-manifold M^n which is a* PL *homeomorphism on each simplex of T,*

L is a subpolyhedron of $P \times [0, 1]$ containing all singularities of g and which is the union of vertical segments of $P \times [0, 1]$. (A *vertical segment* of $P \times [0, 1]$ is a set of the form $x \times [0, 1]$ where $x \in P$.)

C is a closed set in M^n such that $C \cap g(P \times [0, 1]) \subset g(P \times 1) \cup g(L)$, and

U is an open set in M^n containing $g(P \times 1) \cup g(L)$. Then, for each $\epsilon > 0$ there is an engulfing isotopy $H: M^n \times [0, 1] \to M^n$ such that

$H_0 = 1$,

$H_1 g(P \times [0, 1]) \subset U$,

$H_t = 1$ on C, and

for each $x \in M^n$, either $H(x \times [0, 1])$ is a point or there is a point $y \in P$ such that $H(x \times [0, 1])$ lies in the ϵ-neighborhood of $g(y \times [0, 1])$.

Proof. Let L' be a polyhedron in $P \times [0, 1]$ which is the union of vertical segments in $P \times [0, 1]$ and such that $g(L') \subset U$ and L contains no limit points of $(P \times [0, 1]) - L'$. If we add $g(L')$ to C, replace $P \times [0, 1]$ by $\mathrm{Cl}((P \times [0, 1]) - L')$, and replace L by

$$\mathrm{Cl}((P \times [0, 1]) - L') \cap L',$$

we see that we can assume with no loss of generality that g has no singularities. Hence, we suppose $g: P \times [0, 1] \to M^n$ is a homeomorphism and $P \times 0 = P$.

Apply Exercise **4.10.1** to obtain a positive number δ such that any connected set which lies in the δ-neighborhood of the union of two $g(x \times [0, 1])$'s actually lies in the ϵ-neighborhood of one of them.

Let T_1 be a subdivision of T such that L is the union of simplexes of T_1 and g is a linear homeomorphism on each simplex of T_1 that takes it onto a set in M^n with diameter less than δ. Let T^2 be a subdivision of T^1 that is **cylindrical**—that is, the vertical projection of $P \times [0, 1]$ onto $P \times 0$ sends each simplex of T^2 linearly onto a simplex of T^2.

We now prove the lemma by induction on the dimension of P. If this dimension is 0, we push out in M^n along mutually exclusive polygonal arcs in $g(P \times [0, 1])$. If $\dim P = p > 0$, we let P^{p-1} be the $(p - 1)$-skeleton of T^2 restricted to $P \times 0 = P$ and let $\mathring{\sigma}_1, \mathring{\sigma}_2, \ldots, \mathring{\sigma}_m$ be the open p-simplexes of T^2 in $P \times 0 = P$ which miss L. Let $V_1, V_2, \ldots, V_m$ be mutually exclusive open sets in M^n such that $V_i \cap g(P \times [0, 1]) = g(\mathring{\sigma} \times [0, 1])$, $V_i \cap C = \emptyset$ and V_i lies in the δ-neighborhood of $g(x_i \times [0, 1])$, where x_i is the barycenter of $\mathring{\sigma}_i$.

It follows by induction on p that there is an engulfing isotopy $H: M^n \times [0, \frac{1}{2}] \to M^n$ such that $H_0 = 1$, $H^t = 1$ on C,

$$g(P^{p-1} \times [0, 1]) \subset H_{1/2}(U)$$

and if $x \in M^n$, either $H(x \times [0, \frac{1}{2}])$ is a point or there is a point $y \in P$ such that $H(x \times [0, \frac{1}{2}])$ lies in a δ-neighborhood of $g(y \times [0, 1])$. The engulfing $H: M^n \times [0, \frac{1}{2}] \to M^n$ is extended to $H: M^n \times [\frac{1}{2}, 1] \to M^n$ by pushing down through the V_i in a manner similar to the one indicated in Exercise **1.6.12**.

If R is an r-polyhedron and $H: R \times [0, 1] \to M$ is a map into the manifold M, then a polyhedron $L \subset R \times [0, 1]$ is called a **shadow for** H **relative to the set** $X \subset R \times [0, 1]$ if

$\dim L \leqslant r - 1$,
L is the union of vertical segments of $R \times [0, 1]$,
$X \subset L$, and
if $x, y \in R \times [0, 1]$ with $H(x) = H(y)$, then $x \in L$ if and only if $y \in L$.

Shadow-Building Lemma 4.10.2. *Let M^n be a PL n-manifold without boundary, P be an $(n-3)$-polyhedron (not necessarily compact), and Q be a subpolyhedron of P such that $R = \mathrm{Cl}(P - Q)$ is a compact r-subpolyhedron of P. Suppose that T is a triangulation of $(Q \times 0) \cup (R \times [0, 1])$ and $H: (Q \times 0) \cup (R \times [0, 1]) \to M^n$ such that*

$H(x \times 0) = x$,
H is a PL homeomorphism on each simplex of T,
H is in general position with respect to T in the sense that if $\mathrm{Int}\, \sigma^i$ and $\mathrm{Int}\, \sigma^j$ are different open simplexes of T with dimensions i, j, respectively, then $\dim H(\sigma^i) \cap H(\sigma^j) \leqslant i + j - n$.

Let X_1 denote the set of singularities of H in $R \times [0, 1]$ resulting from considering pairs of elements of T at least one of which is of dimension less than $n - 2$, and let T^{n-3} denote the $(n-3)$-skeleton of T. Then for each $\epsilon > 0$ there is a PL ϵ-homeomorphism g of $(Q \times 0) \cup (R \times [0, 1])$ onto itself such that $g \mid T^{n-3} \cup X_1 = 1$ and such that there is a shadow L for $Hg \mid R \times [0, 1]$ relative to the set of singularities of Hg in $R \times [0, 1]$ resulting from considering pairs of simplexes of T such that at least one member of the pair has dimension less than $n - 2$.

Proof. Let X denote the singular set of H. Unless $r = n - 3$, we can let $g = 1$ and L be the union of all vertical segments through the part of X which lies in $R \times [0, 1]$. Hence, suppose $r = n - 3$.

The first approximation of L is the set L_1 which is the union of all vertical segments through X_1. Then, L_1 contains X_1 and is of dimension less than or equal to $r - 1$. Unfortunately, it may contain many points of $X - X_1$ so we shall move some of these off L_1 before getting our second approximation L_2.

Since $\dim L_1 \leqslant n - 4$ and $\dim X \leqslant n - 4$, it follows from the general position techniques of proof of Theorem **1.6.10** (see Theorem 15 of [Zeeman, 1]) that there is a small PL homeomorphism g_1 of $(Q \times 0) \cup (R \times [0, 1])$ onto itself that is the identity on $T^{n-3} \cup X_1$ such that $\dim L_1 \cap g_1(X - X_1) \leqslant n - 6$. Let $X_2 \supset X_1$ be the set of all points x of $g_1(X)$ such that for some point $y \in L_1 \cap g_1(X)$, $Hg_1^{-1}(x) = Hg_1^{-1}(y)$. To see that $\dim (X_2 - X_1) \leqslant n - 6$ see Remark **4.10.1**. Let L_2 be the union of all vertical segments in $R \times [0, 1]$ through points of X_2. Note that $L_1 \subset L_2$ and $\dim(L_2 - L_1) \leqslant n - 5$.

Let g_2 be a small PL homeomorphism of $(Q \times 0) \cup (R \times [0, 1])$ onto itself that equals the identity on $T^{n-3} \cup X_1 \cup g_1(L_1 \cup (X_2 - X_1))$ and such that $\dim(L_2 - L_1) \cap g_2(g_1(X - X_2)) \leqslant n - 9$. Let $X_3 \supset X_2$ be the set of all points x of $g_2 g_1(X)$ such that for some point $y \in L_2 \cap g_2 g_1(X)$,

$Hg_1^{-1}g_2^{-1}(x) = Hg_1^{-1}g_2^{-1}(y)$. Let L_3 be the union of all vertical segments in $R \times [0, 1]$ through points of X_3. Note that $\dim(L_3 - L_2) \leqslant n - 6$.

We continue defining X_i, L_i, and g_i until we have reduced the dimension of $X_{j+1} - X_j$ to a negative. Then $g = g_1^{-1}g_2^{-1} \cdots g_j^{-1}g_{j+1}^{-1}$ and $L = L_j$. If we have taken care that the g_i are sufficiently close to the identity, we will have that g is an ϵ-homeomorphism.

REMARK 4.10.1. We now show why $\dim(X_2 - X_1) \leqslant n - 6$. Since any point of $X - X_1$ is in the interior of an $(n - 2)$-simplex of T, no point of $X - X_1$ can hit a point of X_1 under H. Hence, the only points of $g_1(X)$ which hit points of $L_1 \cap g_1(X)$ under Hg^{-1}, in fact hit points of $L_1 \cap g_1(X - X_1)$ under Hg^{-1}. Since $\dim L_1 \cap g_1(X - X_1) \leqslant n - 6$ and Hg^{-1} is nondegenerate, the set $X_2 - X_1$ of points which hit points of $L_1 \cap g_1(X - X_1)$ under Hg^{-1} also has dimension $\leqslant n - 6$.

Proof of Theorem 4.10.1. The proof is by induction on r. In the case $r = 0$, mutually exclusive polygonal arcs are run from points of R to U so that each of these polygonal arcs lies in an element of $\{f^{-1}(G_\beta')\}$. Then, U is pushed out near these polygonal arcs. Suppose the theorem is true for nonnegative integers less than r. We will now show that it is true for r if $r \leqslant n - 3$.

Let $k = \dim Y$ and let $\{G_\alpha''\}$ be an open covering of Y such that any connected subset of Y that lies in the sum of $k + 2$ elements of $\{G_\alpha''\}$ lies in one element of $\{G_\beta'\}$. For $i = 1, 2, \ldots, k + 1$, let $\{G_{\alpha_i}^i\}$ be a discrete collection of open sets in Y such that $\{G_{\alpha_i}^i\}$ refines $\{G_\alpha''\}$ and $\{G_\alpha'''\} = \{G_{\alpha_1}^1\} \cup \{G_{\alpha_2}^2\} \cup \cdots \cup \{G_{\alpha_{k+1}}^{k+1}\}$ covers Y. (That such collections of open sets exist, follows from the k-dimensionality of Y (see Theorem VI of [Hurewicz and Wallman, 1])).

Let $H^1\colon P^k \times [0, 1] \to M$ be a homotopy pulling R in U along $f^{-1}\{G_\alpha'''\}$, that is, $H_0^1 = 1$, $H_1^1(P^k) \subset U$, $H_t^1 = 1$ on Q and for each point $x \in P^k$, $H^1(x \times [0, 1])$ lies in an element of $f^{-1}\{G_\alpha'''\}$. Apply Part 1 of Theorem **1.6.10** to obtain an approximation H^2 of H^1 and a triangulation T of $(Q \times 0) \cup (R \times [0, 1])$ such that H^2 is a linear homeomorphism on each simplex of T, $H_0^2 = 1$, $H_1^2(P) \subset U$, $H^2(Q \times [0, 1]) \subset U$, and the mesh of T is so fine and H^2 so close to H^1 that if σ is an element of T in $R = R \times 0$, then $H^2(\sigma \times [0, 1])$ lies in an element of $f^{-1}\{G_\alpha'''\}$. It will be convenient for our argument that T restricted to $R \times [0, 1]$ be a cylindrical triangulation, that is, the vertical projection onto $R \times 0$ sends each simplex of T linearly onto a simplex of T.

Apply Part 2 of Theorem **1.6.10** to obtain a PL approximation H^3 of H^2 such that

$H_0^3 = 1$,
$H^3[(R \times 1) \cup ([Q \cap R] \times [0, 1])] \subset U$,

if σ is an element of T in $R = R \times 0$, then $H^3(\sigma \times [0, 1])$ lies in an element of $f^{-1}\{G_\alpha'''\}$,

H^3 is a PL (not necessarily linear) homeomorphism on each simplex of T, and

H^3 is in general position with respect to T in the sense that if Int σ^i, Int σ^j are different open simplexes of T with dimensions i, j respectively, then dimension $H^3(\text{Int } \sigma^i) \cap H^3(\text{Int } \sigma^j) \leqslant i + j - n$.

By applying Shadow Building Lemma **4.10.2**, we may assume that there is a shadow L for $H^3 \mid R \times [0, 1]$ relative to the set of singularities of H^3 in $R \times [0, 1]$ resulting from considering pairs of simplexes of T such that at least one member of the pair has dimension less than $n - 2$.

Let K^{r-2} be the $(r - 2)$-skeleton of T restricted to R and K^{r-1} the $(r - 1)$-skeleton of T restricted to R. Choose an ordering $\sigma_1, \sigma_2, ..., \sigma_{j-2}$ of the r-simplexes of T in $R = R \times 0$ so that there are $k + 2$ integers

$$1 = k_1 < k_2 < \cdots < k_{k+2} = j - 2$$

so that if $k_i < m \leqslant k_{i+1}$ then $H^3(\sigma_m \times [0, 1])$ lies in an element of $\{f^{-1}(G_{\alpha^i}^i)\}$.

It follows by induction on r that there is an engulfing isotopy $H: M^n \times [0, 1/j] \to M^n$ such that

$H_0 = 1$,

$H_t \mid Q \cup H^3((R \times 1) \cup ((Q \cap R) \times [0, 1])) = 1$, and

$H^3(K^{r-2} \times [0, 1] \cup L) \subset H_{1/j}(U)$.

We note that H^3 is a homeomorphism on $(K^{r-1} \times [0, 1]) - L$, so we can apply Engulfing Lemma **4.10.1** to extend $H: M^n \times [0, 1/j] \to M$ to $H: M^n \times [1/j, 2/j] \to M$ so that $H^3(K^{r-1} \times [0, 1]) \subset H_{2/j}(U)$ and for $1/j \leqslant t \leqslant 2/j$, $H_t = H_{1/j}$ on

$$Q \cup H^3(((Q \cap R) \times [0, 1]) \cup (R \times 1) \cup (K^{r-2} \times [0, 1]) \cup L).$$

Now consider σ_1 and all the $(r + 1)$-simplexes of T above it. Suppose there are j_1 of these. By starting at the top and working his way down inside the element of $f^{-1}(G_{\alpha_1}^1)$ containing $H^3(\sigma_1 \times [0, 1])$, one finds in j_1 applications of Lemma **4.10.1**, that $H: M^n \times [0, 2/j] \to M$ can be extended to an isotopy $H: M^n \times [2/j, 3/j] \to M^n$ such that

$$H^3(\text{all } r\text{-simplexes of } T \text{ above } \sigma_1) \subset H_{3/j}(U)$$

and for $2/j \leqslant t \leqslant 3/j$, $H_t = H_{2/j}$ on

$$Q \cup H^3(((Q \cap R) \times [0, 1]) \cup (R \times 1) \cup (K^{r-1} \times [0, 1]) \cup L).$$

[In each application of Lemma **4.10.1**, we cover up the image under H^3 of an $(r + 1)$-simplex of T, but in working down to those beneath it, we do not necessarily keep it covered, since L was not required to contain all singularities of H^3. Notice here also that things might have gone wrong if we had omitted the last requirement in the definition of a shadow. As we pushed across an $H^3(\tau)$, where τ is an $(r + 1)$-simplex, we might have uncovered part of $H^3(L)$ through $H^3(\tau')$ where τ' is another $(r + 1)$-simplex.]

By working his way down the cylinders above $\sigma_2, \sigma_3, \ldots, \sigma_{j-2}$ one at a time, one finds, on repeated applications of Lemma **4.10.1**, that H can be extended to an isotopy $H\colon M^n \times [0, 1] \to M^n$ such that

$H_0 = 1$,
$H_t = 1$ on Q,
$H^3(r\text{-skeleton of } T) \subset H_1(U)$, and
for each $x \in M^n$, there is a $G_\gamma' \in \{G_\beta'\}$ such that $H(x \times [0, 1])$ lies in $f^{-1}(G_\gamma')$.

This H is the required engulfing isotopy.

As mentioned at the first of this section, Bing has proved the following more refined engulfing theorem for codimensions greater than three. The proof of this theorem follows the lines of proof of Theorem **4.10.1** except that one doesn't have to worry much about shadows since they are easy to find in codimensions greater than three.

Radial Engulfing Theorem 4.10.2 (Bing). *Let M^n be a connected* PL *n-manifold without boundary, U an open set in M^n, P an $(n - 3)$-dimensional polyhedron (not necessarily compact), $Q \subset P$ a subpolyhedron in U such that $R = \mathrm{Cl}(P - Q)$ is compact and of dimension $r \leqslant n - 4$. Suppose that $\{A_\alpha\}$ is a collection of sets such that finite r-complexes in M^n can be pulled into U along $\{A_\alpha\}$. Then, for each $\epsilon > 0$, there is an engulfing isotopy $H\colon M^n \times [0, 1] \to M^n$ such that $H_0 = 1$, $H_t \mid Q = 1$, $P \subset H_1(U)$, and for each $x \in M^n$ there are $r + 1$ elements of $\{A_\alpha\}$ such that the track $H(x \times [0, 1])$ lies in the ϵ-neighborhood of the union of these $r + 1$ elements.*

Before stating a codimension three form of the above theorem, let us make a definition. The **double ϵ-neighborhood of an element A_α** of a collection $\{A_\alpha\}$ of sets in M^n is defined as follows:

$$N_\epsilon^2(A_\alpha) = \{x \in M^n \mid x \in N_\epsilon(A_\beta, M^n) \quad \text{for some} \quad A_\beta \quad \text{which}$$
$$\text{intersects} \quad N_\epsilon(A_\alpha, M^n)\}.$$

Wright [1] in generalizing Theorem **4.10.2** to the case $r = n - 3$ employed Zeeman's piping lemma (discussed in Section **4.6**) and piping apparently necessitates the use of the double ϵ-neighborhood. Certainly the theorem would be more pleasing if this concept could be avoided.

Radial Engulfing Theorem 4.10.3 (Wright). *Assume the hypotheses of Theorem* **4.10.2** *except allow r to be $n - 3$. Then, for each $\epsilon > 0$, there is an engulfing isotopy $H: M^n \times [0, 1] \to M^n$ such that $H_0 = 1$, $H_t \mid Q = 1$, $P \subset H_1(U)$, and for each $x \in M^n$ there are $r + 2$ elements of $\{A_\alpha\}$ such that the track $H(x \times [0, 1])$ lies in the ϵ-neighborhood of the sum of some $r + 1$ of these elements and the double ϵ-neighborhood of the remaining element.*

For an indication of how to improve Theorem **4.10.3** see [Edwards and Glaser, 1].

4.11. THE PL APPROXIMATION OF STABLE HOMEOMORPHISMS OF E^n

A homeomorphism h of a connected manifold M is said to be **stable** if $h = h_m h_{m-1} \cdots h_1$, where h_i is a homeomorphism of M which is the identity on some nonvoid open subset U_1 of M, $i = 1, 2, \ldots, m$. The **n-dimensional stable homeomorphism conjecture** is the conjecture that every orientation-preserving homeomorphism of S^n (equivalently E^n) is stable. In [Brown and Gluck, 1], it is shown that the n-dimensional stable homeomorphism conjecture implies the n-dimensional annulus conjecture (given in Remark **1.8.3**) and that the annulus conjecture in all dimensions $\leqslant n$ implies the stable homeomorphism conjecture in all dimensions $\leqslant n$. Kirby, Siebenmann, and Wall [1] have established the n-dimensional stable homeomorphism conjecture for $n \geqslant 5$ (hence the n-dimensional annulus conjecture for $n \geqslant 5$). The stable homeomorphism conjecture follows for $n = 3$ from [Bing, 4], [Bing, 5], or [Moise, 2] and for $n = 2$ from the classical Schoenflies Theorem. The purpose of this section is to show that stable homeomorphisms of E^n, $n \geqslant 5$, can be approximated by PL homeomorphisms. (As stated in the last section, this was first proved in [Connell, 1] for $n \geqslant 7$ and extended to $n = 5, 6$ by Bing [10].) Since the n-dimensional stable homeomorphism conjecture is now known for $n \geqslant 5$, it follows that any homeomorphism of E^n, $n \geqslant 5$, can be approximated by a PL homeomorphism. It is also known that any homeomorphism of E^3 can be approximated by a piecewise linear one (see [Bing, 5] or [Moise, 3]).

In this section we shall continue our policy of shunning abstract triangulations and using only rectilinear triangulations, although Connell's original proof does use abstract triangulations. The proof of the main theorem (Theorem **4.11.1**) of this section (which we are about to state) will be presented "backwards" in the sense that it will first be given modulo a lemma (Lemma **4.11.1**) and then that lemma will be proved modulo another lemma (Lemma **4.11.2**) and finally a proof, based on results of the preceding section, will be given for the latter lemma. We feel that this order of presentation motivates the proof.

Theorem 4.11.1 (Connell and Bing). *If* $g: E^n \to E^n$, $n \geqslant 5$, *is a stable homeomorphism and* $\epsilon(x): E^n \to (0, \infty)$ *is a continuous function, then there exists a* PL *homeomorphism* $f: E^n \twoheadrightarrow E^n$ *such that*

$$\text{dist}(f(x), g(x)) < \epsilon(x) \qquad \textit{for} \quad x \in E^n.$$

Before giving the proof of Theorem **4.11.1**, let us state a preliminary lemma.

Lemma 4.11.1. *Let* $g: E^n \twoheadrightarrow E^n$, $n \geqslant 5$, *be an arbitrary (topological) homeomorphism such that* $g \mid O$ *is* PL (*where* $O = O_1{}^n$ *as defined in the last section*). *Let* $h: O \twoheadrightarrow E^n$ *be a homeomorphism such that* $h(0) = 0$, $\theta(h(x), x) = 0$ *for* $x \in O$ *and if* $0 \leqslant r \leqslant 1$, *there exists a number* $\mu(r) > r$ *such that* $h((\text{Cl}\, O_r{}^n) - O_r{}^n) = \text{Cl}\, O^n_{\mu(r)} - O^n_{\mu(r)}$. *Then, if* $\epsilon(x): O \to (0, \infty)$ *is continuous, there is a* PL *homeomorphism* $f: g(O) \twoheadrightarrow E^n$ *such that* $\text{dist}(fg(x), gh(x)) < \epsilon(x)$ *for* $x \in O$.

Proof of Theorem 4.11.1 Modulo Lemma 4.11.1. Since g is stable, there exist homeomorphisms $g_1, g_2, \ldots, g_m$ and nonvoid open sets $U_1, U_2, \ldots, U_m$ such that $g_i \mid U_i = 1$ for $i = 1, 2, \ldots, m$ and $g = g_m g_{m-1} \cdots g_1$. If each g_i can be approximated by a PL homeomorphism, then clearly g can also. Thus, it may be assumed that there exists a nonvoid open set U such that $g \mid U = 1$. We now see that it will more than suffice to prove the following proposition.

Proposition 4.11.1. *If* $g: E^n \to E^n$, $n \geqslant 5$, *is a homeomorphism which is* PL *on some nonempty open set* U *and* $\epsilon(x): E^n \to (0, \infty)$ *is a continuous function, then there exists a* PL *homeomorphism* $f: E^n \twoheadrightarrow E^n$ *such that* $\text{dist}(f(x), g(x)) < \epsilon(x)$ *for* $x \in E^n$.

At first glance, Proposition **4.11.1** appears to be a stronger fact than we need. However, in view of the following exercise, it is not. We prove Proposition **4.11.1** directly because it causes us no extra trouble.

EXERCISE 4.11.1. Show that an orientation preserving homeomorphism h of a connected PL manifold M which is PL on some nonempty open set U is stable. (Hint: First show that a homeomorphism of a connected manifold M which agrees with a stable homeomorphism of M on a nonempty open set is stable.)

Proof of Proposition 4.11.1 Modulo Lemma 4.11.1. For convenience suppose $U \supset O_1{}^n = O$. Let $h: O \twoheadrightarrow E^n$ be a homeomorphism as in the statement of Lemma **4.11.1**, that is, $h(x) = \mu(\| x \|)x$. By applying Lemma **4.11.1**, we can obtain a PL homeomorphism $f_2: g(O) \twoheadrightarrow E^n$ such that $f_2 g \mid O$ approximates gh very closely. We can also apply Lemma **4.11.1** to obtain a PL homeomorphism $f_1: O \twoheadrightarrow E^n$ such that f_1 approximates h very closely. (To get f_1, let the g of Lemma **4.11.1** be the identity.) Then, $f = f_2 g f_1^{-1}: E^n \twoheadrightarrow E^n$ is the PL homeomorphism we seek. It is easy to see that by making f_2 approximate gh closely enough and f_1 approximate h closely enough, we make $f_2 g f_1^{-1}$ $\epsilon(x)$-approximate g.

Before proving Lemma **4.11.1**, let us state another lemma.

Lemma 4.11.2. *Let $g: E^n \twoheadrightarrow E^n$ be an arbitrary (topological) homeomorphism and a, b, and ϵ numbers with $0 < a < b$ and $\epsilon > 0$. Then, there is a PL homeomorphism $f_*: E^n \twoheadrightarrow E^n$ such that $f_* \mid g(O^n_{a-\epsilon}) = 1$, $f_* \mid g({}_{b+\epsilon}O^n) = 1$, $f_*(g(O_a{}^n)) \supset g(O_b{}^n)$ and $\theta\{g^{-1}(f_*(y)), g^{-1}(y)\} < \epsilon$ for all $y \in E^n$.*

Proof of Lemma 4.11.1 Modulo Lemma 4.11.2. First, we claim that given $\epsilon'(x): O \to E^n$, it will suffice to find a PL homeomorphism $f: g(O) \twoheadrightarrow E^n$ such that $\text{dist}(g^{-1}fg(x), h(x)) < \epsilon'(x)$ for $x \in O$. We can ensure by choosing $\epsilon'(x)$ small enough that $\text{dist}(y, z) < \epsilon'(h^{-1}(z))$, $y, z \in E^n$, implies $\text{dist}(g(y), g(z)) < \epsilon(h^{-1}(z))$ and so the claim follows by setting $y = g^{-1}fg(x)$ and $z = h(x)$.

Thus, the proof calls for a PL expansion f of $g(O)$ onto E^n such that $g^{-1}fg \mid O: O \twoheadrightarrow E^n$ is nearly radial. The difficulty arises that after a sequence of expansions via Lemma **4.11.2**, an "angle error" may accumulate. This is overcome, however.

Let $\delta(w): E^n \to (0, \infty)$ be a continuous function such that if $v, w \in E^n$, $-\delta(w) < \| v \| - \| w \| < \delta(w)$ and $\theta\{v, w\} < \delta(w)$, then $\text{dist}(v, w) < \epsilon'(h^{-1}(w))$. Let $0 = r_0 < r_1 < r_2 \cdots$ be an increasing sequence of numbers such that $r_n \to 1$ as $n \to \infty$ and $\mu(r_{i+2}) - \mu(r_i) < \min \delta(w)$ for $w \in \text{Cl}(O_{\mu(r_{i+2})})$ (denote this min by δ_i).

It follows from Lemma **4.11.2**, that there exists a PL homeomorphism

$f_1: E^n \twoheadrightarrow E^n$ such that $f_1 \mid g(O_{r_0}) = 1$, $f_1 \mid g({}_{\mu(r_2)}O) = 1$, $f_1(g(O_{r_1})) \supset g(O_{\mu(r_2)})$ and $\theta\{g^{-1}(f_1(y)), g^{-1}(y)\} < \delta_1/2$.

Again applying Lemma **4.11.2**, there is a PL homeomorphism $f_2: E^n \twoheadrightarrow E^n$ such that $f_2 \mid f_1 g(O_{r_1}) = 1, f_2 \mid f_1 g({}_{\mu(r_3)}O) = 1, f_2(f_1 g(O_{r_2}) \supset f_1 g(O_{\mu(r_2)})$, and $\theta\{(f_1 g)^{-1}(f_2(y)), (f_1 g)^{-1}(y)\} < \delta_2/2$.

In general suppose $f_1, f_2, \ldots, f_{k-1}$ have been defined. Let the f_* of Lemma **4.11.2** be $f_{k-1}f_{k-2} \cdots f_1 g$. See Fig. **4.11.1**. Then, there is a PL

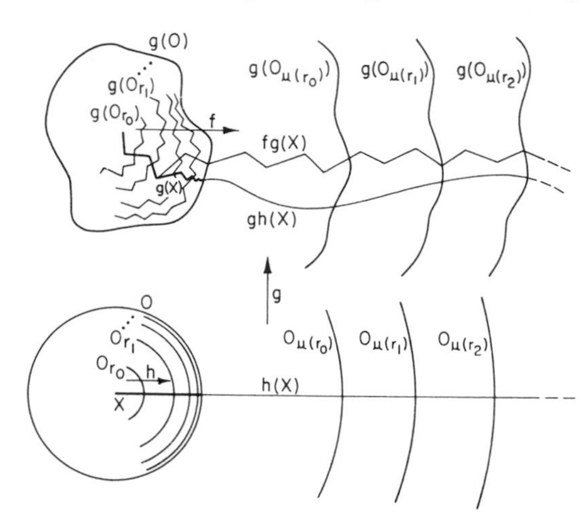

Figure 4.11.1

homeomorphism $f_k: E^n \twoheadrightarrow E^n$ such that

(1) $f_k \mid f_{k-1}f_{k-2} \cdots f_1 g(O_{r_{k-1}}) = 1$,

(2) $f_k \mid f_{k-1}f_{k-2} \cdots f_1 g({}_{\mu(r_{k+1})}O) = 1$,

(3) $f_k(f_{k-1}f_{k-2} \cdots f_1 g(O_{r_k})) f_{k-1}f_{k-2} \cdots f_1 g(O_{\mu(r_k)})$,

(4) $\theta\{(f_{k-1}f_{k-2} \cdots f_1 g)^{-1} f_k(z), (f_{k-1}f_{k-2} \cdots f_1 g)^{-1}(z)\} < \delta_k/2$ for $z \in E^n$.

From (2) follows

(2′) $f_k f_{k-1} \cdots f_1 g \mid ({}_{\mu(r_{k+1})}O) = 1$.

Suppose (2′) is true for $k - 1$, that is, suppose $f_{k-1} \cdots f_1 g \mid_{\mu(r_k)} O = 1$. Then, since ${}_{\mu(r_{k+1})}O \subset {}_{\mu(r_k)}O$, $f_{k-1}f_{k-2} \cdots f_1 g \mid_{\mu(r_{k+1})} O = 1$, and thus by (2), $f_k f_{k-1} \cdots f_1 g \mid_{\mu(r_{k+1})} O = 1$. Therefore, (2′) follows from (2) by induction.

Define $f: g(O) \twoheadrightarrow E^n$ by $f = \cdots f_3 f_2 f_1 \mid g(O)$. This f will satisfy the conclusion of the lemma. First it will be shown that f is well defined. Let $x \in g(O)$. Then there is an integer k such that $x \in g(O_{r_k})$. From (1), $f_s \mid f_{s-1} \cdots f_1 g(O_{r_{s-1}}) = 1$ for $s = 1, 2, \ldots$ and since $x \in g(O_{r_s})$ when $s > k$, $f(x) = f_k \cdots f_1(x)$ and thus f is well defined. Since each f_i is PL and since f is defined by a finite number of the f_i on any compact subset of $g(O)$, it is clear that f is a PL homeomorphism.

It remains to be shown that $|g^{-1}f(x) - h(x)| < \epsilon'(x)$ for $x \in O$.

Observation A_1. $f(g(O_{r_k})) \supset g(O_{\mu(r_k)})$ for $k = 1, 2, \ldots$. To see this, note that $f(g(O_{r_k})) = f_k f_{k-1} \cdots f_1(g(O_{r_k}))$ which by (3), $\supset f_{k-1}f_{k-2} \cdots f_1(g(O_{\mu(r_k)}))$, and it follows from (2′) that this is equal to $g(O_{\mu(r_k)})$.

Observation A_2. $f(g(O_{r_{k+1}})) \subset g(O_{\mu(r_{k+2})})$ for $k = 0, 1, 2 \ldots$. To see this, note that

$$f((g(O_{r_{k+1}})) = f_{k+1}f_k \cdots f_1(g(O_{r_{k+1}})) \subset f_{k+1}f_k \cdots f_1(g(O_{\mu(r_{k+2})}))$$

and this is equal to $g(O_{\mu(r_{k+2})})$ because $f_{k+1}f_k \cdots f_1$ is the identity on $O_{\mu(r_{k+2})}$.

Observation A. If $x \in O_{r_{k+1}} - O_{r_k}$, then

$$-\delta_k < \| g^{-1}fg(x)\| - \| h(x)\| < \delta_k .$$

It follows from A_1 and A_2 that $fg(x) \in g({}_{\mu(r_k)}O) \cap g(O_{\mu(r_{k+2})})$, hence $g^{-1}fg(x) \in {}_{\mu(r_k)}O \cap O_{\mu(r_{k+2})}$ from which it follows that $\mu(r_k) \leqslant \| g^{-1}fg(x)\| < \mu(r_{k+2})$. Also, $\mu(r_k) < \| h(x)\| < \mu(r_{k+1})$. Thus, the absolute value of $\| g^{-1}fg(x)\| - \| h(x)\|$ is $< \mu(r_{k+2}) - \mu(r_k)$ which is $< \delta_k$, and the result follows.

Observation B_1. If $y \in {}_{r_k}O$ then $\theta\{g^{-1}(f_kf_{k-1} \cdots f_1)g(y), y\} < \delta_k/2$. According to A_1, $f_kf_{k-1} \cdots f_1 g(y) \in g({}_{\mu(r_k)}O)$, hence

$$g^{-1}(f_kf_{k-1} \cdots f_1)g(y) \in {}_{\mu(r_k)}O.$$

By (2′), $f_{k-1}f_{k-2} \cdots f_1 g \mid ({}_{\mu(r_k)}O) = 1$. Therefore,

$$\begin{aligned}\theta\{g^{-1}(f_kf_{k-1} \cdots f_1)g(y), y\} &= \{(f_{k-1} \cdots f_1 g)^{-1}(f_k \cdots f_1)g(y), y\}\\ &= \theta\{(f_{k-1} \cdots f_1 g)^{-1}(f_k \cdots f_1)g(y), (f_{k-1} \cdots f_1 g)^{-1}(f_{k-1} \cdots f_1 g)(y)\}\end{aligned}$$

which is $< \delta_k/2$ by (4) as can be seen by setting z of (4) to be $f_{k-1} \cdots f_1 g(y)$. This shows B_1.

Observation B_2. $g^{-1}(f_k \cdots f_1)^{-1}(f_{k+1} \cdots f_1)g({}_{r_k}O) \subset ({}_{r_k}O)$. This will be true if $(f_{k+1} \cdots f_1)g({}_{r_k}O) \subset (f_k \cdots f_1)g({}_{r_k}O)$, which follows easily from (1), $f_{k+1} \mid f_k \cdots f_1 g(O_{r_k}) = 1$.

Observation B. If $x \in O_{r_{k+1}} - O_{r_k}$, then $\theta\{g^{-1}fg(x), x\} < \delta_k$. Let $y = g^{-1}(f_k \cdots f_1)^{-1}(f_{k+1} \cdots f_1)g(x)$. By B_2, $y \in {}_{r_k}O$. Substitute y in the inequality of B_1 and obtain

$$\theta\{g^{-1}(f_{k+1}f_k \cdots f_1)g(x), y\} = \theta\{g^{-1}fg(x), y\} < \delta_k/2.$$

Now,

$$\theta\{y, x\} = \theta\{g^{-1}(f_k \cdots f_1)^{-1}(f_{k+1} \cdots f_1)g(x), x\}$$
$$= \theta\{(f_k \cdots f_1 g)^{-1} f_{k+1}(f_k \cdots f_1 g)(x), (f_k \cdots f_1 g)^{-1}(f_k \cdots f_1 g)(x)\}$$

which is $< \delta_{k+1}/2$ by (4). Now, by the triangle inequality,

$$\theta\{g^{-1}fg(x), x\} < (\delta_k/2) + (\delta_k/2) \leqslant \delta_k ,$$

and B follows.

Conclusion. If $x \in O$, then there is an integer k such that $x \in O_{r_{k+1}} - O_{r_k}$. By A, $-\delta_k < \| g^{-1}fg(x)\| - \| h(x)\| < \delta_k$ and by B, $\theta\{ g^{-1}fg(x), x\} = \theta\{ g^{-1}fg(x), h(x)\} < \delta_k$. In the definition of $\delta(x)$ at the beginning of the proof, let $v = g^{-1}fg(x)$ and $w = h(x)$, and note that $\delta(h(x)) \geqslant \delta_k$ because $\| h(x)\| < \mu(r_{k+1})$. Thus, from the definition of $\delta(w)$, $\text{dist}(v, w) < \epsilon^{-1}(h^{-1}(w))$ or $\text{dist}(g^{-1}fg(x), h(x)) < \epsilon'(h(^{-1}(x))) = \epsilon'(x)$. This completes the proof.

We will now complete the proof of Theorem **4.11.1** by giving a proof of Lemma **4.11.2**. The proof employs a trick of Stallings which we have already used several times in this chapter. In carrying out the trick we will use Corollaries **4.10.1** and **4.10.2** of the previous section.

Proof of Lemma 4.11.2. Let T be an arbitrary (rectilinear) triangulation of E^n and let T_1 be a subdivision of T such that if v is a simplex of T_1 which intersects $g(O_{b+\epsilon})$ and $x, y \in v \subset E^n$, then $\text{dist}(g^{-1}(x), g^{-1}(y)) < \epsilon/3$ and $\theta\{ g^{-1}(x), g^{-1}(y)\} < \epsilon/3$. Let J be the simplicial neighborhood (in T_1) of $g((_{a-(2/3)\epsilon}O) \cap (\text{Cl}(O_{b+(2/3)\epsilon})))$ and let J^2 be the 2-skeleton of J.

Now $n - \dim J^2 \geqslant 5 - 2 = 3$, and so by Corollary **4.10.1** there is a PL homeomorphism $h_1\colon E^n \twoheadrightarrow E^n$ such that $h_1 \mid g(O_{a-(\epsilon/3)}) = 1$, $h_1 \mid g(_{b+\epsilon}O) = 1$, $h_1(g(O_a)) \supset J^2$ and $\theta\{ g^{-1}(h_1(x)), g^{-1}(x)\} < \epsilon/3$ for all $x \in E^n$.

Let L be the subcomplex of the first barycentric subdivision of J which is maximal with respect to the property of not intersecting J^2. Now $\dim L = n - (\dim J^2 + 1) = n - 3$ and so by Corollary **4.10.2**, there is a PL homeomorphism $h_2\colon E^n \twoheadrightarrow E^n$ such that $h_2 \mid g(_{b+(\epsilon/3)}O) = 1$, $h_2 \mid g(O_{a-\epsilon}) = 1$, $h_2(g(\text{Cl}(_bO))) \supset L$, and $\theta\{ g^{-1}(h_2(x)), g^{-1}(x)\} < \epsilon/3$ for all $x \in E^n$.

If A is an n-simplex of J, A is the join of $A \cap J^2$ and $A \cap L$ and $A \cap J^2 \subset h_1(g(O_a))$ and $A \cap L \subset h_2(g(\text{Cl}(_bO)))$. Thus by pushing up this join structure in each simplex of J, we obtain a homeomorphism $h_3\colon E^n \twoheadrightarrow E^n$ such that $h_3 \mid g(O_{a-\epsilon}) = 1$, $h_3 \mid g(_{b+\epsilon}O) = 1$, for each n-simplex u of J, $h_1(g(O_a)) \cup h_3h_2(g(\text{Cl}(_bO)) \supset u$, and for each simplex

v of T_1, $h_3(v) = v$. The triangulation T_1 was chosen fine enough that $h_3(v) = v$ implies $\theta\{g^{-1}h_3(x), g^{-1}(x)\} < \epsilon/3$ for all $x \in E^n$.

Let $h_4 = h_3h_2$. We will show that $h_1(g(O_a)) \cup h_4(g({}_bO)) = E^n$. Now h_3 was chosen so that $h_1(g(O_a)) \cup h_4(g({}_bO)) \supset J$ and since $h_1 \mid g(O_{a-(\epsilon/3)}) = 1$, it clearly contains $g(O_{a-(\epsilon/3)}) \cup J \supset g(O_{b+(2/3)\epsilon})$. It remains to show that $h_3h_2({}_bO) \supset E^n - g(O_{b+(2/3)\epsilon})$. Let v be a simplex of T_1 in $E^n - g({}_{b+(2/3)\epsilon}O)$. Then, $v \subset g({}_{b+(\epsilon/3)}O)$ and thus $h_2 \mid v = 1$. Since $h_3(v) = v$, $h_3h_2(v) = v$ and therefore $h_3h_2(g({}_bO)) \supset v$. This shows that $h_1(g(O_a)) \cup h_4(g({}_bO)) = E^n$.

This gives $h_1(g(O_a)) \supset h_4(g(E^n - {}_bO))$ which implies

$$h_4^{-1}h_1(g(O_a)) \supset E^n - {}_bO = O_b\,.$$

Let $f_* = h_4^{-1}h_1$, and note that $f_* \mid g(O_{a-\epsilon}) = 1$, $f_* \mid g({}_{b+\epsilon}O) = 1$ and $\theta\{g^{-1}(f_*(x)), g^{-1}(x)\} < \epsilon$ and that f_* is a PL homeomorphism. Thus, f_* satisfies the conclusion of the lemma.

Some of the exercises and remarks which follow establish Diagram **4.11.1** for homeomorphisms of E^n.

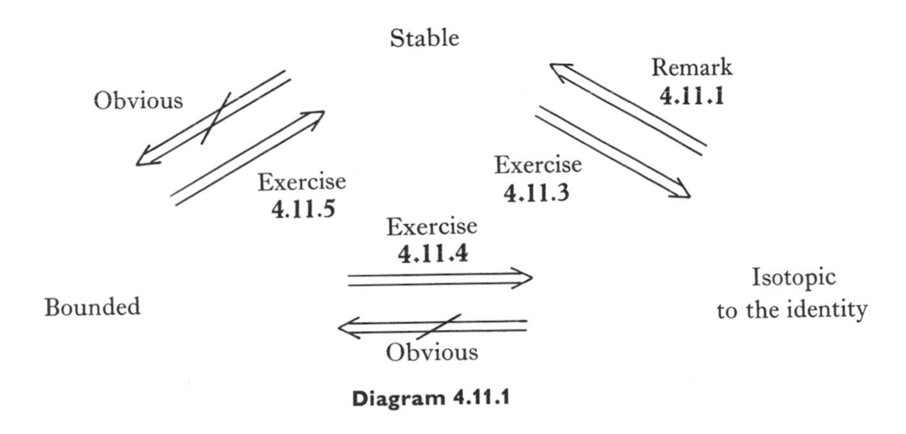

Diagram 4.11.1

EXERCISE 4.11.2. Show that every homeomorphism of I^n onto itself which restricts to the identity on ∂I^n, is isotopic to the identity. (This was first proved in [Alexander, 4]. For a generalization of this result see [Rushing, 9].)

EXERCISE 4.11.3. Show that every stable homeomorphism of S^n and E^n is isotopic to the identity. (Hint: Use Exercise **4.11.2**.)

EXERCISE 4.11.4. Show that every bounded homeomorphism of E^n is isotopic to the identity. (Hint: This isotopy is similar to the one of Exercise **4.11.2**) (This result was first proved in Theorem 1 of [Kister, 1].)

EXERCISE 4.11.5. Show that bounded homeomorphisms of E^n are stable. (This was first shown in Lemma 5 of [Connell, 1].)

EXERCISE 4.11.6. Suppose that g and h are homeomorphisms from E^n onto itself such that for some $M > 0$, $\text{dist}(g(x), h(x)) < M$ for $x \in E^n$. Then if g is stable, h is also stable. (Hint: Use Exercise **4.11.5**.) (This was first proved in Theorem 5 of [Connell, 1].)

REMARK 4.11.1. In [Kirby, 3] it was announced that a homeomorphism of E^n which is isotopic to the identity is stable. Later, as mentioned at the first of this section, it was announced in [Kirby *et al.*, 1] that every (orientation preserving) homeomorphism of E^n, $n \geqslant 5$, is stable.

REMARK 4.11.2. It should be pointed out that the analog of Theorem **4.11.1** for S^n is established in [Connell, 1].

4.12. TOPOLOGICAL ENGULFING

The first engulfing theorem for topological manifolds was given in [Newman, 2], and it was proved by methods which do not depend on the theory of PL manifolds. A weaker topological engulfing theorem was independently proved by Connell [2]. Connell's proof did use PL theory. In fact, it is fair to say that the main idea of Connell's proof is to use the techniques of PL engulfing locally. Connell did not seem to worry about getting the most general topological engulfing theorem possible, but was only interested in obtaining a theorem sufficient to prove a certain topological H-cobordism theorem and the topological Poincaré theorem. (These results will be discussed in the next section.) Another topological engulfing theorem is proved in [Lees, 2]. In this section we will adapt Connell's techniques to prove a topological engulfing theorem which would be the topological analog of Stallings' (PL) Engulfing Theorem **4.2.1** if the monotonic r-connectivity hypothesis (which we are about to define) were replaced by a r-connectivity hypothesis.

The pair (M, U) is **monotonically r-connected** if given any compact subset C_1 of U, there exists a closed (rel M), proper subset C_2 of U containing C_1 such that $(M - C_2, U - C_2)$ is r-connected.

Topological Engulfing Theorem 4.12.1. *Let M^n be a connected topological n-manifold without boundary, U an open set in M^n, P^k, $k \leqslant n - 3$, a possibly noncompact polyhedron, $f\colon P^k \to M$ a closed, locally tame* *embedding, and Q a (possibly noncompact) subpolyhedron of P^k such that $f(Q) \subset U$ and $R = \mathrm{Cl}(P - Q)$ is a compact r-subpolyhedron of P. Let (M, U) be monotonically r-connected. Then, given any compact subset C of U, there is a compact set $E \subset M^n$ and an ambient isotopy e_t of M^n such that $f(P) \subset e_1(U)$ and*

$$e_t \,|(M - E) \cup f(Q) \cup C = 1 \,|(M - E) \cup f(Q) \cup C.$$

Recall that in Section **3.6**, we defined an embedding $f\colon P^k \to M^n$ of the (possibly infinite) k-polyhedron P^k into the interior of the topological n-manifold M^n to be **locally tame** if there is a triangulation of P^k such that for each point $x \in P^k$ there is an open neighborhood U of $f(x)$ in M^n and a homeomorphism $h_U\colon U \to E^n$ such that $h_U f$ is PL on some neighborhood of x. [Without loss of generality, we may assume that $h_U f$ is PL on $f^{-1}(U \cap f(P))$ and we will do so in this section.]

Proof of Theorem 4.12.1. In order to make the proof more transparent, we will give the proof for the case $m \leqslant n - 4$. The case $m = n - 3$ contains an added difficulty similar to the one encountered in the proof of Stallings' engulfing theorem **4.2.1** in that case. This difficulty may be handled in a way completely analogous to the way the problem was handled in the proof of Stallings' engulfing theorem or alternately the way this problem was handled in the section on radial engulfing (see note at end of Case 1).

Without loss of generality we may assume that $C \cap f(R) = \emptyset$ and that there is a closed proper subset C_2 of U containing C such that $(M - C_2, U - C_2)$ is r-connected and such that $C_2 \cap f(R) = \emptyset$.

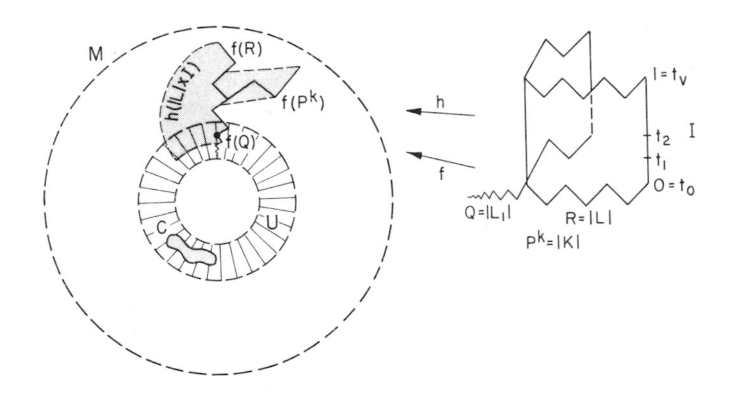

Figure 4.12.1

Let $h: R \times I \to M - C$ be a continuous function satisfying

(a) $h(x, 1) = f(x)$ for $x \in R$,
(b) $h(x, t) = f(x)$ for $x \in Q \cap R$ and $t \in [0, 1]$,
(c) $h(x, 0) \in U$ for $x \in R$.

Such an h exists because

$$\pi_i(M - C_2, U - C_2) = 0, \qquad i = 1, 2, ..., r.$$

Since $f(P^k) \cap h(R \times I)$ is compact and since f is locally tame there are compact subpolyhedra $X_i \subset P^k$, $i = 1, ..., u$, and sets $U_i \subset M - C$, $i = 1, ..., u$, and homeomorphisms $h_{U_i}: U_i \twoheadrightarrow E^n$ such that $h_{U_i} f \mid f^{-1}(U_i \cap f(P))$ is PL and $\bigcup_{i=1}^u \text{Int } X_i \supset f(P^k) \cap h(R \times I)$, where Int X_i denotes point-set interior of X_i relative to P^k. Let $\delta > 0$ be such that

$$N_{2\delta}(f(X_i), M) \subset U_i$$

$i = 1, ..., v$. Let $h_{U_i}: U_i \twoheadrightarrow E^n$, $i = v + 1, v + 2, ..., w$ be homeomorphisms where the U_i are open subsets of $M - C$ such that

$$\bigcup_{i=1}^{w} U_i \supset h(R \times I).$$

Let K be a triangulation of P, with L_1 and L the induced triangulations of Q and R, such that f is an a simplicial homeomorphism on K. Let $\sigma_1{}^k, \sigma_2{}^k, ..., \sigma_{q(k)}^k$ be the k-simplexes of L for $i = 0, 1, ..., r$. Finally, let $0 = t_0 < t_1 < \cdots < t_v = 1$ be a partition of $[0, 1]$. If the triangulation K and the partition $t_0 < t_1 < \cdots < t_v$ are fine enough, then $h: |L| \times I \to M$ will satisfy Property P below.

Definition. *A* continuous function $h: |L| \times I \to M - C$ has **Property P** provided

(1) $h(x, 1) = f(x)$ for $x \in |L|$,

(2) $h((|L_1| \cap |L|) \times [0, 1]) \subset U - C$,

(3) $h(\sigma_m{}^k \times [t_{a-1}, t_a]) \subset U_i$ for some $i = i(k, m, a)$ when $0 \leqslant k \leqslant r$, $1 \leqslant m \leqslant q(k)$ and $1 \leqslant a \leqslant v$.

(4) $\bigcup_{i=1}^u \text{Int } X_i \supset f(P^k) \cap h(R \times I)$,

(5) $\text{diam}(h(\sigma_m{}^k \times [t_{a-1}, t_a]) < \delta$ for $0 \leqslant k \leqslant r$, $1 \leqslant m \leqslant q(k)$ and $1 \leqslant a \leqslant v$.

Now suppose that the triangulation K and the partition $t_0 < t_1 < \cdots < t_v$, are given so that h satisfies Property P. For the remainder of this proof, the simplexes σ_m^k, the partition $t_0 < t_1 < \cdots < t_v$ and the function $(k, m, a) \to i(k, m, a)$ are fixed. The statement that $\alpha\colon |L| \times I \to M$ satisfies Property P means with respect to this fixed data. Notice that if α satisfies Property P and $\beta\colon |L| \times I \to M$ has $\beta(x, 1) = f(x, 1)$ and β is a close enough approximation to α, then β will also satisfy Property P.

Definition. For

$$0 \leqslant k \leqslant r, \qquad 1 \leqslant m \leqslant q(k), \qquad 1 \leqslant a \leqslant v, \qquad X(k, m, a) \subset (|L| \times I)$$

is defined by

$$\begin{aligned} X(k, m, a) = {} & |L| \times 0 \cup (|L| \cap |L_1| \times [0, 1]) \\ & \cup |L| \times [0, t_{a-1}] \cup \{\sigma_t^s \times [t_{a-1}, t_a]\colon s \leqslant k, \quad 1 \leqslant t \leqslant q(s)\} \\ & \cup \{\sigma_t^k \times [t_{a-1}, t_a]\colon 1 \leqslant t \leqslant m\}. \end{aligned}$$

Inductive Hypothesis (k, m, a) = IH(k, m, a). *There exists a continuous function*

$$\alpha_{(k,m,a)}\colon |L| \times I \to M - C,$$

which satisfies Property P and a homeomorphism

$$H_{(k,m,a)}\colon M \to M,$$

which can be realized by an ambient isotopy e_t of M^n satisfying

(1) *For some compact subset $E \subset M$*

$$e_t |(M - E) \cup f(Q) \cup C = 1 |(M - E) \cup f(Q) \cup C,$$

(2) $H_{(k,m,a)}(U) = e_1(U) \supset f(Q) \cup \alpha_{(k,m,a)}(X(k, m, a))$.

The purpose of the proof is to show that $IH(r, q(r), v)$ is true.

Fact 1. $IH(0, 1, 1)$ is true.

Fact 2. $IH(k, m - 1, a) \Rightarrow IH(k, m, a)$ for $0 \leqslant k \leqslant r, 2 \leqslant m \leqslant q(k)$, $1 \leqslant a \leqslant v$.

Fact 3. $IH(k, q(k), a) \Rightarrow IH(k + 1, 1, \alpha)$ for $0 \leqslant k \leqslant r, 1 \leqslant a \leqslant v$.

Fact 4. $IH(r, q(r), a) \Rightarrow IH(0, 1, a + 1)$ for $1 \leqslant a \leqslant v$.

The proof of Fact 2 is presented in detail. The proof of Facts 1, 3, and 4 require only trivial modifications and are not included.

Suppose $0 \leqslant k \leqslant r$, $2 \leqslant m \leqslant q(k)$, $1 \leqslant a \leqslant v$, and $IH(k, m-1, a)$ is true. For simplicity of notation let

$$H = H_{(k,m-1,a)}: M \to M \qquad \text{and} \qquad \alpha = \alpha_{(k,m-1,a)}: |L| \times I \to M - C_2 .$$

Then, α has Property P and H can be realized by an ambient isotopy e_t of M^n satisfying

(1) For some compact subset $E \subset M$,

$$e_t \,|(M - E) \cup f(Q) \cup C = 1 \,|(M - E) \cup f(Q) \cup C,$$

(2) $H(U) = e_1(U) \supset f(Q) \cup \alpha(X(k, m-1, a))$.

Proof of Fact 2

Case 1. $(\alpha(\sigma_m{}^k \times [t_{a-1}, t_a]) \cap f(P^k) = \emptyset)$. Let W_1, W_2, W_3 be open subsets of M with $\alpha(\sigma_m{}^k \times [t_{a-1}, t_a]) \in W_1$, $\text{Cl}(W_1) \subset W_2$, $\text{Cl}(W_2) \subset W_3$, $\text{Cl}(W_3) \subset U_{i(k,m,a)}$ and $W_3 \cap f(P^k) = \emptyset$. Let Z be a subpolyhedron of $|L| \times I$ with $\alpha^{-1}(W_2) \subset Z \subset \alpha^{-1}(W_3)$. Now by a general position argument (see Part d of Section **1.6**) there is a continuous

$$\alpha_{(k,m,a)} = \beta: |L| \times I \to M - C,$$

which satisfies Property P and

(1) $\beta(\sigma_m{}^k \times [t_{a-1}, t_a]) \subset W_1$,
(2) $\beta^{-1}(W_1) \subset \alpha^{-1}(W_2) \subset Z$,
(3) $\beta \mid \alpha^{-1}(M - W_3) = \alpha \mid \alpha^{-1}(M - W_3)$,
(4) $h_{U_{i(k,m,a)}} \beta \mid Z$ is PL and in general position.

In particular, if

$$S = \text{Cl}\{x \in \sigma_m{}^k \times [t_{a-1}, t_a]\text{: there is } y \in Z \quad \text{with} \quad x \neq y, \quad \beta(x) = \beta(y)\},$$

then $\dim S \leqslant 2(r+1) - n \leqslant (n-4) + r + 2 - n = r - 2$. In addition, it is assumed that β approximates α closely enough that

$$H(U) \supset f(Q) \cup \beta(X(k, m-1, a)). \qquad \text{[See (2) above.]}$$

Let $\pi: \sigma_m{}^k \times [t_{a-1}, t_a] \to \sigma_m{}^k$ be the projection. Since β has Property P,

$$\beta(\pi(S) \times [t_{a-1}, t_a]) \subset U_{i(k,m,a)} .$$

Since $g_{U_{i(k,m,a)}} \beta(\pi(S) \times [t_{a-1}, t_a])$ is a polyhedron in E^n of dimension $\leqslant r-1$, the inductive hypothesis on r may be applied. (Note that if

$2(r + 1) < n$, then no induction on r is necessary.) In applying the induction,

$$(H(U), C \cup \beta(X(k, m - 1, a)), \beta(\pi(S) \times [t_{a-1}, t_a]) \cup f(Q))$$

corresponds to

$$(U, C, f(P)).$$

Then, there is a homeomorphism $G_1: M \twoheadrightarrow M$ which can be realized by an ambient isotopy e_t satisfying

(a) For some compact set $E \in M$

$$e_t \,|(M - E) \cup f(Q) \cup C \cup \beta(X(k, m - 1, a)) \\ = 1 \,|(M - E) \cup f(Q) \cup C \cup \beta(X(k, m - 1, a)).$$

(b) $G_1(H(U)) \supset \beta(\pi(S) \times [t_{a-1}, t_a])$.

Now since $\sigma_m{}^k \times [t_{a-1}, t_a]$ collapses to

$$(X(k, m - 1, a) \cap (\sigma^k{}_m \times [t_{a-k}, t_a])) \cup (\pi(S) \times [t_{a-1}, t_a]) \\ = (\sigma_m{}^k \times t_{a-1}) \cup (\pi(S) \cup \partial\sigma_m{}^k) \times [t_{a-1}, t_a],$$

by Exercise **1.6.12**, there is a homeomorphism $G_2: M \to M$ which can be realized by an ambient isotopy e_t such that

(A) $e_t(x) = x$ for $x \in (M - W_1) \cup f(Q) \cup C \cup \beta(X(k, m - 1, a))$,

(B) $G_2G_1H(U) \supset f(Q) \cup \beta(X(k, m, a))$.

The homeomorphism $H_{(k,m,a)}$ is given by

$$H_{(k,m,a)} = G_2G_1H = G_2G_1H_{(k,m-1,a)}\,.$$

$H_{(k,m,a)}$ and $\alpha_{(k,m,a)} = \beta$ satisfy $IH(k, m, a)$. [Note: The changes necessary for the case $r = n - 3$ are almost identical to the changes necessary in the PL case. The inductive hypothesis $IH(k, m - 1, a)$ would require covering only the m-skeleton of $\alpha_{(k,m,a)}(X(k, m - 1, a))$, that is, the $(m + 1)$-cells need not be contained in $H_{(k,m-1,a)}(U)$. The singular set S would be defined by intersections of $\alpha(\sigma_m{}^k \times [t_{a-1}, t_a])$ with $\alpha(Z^r)$, where Z^r is the r-skeleton of Z.]

Case 2. $(\alpha(\sigma_m{}^k \times [t_{a-1}, t_a]) \cap f(P^k) \neq \emptyset)$. This case is similar to Case 1 except that we are sure to use a g_{U_i} where $1 \leqslant i \leqslant u$. Note that Case 2 always holds when $a = v$.

Since $\alpha(\sigma_m{}^k \times [t_{a-1}, t_a]) \cap f(P^k) \neq \emptyset$ it follows that for some i such that $1 \leqslant i \leqslant u$, $\alpha(\sigma_m{}^k \times [t_{a-1}, t_a]) \cap f(X_i) \neq \emptyset$. But since $N_{2\delta}(f(X_i), M) \subset U_i$ and $\text{diam}(\alpha(\sigma_k{}^m \times [t_{a-1}, t_a]) < \delta$ it follows that

$$\alpha(\sigma_m{}^k \times [t_{a-1}, t_a]) \subset N_{2\delta}(f(X_i), M) \subset U_i .$$

Let W_1, W_2, W_3 be open subsets of M with

$$\alpha(\sigma_k{}^m \times [t_{a-1}, t_a]) \subset W_1, \qquad \text{Cl}(W_1) \subset W_2, \qquad \text{Cl}(W_2) \subset W_3,$$

and

$$\text{Cl}(W_3) \subset U_i .$$

Let Z be a subpolyhedron of $|L| \times I$ with $\alpha^{-1}(W_2) \subset Z \subset \alpha^{-1}(W_3)$. Since $1 \leqslant i \leqslant u$, we have that $h_{U_i} f \mid f^{-1}(U_i \cap f(P))$ is PL and so by a relative general position approximation argument, there is a continuous

$$\alpha(k, m, a) = \beta: |L| \times I \to M - C,$$

which satisfies Property P and

(1) $\beta(\sigma_m{}^k \times [t_{a-1}, t_a]) \subset W_1$,
(2) $\beta^{-1}(W_1) \subset \alpha^{-1}(W_2) \subset Z$,
(3) $\beta \mid \alpha^{-1}(M - W_3) = \alpha \mid \alpha^{-1}(m - W_3)$,
(4) $h_{U_i}\beta \mid Z: Z \to E^n$ is PL and in general position relative to $f(Q)$.

In particular, if $S = \text{Cl}\{x \in \sigma_m{}^k \times [t_{a-1}, t_a]$: (there is a $y \in Z$, $y \neq x$, $\beta(x) = \beta(y)$) or (there is $w \in Q - R$ with $\beta(x) = f(w)$)$\}$ then

$$\dim S \leqslant 2(r + 1) - n \leqslant n - 4 + r + 2 - n = r - 2.$$

The remainder of the proof is a repeat from Case 1.

4.13. TOPOLOGICAL *H*-COBORDISMS AND THE TOPOLOGICAL POINCARÉ THEOREM

This section is organized as follows: We first give a few definitions, next we state the main results, then we discuss these results and related work and finally we give the proofs. A **cobordism** of dimension n is a triple $(M; A, B)$ where M is a compact (topological) n-manifold with boundary components A and B. A cobordism $(M; A, B)$ is called an ***H*-cobordism** if A and B are deformation retracts of M. If $(M; A, B)$ and $(N; B, C)$ are two cobordisms with a common boundary part B, we define the **composed cobordism** $(M; A, B) \cdot (N; B, C)$ to be $(M \cup N; A, C)$ where $M \cup N$ is the union of M and N pasted together along B. The cobordism $(M; A, B)$ is a **product cobordism** if

$$(M; A, B) \approx (A \times [0, 1], A \times 0, A \times 1).$$

The cobordism $(M; A, B)$ is **invertible** if there exists a cobordism $(N; B, A)$ so that the composed cobordisms $(M; A, B) \cdot (N; B, A)$ and $(N; B, A) \cdot (M; A, B)$ are product cobordisms (see Fig. **4.13.1**).

Figure 4.13.1

Theorem 4.13.1. *Let $(M; A, B)$ be a connected, n-dimensional, $n \geqslant 5$, (topological) cobordism such that $\pi_i(M, A) = \pi_i(M, B) = 0$ for $i = 1, 2, \ldots, n - 3$. Let*

$$g_A\colon A \times [0, 1] \to M - B$$

and

$$g_B\colon B \times [0,1] \to M - A$$

be embeddings with $g_A(x, 0) = x$ for all $x \in A$ and $g_B(y, 0) = y$ for $y \in B$ such that $g_A(A \times [0, 1]) \cap g_B(B \times [0, 1]) = \emptyset$. Then, if b is a number, $0 < b < 1$, there exist homeomorphisms $f_A\colon M \to M$ and $f_B\colon M \to M$ such that

$$f_A \mid g_A(A \times [0, 1 - b]) \cup g_B(B \times [0, 1 - b]) = 1,$$

$$f_B \mid g_A(A \times [0, 1 - b]) \cup g_B(B \times [0, 1 - b]) = 1,$$

and

$$f_A(g_A(A \times [0, 1))) \cup f_B(g_B(B \times [0, 1))) = M.$$

Also, there is a homeomorphism $H\colon g_A(A \times [0, 1)) \twoheadrightarrow M - B$.

Remark 4.13.1. The embeddings g_A and g_B in the hypothesis of Theorem **4.13.1** always exist by Theorem **1.7.4**.

Remark 4.13.2. The f_A and f_B will actually be constructed so that they are isotopic to the identity.

Corollary 4.13.1 (Weak H-cobordism theorem). *If $(M; A, B)$ is an n-dimensional* (topological) *H-cobordism, $n \geqslant 5$, then*

$$M - B \approx A \times [0, 1).$$

Corollary 4.13.2 (Topological Poincaré theorem). *If Y is a compact topological n-manifold, $n \geqslant 5$, without boundary, which has the homotopy type of S^n, then Y is homeomorphic to S^n.*

Corollary 4.13.3. *For $n \geqslant 5$, every n-dimensional (topological) H-cobordism $(M; A, B)$ is invertible.*

It is shown in [Stallings, 1], Theorem 4, that in the case that the M of Corollary **4.13.1.** is a PL manifold, one can conclude that $M - B \overset{\text{PL}}{\approx} A \times [0, 1)$. The, now classical, ***H*-cobordism theorem** states that if $(M; A, B)$ is an H-cobordism of dimension greater than 5 such that M is a PL manifold and such that A and B are simply connected, then M is PL homeomorphic to $A \times [0, 1]$ and (consequently) A is PL homeomorphic to B. This was first proved by Smale [1, 2] in the differential category and another good proof appears in [Milnor, 2]. If one omits the restriction that A and B be simply connected, the theorem becomes false (see [Milnor, 1]). But it will remain true if we at the same time assume that the inclusion of A (or B) into M is a simple homotopy equivalence in the sense of Whitehead. This generalization, called the ***s*-cobordism theorem**, is due to Mazur [3], Barden [1], and Stallings.

The remainder of this section is based primarily on [Connell, 2]. For other work related to this section see [Lees, 1], [Stallings, 1], [Newman, 2], and [Siebenmann, 1]. (The proof of Corollary **4.13.3** given here appears in [Lees, 1] and is implied in [Siebenmann, 1].)

Before proving the theorem, let us prove the corollaries. Of course, Corollary **4.13.1** is immediate.

Proof of Corollary 4.13.2. Let B^n and $B_1{}^n$ be disjoint, locally flat topological n-cells in Y and $p \in \text{Int } B_1{}^n$. Then, $Y - B_1{}^n$ is homeomorphic to $Y - p$ by Corollary **1.8.2.** It follows from Corollary **4.13.1** and the fact that

$$Y - (\text{Int } B^n \cup \text{Int } B_1{}^n)$$

is a topological H-cobordism (see Exercise **4.13.1**) that $Y - B_1{}^n$ is homeomorphic to E^n. Thus, $Y - p$ is homeomorphic to E^n and Y is homeomorphic to S^n.

EXERCISE 4.13.1. Show that $Y - (\text{Int } B^n \cup \text{Int } B_1{}^n)$ in the above proof is indeed a topological H-cobordism.

Proof of Corollary 4.13.3. (This proof is taken from Theorem 2 of [Lees, 1].) By Theorem **4.13.1** there are embeddings

$$h_A: A \times [0, 2] \rightarrow M - B$$

and

$$h_B: B \times [0, 2] \to M - A$$

with $h_A(x, 0) = x$ for all $x \in A$ and $h_B(x, 0) = x$ for all $x \in B$ such that $h_A(A \times [0, 1)) \cup h_B(B \times [0, 1)) = M$. Let $N_A = h_A(A \times [0, 1])$ and $N_B = h_B(B \times [0, 1])$. Let $M' = N_A \cap N_B$. Then,

$$(M'; \mathrm{Fr}_M(N_B), \mathrm{Fr}_M(N_A))$$

is an h-cobordism. ($\mathrm{Fr}_X\, Y$ denotes the frontier of Y relative to X, see Fig. **4.13.2**). Note that $\mathrm{Bd}\, M' = \mathrm{Fr}_M(N_B) \cup \mathrm{Fr}_M(N_A) \approx A \cup B$.

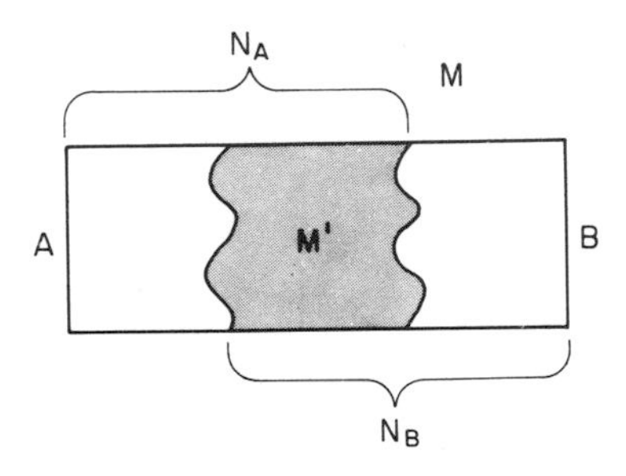

Figure 4.13.2

Now,

$$M' \cup_{\mathrm{id}(B)} M \approx M' \cup (M - N_A) \approx N_B \approx B \times [0, 1]$$

and

$$M' \cup_{\mathrm{id}(A)} M \approx M' \cup (M - N_B) \approx N_A \approx A \times [0, 1].$$

Before proving Theorem **4.13.1**, let us establish a couple of preliminary lemmas.

Engulfing Lemma 4.13.1. *Let $(M; A, B)$ be a connected, n-dimensional $n \geqslant 5$, (topological) cobordism such that $\pi_i(M, A) = \pi_i(M, B) = 0$ for $i = 1, 2, \ldots, n - 3$. Let $g_A: A \times [0, 1] \to M - B$ be an embedding with $g_A(x, 0) = x$ for all $x \in A$. Suppose $K \subset E^n$ is a compact k-polyhedron, $k \leqslant n - 3$, $h: E^n \to \mathrm{Int}\, M$ is a topological embedding, and ϵ is a number with $0 < \epsilon < 1$. Then, there is a homeomorphism $H: M \to M$ satisfying*

(1) *$H(x) = x$ for $x \in B \cup g_A(A \times [0, 1 - \epsilon])$, and*

(2) *$H(g_A(A \times [0, 1))) \supset h(K)$.*

Proof. This lemma follows by applying Theorem **4.12.1**, where

$$[M - (g_A(A \times [0, 1 - \epsilon] \cup B), h^{-1}(h(K) - (h(K) \cap g_A(A \times [0, 1 - \epsilon]))),$$
$$h\,|(h^{-1}(h(K) - (h(K) \cap g_A(A \times [0, 1 - \epsilon])))), g_A(A \times (1 - \epsilon, 1))]$$

corresponds to (M, P^k, f, u) of that theorem. The collar structure very easily gives the required monotonic connectivity (see Fig. 4.13.3).

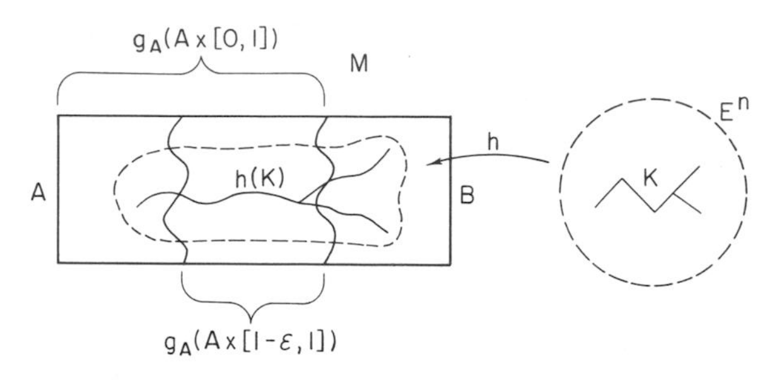

Figure 4.13.3

Lemma 4.13.2. *Let $(M; A, B)$ be a connected, n-dimensional, $n \geqslant 5$, (topological) cobordism such that $\pi_i(M, A) = \pi_i(M, B) = 0$ for $i = 1, 2, \ldots, n-3$. Let*

$$g_A\colon A \times [0, 1] \to M - B \qquad \textit{and} \qquad g_B\colon B \times [0, 1] \to M - A$$

be embeddings with $g_A(x, 0) = x$ for all $x \in A$ and $g_B(y, 0) = y$ for all $y \in B$. Suppose b is a number with $0 < b < 1$,

$$g_A(A \times [0, 1]) \subset M - g_B(B \times [0, 1-b]),$$

$$g_B(B \times [0, 1]) \subset M - g_A(A \times [0, 1-b]),$$

and $h\colon E^n \to \operatorname{Int} M$ is a topological embedding. Then, for any number a with $0 < a < b$, there are homeomorphisms $f_A\colon M \to M$ and $f_B\colon M \to M$ with

$$f_A \mid g_A(A \times [0, 1-a]) \cup g_B(B \times [0, 1-b]) = 1,$$

$$f_B \mid g_B(B \times [0, 1-a]) \cup g_A(A \times [0, 1-b]) = 1,$$

and

$$f_A g_A(A \times [0, 1)) \cup f_B g_B(B \times [0, 1)) \supset h(I^n).$$

Proof. Let T be a triangulation of E^n which triangulates I^n as a subcomplex. Let X be the subcomplex of T composed of all closed simplexes $\sigma \subset I^n$ with

$$h(\sigma) \cap \{M - [g_A(A \times [0, 1-a/2]) \cup g_B(B \times [0, 1-a/2])]\} \neq \emptyset$$

and let Y be the simplicial neighborhood of X in T (in all of E^n). Suppose that the triangulation T is fine enough that

$$h(Y) \subset \{M - [g_A(A \times [0, 1 - 3a/4] \cup g_B(B \times [0, 1 - 3a/4])]\}$$

(see Fig. **4.13.4**).

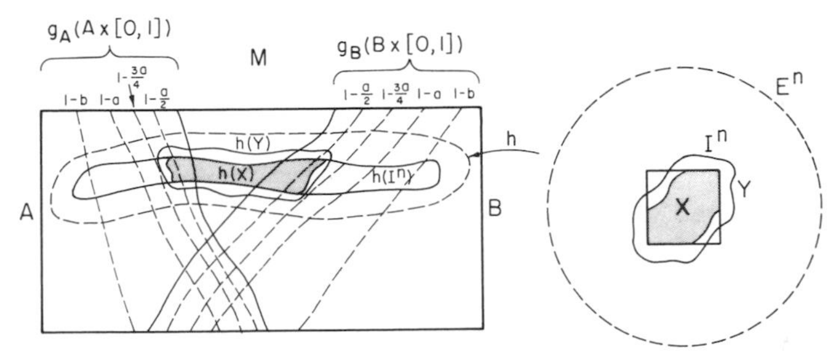

Figure 4.13.4

Let $\Delta > 0$ be such that

$$N_{3\Delta}(h(X), M) \subset h(Y)$$

and

$$N_\Delta(g_A(A \times [0, 1 - a/2]), M) \subset g_A(A \times [0, 1 - a/4]).$$

Let T_1 be a subdivision of T such that for any simplex σ_1 of T_1, $\operatorname{diam}(h(\sigma_1)) < \Delta$. Let X_1 and Y_1 be the sets X and Y under the triangulation T_1. Let K be the $(n - 3)$-skeleton of Y_1 and K_* be the maximal complex of a first derived of Y_1 which does not intersect K. Then, $\dim K_* \leqslant n - 3$. Now apply Lemma **4.13.1** to the H-cobordism $M - g_B(B \times [0, 1 - b])$ and obtain a homeomorphism

$$f_A'\colon M - g_B(B \times [0, 1 - b)) \to M - g_B(B \times [0, 1 - b))$$

such that $f_A'(x) = x$ for $x \in g_B(B \times (1 - b)) \cup g_A(A \times [0, 1 - a/4])$ and

$$f_A'(g_A(A \times [0, 1))) \supset h(K).$$

Extend f_A' to M by the identity on $g_B(B \times [0, 1 - b])$. In the same manner, apply Lemma **4.13.1** to the H-cobordism $M - g_A(A \times [0, 1 - b))$ and obtain a homeomorphism $f_B\colon M \to M$ satisfying

(1) $f_B(x) = x$ for $x \in g_A(A \times [0, 1 - b]) \cup g_B(B \times [0, 1 - a/4])$,
(2) $f_B(g_B(B \times [0, 1))) \supset h(K_*)$.

Statement A. There exists a homeomorphism $f_A'': M \to M$ such that

(a) $f_A''(x) = x$ for $x \in M - h(Y_1) \supset g_A(A \times [0, 1 - 3a/4]) \cup g_B(B \times [0, 1 - 3a/4])$,

(b) $f_A'' f_A'(g_A(A \times [0, 1))) \cup f_B(g_B(B \times [0, 1))) \supset h(X_1)$,

(c) $\text{dist}(f_A''(x), x) < \Delta$ for any $x \in M$.

Statement B. The proof of Lemma **4.13.2** is completed by setting $f_A = f_A'' f_A'$.

Proof of Statement B Assuming Statement A. It must be shown that if $p \in I^n$ then

$$h(p) \in f_A'' f_A'(g_A(A \times [0, 1))) \cup f_B(g_B(B \times [0, 1))).$$

If $p \in X_1$, then this follows from Statement A (b). Now suppose $p \in I^n - X_1$. Then it follows from the definition of X that

$$h(p) \in g_A(A \times [0, 1 - a/2]) \cup g_B(B \times [0, 1 - a/2]).$$

Case 1. ($h(p) \in g_B(B \times [0, 1 - a/2])$.) Since

$$f_B \mid g_B(B \times [0, 1 - a/2]) = 1,$$

it follows that

$$h(p) \in f_B(g_B(B \times [0, 1)))$$

and this case is immediate.

Case 2. ($h(p) \in g_A(A \times [0, 1 - a/2])$.) The sequence of facts

(a) $f_A' \mid g_A(A \times [0, 1 - a/4]) = 1$.

(b) $N_\Delta(g_A(A \times [0, 1 - a/2]), M) \subset g_A(A \times [0, 1 - a/4])$, and

(c) $\text{dist}(f_A''(x), x) < \Delta$ for $x \in M$

imply that $h(p) \in f_A'' f_A'(g_A(A \times [0, 1)))$. This completes the proof of Statement *B*.

Proof of Statement A. This proof is just an application of what has now become the old familiar trick of Stallings first given in the proof of Theorem **4.4.1**. (This is the fifth time that we have used this trick in this chapter, for it was used in the proof of Theorem **4.4.1**, Lemma **4.5.5**, Theorem **4.8.1** and Lemma **4.11.2**. We will use it again in Section **5.2**.)

Proof of Theorem 4.13.1. Let each of $h_1, h_2, \ldots, h_k \colon E^n \to \operatorname{Int} M$ be a topological embedding with

$$\bigcup_{1 \leqslant i \leqslant k} h_i(I^n) \cup g_A(A \times [0, 1-b]) \cup g_B(B \times [0, 1-b]) = M.$$

INDUCTIVE HYPOTHESIS (i) = IH(i), $i = 1, 2, \ldots, k$. There exist homeomorphisms $f_A{}^i$ and $f_B{}^i \colon M \to M$ such that each of $f_A{}^i$ and $f_B{}^i$ restricted to

$$g_A(A \times [0, 1-b]) \cup g_B(B \times [0, 1-b])$$

equals the identity, and

$$f_A{}^i(g_A(A \times [0, 1))) \cup f_B{}^i(g_B(B \times [0, 1))) \supset \bigcup_{1 \leqslant t \leqslant i} h_t(I^n).$$

The proof involves showing IH(k) is true and setting $f_A = f_A{}^k$ and $f_B = f_B{}^k$. IH(1) follows immediately from Lemma **4.13.2**. Suppose IH(i) is true for some i, $1 \leqslant i \leqslant k$, and show IH($i+1$) true. The collars of Lemma **4.13.2** will be

$$f_A{}^i g_A(A \times [0, 1]) \subset M - g_B(B \times [0, 1-b]) = M - f_B{}^i g_B(B \times [0, 1-b])$$

and

$$f_B{}^i g_B(B \times [0, 1]) \subset M - g_A(A \times [0, 1-b]) = M - f_A{}^i g_A(A \times [0, 1-b]).$$

Now there is a number a, $0 < a < b$ with

$$f_A{}^i g_A(A \times [0, 1-a]) \cup f_B{}^i g_B(B \times [0, 1-a]) \supset \bigcup_{1 \leqslant t \leqslant i} h_t(I^n).$$

By Lemma **4.13.2**, there exist homeomorphisms α_A and $\alpha_B \colon M \to M$ with

$$\alpha_A \mid f_A{}^i g_A(A \times [0, 1-a]) \cup f_B{}^i g_B(B \times [0, 1-b]) = 1,$$

$$\alpha_B \mid f_B{}^i g_B(B \times [0, 1-a]) \cup f_A{}^i g_A(A \times [0, 1-b]) = 1,$$

and

$$\alpha_A f_A{}^i g_A(A \times [0, 1)) \cup \alpha_B f_B{}^i g_B(B \times [0, 1)) \supset h_{i+1}(I^n).$$

The induction is completed by setting

$$f_A^{i+1} = \alpha_A f_A{}^i \colon M \to M \qquad \text{and} \qquad f_B^{i+1} = \alpha_B f_B{}^i \colon M \to M.$$

This completes the proof of the first part of Theorem **4.13.1**.

Note that $f_B^{-1}f_A\colon M \to M$ satisfies

$$f_B^{-1}f_A \mid g_A(A \times [0, 1 - b]) = 1$$

and

$$f_B^{-1}f_A(g_A(A \times [0, 1))) \cup g_B(B \times [0, 1)) = M.$$

Thus, the existence of the homeomorphism $H\colon g_A(A \times [0, 1)) \to M - B$ follows by making a countable number of applications of the first part of the theorem.

EXERCISE 4.13.2. Show how to rigorously define the homeomorphism H just mentioned.

4.14. INFINITE ENGULFING

Upon a moment's reflection, one sees that the direct analog of Stallings' Engulfing Theorem **4.2.1** does not hold for infinite polyhedra. For instance, if we take the ambient manifold M^n to be E^4 and the open set U to be the open unit ball in E^4, then we cannot engulf E^1_+ with U. However, engulfing theorems for infinite polyhedra can be proved. In particular, Černavskiĭ [5, 6] established specialized engulfing theorems for infinite polyhedra and used them to prove some very interesting results. The purpose of this section is to prove a corollary (Theorem **4.14.1**) of a general infinite engulfing theorem stated at the end of this section. A special case of Theorem **4.14.1** will suffice for our development in Section **5.2** of a version of some work of Černavskiĭ. (We will also mention in Section **5.2** another result of Černavskiĭ which requires the more general infinite engulfing theorem stated at the end of this section.) Although the statement of Theorem **4.14.1** is more general than Lemma 4.1 of [Cantrell and Lacher, 1], the proof of Theorem **4.14.1** given here follows their modification of techniques of [Černavskiĭ, 6] very closely. After noticing that Cantrell and Lacher's proof generalized to prove Theorem **4.14.1**, we saw how to construct a classical type proof, which uses Theorem **4.14.1**, of a generalization (Theorem **4.14.2**) of "the basic lemma" of [Černavskiĭ, 6].

Before stating our first infinite engulfing theorem, let us make a couple of definitions. Let P and U be subspaces of a metric space M. Then we say that P **tends** to U if given $\epsilon > 0$ there is a compact subset C of P such that $\operatorname{dist}(x, U) < \epsilon$ for any $x \in P - C$. We will call a manifold M **uniformly locally *p*-connected** (p-ULC) if given $\epsilon > 0$ there is a $\delta > 0$ for which every mapping $f\colon S^p \to M$ such that $\operatorname{diam} f(S^p) < \delta$ is null-homotopic through a homotopy whose track has diameter less than ϵ. If M is p-ULC for $0 \leqslant p \leqslant k$, then we say that M is $\mathbf{ULC}^k$.

Weak Infinite Engulfing Theorem 4.14.1. *Let M^n be connected* PL *n-manifold without boundary, U an open set in M, P^k a (possibly infinite) polyhedron of dimension k at most $n - 3$ which is contained in M^n (P not necessarily closed in M^n) and $Q \subset U$ a (possibly infinite) polyhedron of dimension at most $n - 3$ such that $(\bar{Q} - Q) \cap P = \emptyset$ and $(\bar{P} - P) \cap Q = \emptyset$. We let $\tilde{\infty}$ denote a closed subset of $M - (P \cup Q)$ containing $(\bar{P} - P) \cup (\bar{Q} - Q)$. Suppose that P tends to U, that $U - \tilde{\infty}$ is* ULC^{k-1} *and that $M^n - \tilde{\infty}$ is* ULC^k. *Then, given a compact subset C of P and given $\delta > 0$, there exists an ambient ϵ-isotopy e_t of M^n such that $e_0 = 1$, such that*

$$e_t \,|(M - N_\epsilon(P - C, M)) \cup Q = 1 \,|(M - N_\epsilon(P - C, M)) \cup Q$$

and such that $e_1(U)$ contains all of P except some compact subset. Furthermore, for each $\delta > 0$, there exists a compact subset K of $M - \tilde{\infty}$ such that $e_t \mid M - K$ is a δ-isotopy. (See Fig. **4.14.1.**)

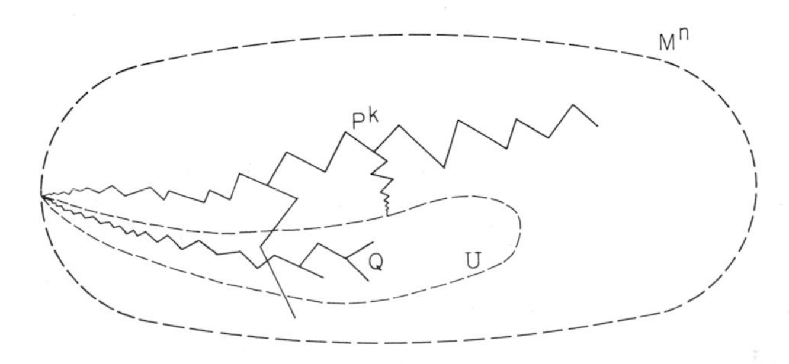

Figure 4.14.1

By using the above infinite engulfing theorem and Stallings' engulfing theorem, we easily obtain the following corollary.

Corollary 4.14.1. *Assume the hypotheses of the previous theorem and in addition assume that the pair $(M - \tilde{\infty}), U - \tilde{\infty})$ is k-connected. Then, there is an ambient isotopy e_t of M such that $P \subset e_1(U)$ and $e_t \mid Q = 1 \mid Q$. Furthermore, for each $\delta > 0$, there exists a compact subset K of $M - \tilde{\infty}$ such that $e_t \mid M - K$ is a δ-isotopy.*

Remark 4.14.1. Note that $(\bar{P} - P) \cup (\bar{Q} - Q)$ is a closed subset of U because P and Q have locally finite triangulations.

Proof of Theorem 4.14.1. The proof of the theorem will be by induction on k, the dimension of P.

In the case $k = 0$, P is a countable set of points. Let $P = \{x_1, x_2, ...\}$. (Ignore points of P on Q.) Notice that $\text{dist}(x_i, U)$ tends to zero as i increases. It is easy to construct a sequence $J_1, J_2, ...$ of polygonal arcs such that

(1) The J_i are pairwise disjoint and all lie in $M - \widetilde{\infty}$: no J_i intersects Q,
(2) One end-point of J_i is x_i and the other lies in U, and
(3) $\text{diam } J_i \to 0$ as $i \to \infty$.

(By our codimension restriction, $n \geqslant 3$, and so (1) is possible by general position.) Let j be an integer such that $\text{diam } J_i < \epsilon/2$ and $x_i \notin C$ whenever $i > j$; and for each $i > j$, let N_i be a regular neighborhood of J_i such that $\text{diam } N_i < 2 \text{ diam } J_i$. The N_i are chosen pairwise disjoint, and each N_i is in $M - (\widetilde{\infty} \cup Q)$. For each $i > j$ there is an ϵ-homeomorphism h_i of N_i onto itself which is the identity on Bd N_i and which maps $N_i \cap U$ onto a set which contains x_i. Since $\text{diam } N_i \to 0$ as $i \to \infty$, we define h to be h_i on N_i for $i > j$, and let h be the identity otherwise.

To do the inductive step, we assume Theorem **4.14.1** for $k < p \leqslant n - 3$ and show that Theorem **4.14.1** is true for $k = p$. The inductive step is much more complicated than the above argument for $k = 0$; however, the ideas are essentially the same. We use standard engulfing techniques, and are able to engulf infinitely many simplexes one at a time by using a method similar to the above one.

In beginning the proof of Theorem **4.14.1** for the case $k = p$, we are given P^p, Q, U, C and $\epsilon > 0$. Let T be a triangulation of $P \cup Q$ such that any simplex of T which intersects Q lies in U. (We may also assume that the diameters of the simplexes of T tend to 0.) Now apply our inductive assumption for $k = p - 1$ where

$$(U, |\, T^{p-1} \,|, Q \cup \{\sigma \in T \mid \sigma \subset U\}, \epsilon/4)$$

corresponds to (U, P, Q, ϵ). Then, by induction there exists an $\epsilon/4$-isotopy e_t^1 of M such that $e_0^1 = 1$, such that

$$e_t^1 \mid (M - N_{\epsilon/4}(|\, T^{p-1} \,| - C, M) \cup Q \cup \{\sigma \in T \mid \sigma \subset U\} = 1$$

and such that $e_1^1(U)$ contains all of $|\, T^{p-1} \,|$ except some compact subset. Also, for each $\delta > 0$, there exists a compact subset K of $M - \widetilde{\infty}$ such that $e_t^1 \mid M - K$ is a δ-isotopy. Moreover, any simplex of T which intersects Q must lie in $e_1^1(U)$.

Let $\sigma_1, \sigma_2, ...$ be the p-simplexes of T which do not lie in $e_1^1(U)$.

No σ_i intersects Q. By "throwing away" a finite number of simplexes of T, we can assume that, for each i, $\mathrm{Bd}\,\sigma_i \subset e_1{}^1(U)$ and that $\sigma_i \cap C = \emptyset$.

Consider $M \times [0, 1]$ and identify M with $M \times 0$ (see Fig. **4.14.2**). For each i, let $v_i \in \hat{\sigma}_i \times [0, 1]$, where $\hat{\sigma}_i$ is the barycenter of σ_i. Let $\Delta_i = \sigma_i * v_i$, the join of σ_i and v_i in $\sigma_i \times [0, 1]$. The v_i are chosen so

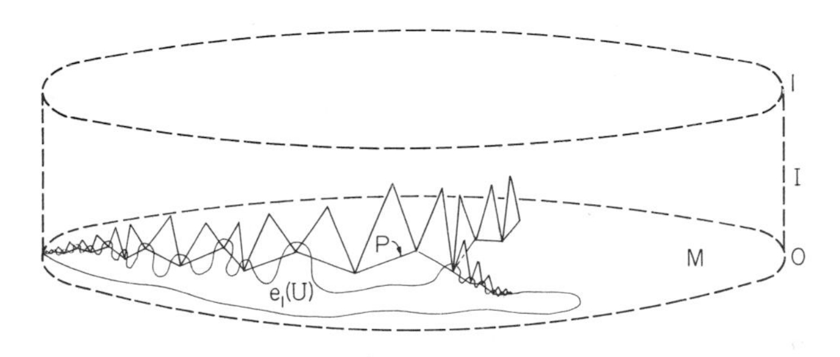

Figure 4.14.2

that the Δ_i are $(p + 1)$-simplexes with diameters tending to zero as i increases. (This is not necessary, but it makes a good picture.) Notice that the sets $\Delta_i - \sigma_i$ are pairwise disjoint.

Let X be the polyhedron $P \cup Q \cup \bigcup_{i=1}^{\infty} \Delta_i$. We want to find a PL map f of X into $M - \tilde{\infty}$ with the following properties:

(1) $f \mid P \cup Q =$ inclusion

(2) $f(v_i * \mathrm{Bd}\,\sigma_i) \subset e_1{}^1(U)$ for large i, and

(3) $\operatorname{diam} f(\Delta_i) \to 0$ as $i \to \infty$.

It is easy to construct f by letting f be the inclusion on $P \cup Q$ and extending f first over each $v_i * \mathrm{Bd}\,\sigma_i$ and then over each Δ_i. In extending f we use the fact that $e_1{}^1(U - \tilde{\infty})$ is ULC^{k-1} and that $M - \tilde{\infty}$ is ULC^k. By applying Theorem **1.6.11** we may throw f into general position staying fixed on $P \cup Q$.

Suppose first that $p \leqslant n - 4$, and let $S(f)$ denote the singular set of f as a map of X into M. Let $S_i(f) = S(f) \cap \Delta_i$, and let Σ_i be the union of all line segments in Δ_i which intersect $S_i(f)$ and are perpendicular to σ_i. Then, $\dim S(f) \leqslant p - 2$, so that $\dim \Sigma_i \leqslant p - 1$. Hence, we can apply the inductive assumption to engulf the $f(\Sigma_i)$ by letting $(e_1{}^1(U),\ \bigcup_i f(\Sigma_i),\ \{\sigma \in T \mid \sigma \subset e_1{}^1(U)\} \cup f(v_i * \mathrm{Bd}\,\sigma_i)_{i\,\mathrm{large}},\ \epsilon/4)$ correspond to (U, P, Q, ϵ). It follows that there is an $\epsilon/4$-isotopy $e_t{}^2$ of M such that $e_0{}^2 = 1$, such that $e_t{}^2 \mid Q = 1$ and such that $e_1{}^2 e_1{}^1(U)$ contains all but a finite part of $|T^{p-1}| \cup \bigcup_i (f(\Sigma_i)) \cup \{\sigma \in T \mid \sigma \subset e_1{}^1(U)\} \cup$

$f(v_i * \text{Bd}\,\sigma_i)_{i\text{large}}$. Again by "throwing away" a few simplexes of $T \mid P$ if necessary, we may assume that $e_1{}^2e_1{}^1(U)$ contains all of $Q \cup \mid T^{p-1} \mid \cup \bigcup_i (f(\Sigma_i))$. If in obtaining $e_t{}^2$ we do not worry about engulfing the $f(\Sigma_i)$ corresponding to simplexes σ_i which hit C and we only engulf $f(\Sigma_i)$ of small diameter, then $e_t{}^2$ has the additional property that $e_t{}^2 \mid M - N_{\epsilon/2}(P - C, M) = 1$. Finally, $e_t{}^2$ has the property that for any $\delta > 0$ there is a compact subset K of $M - \widetilde{\infty}$ such that $e_t{}^2 \mid M - K$ is a δ-isotopy.

For each i, choose a regular neighborhood N_i of $\Sigma_i \cup v_i * \text{Bd}\,\sigma_i$ in Δ_i such that $f(N_i) \subset e_1{}^2e_1{}^1(U)$ and such that Δ_i collapses to N_i. Since N_i is a neighborhood of $S_i(f)$, $f(\Delta_i)$ collapses to $f(N_i)$. Moreover, we can find open sets U_i in $M - \widetilde{\infty}$ such that diam $U_i \to 0$ and $U_i \cap U_j = \emptyset$ for $i \neq j$, and such that $f(\text{Cl}(\Delta_t - N_i)) \subset U_i$. Since the collapse of $f(\Delta_i)$ onto $f(N_i)$ takes place in U_i, we can use Exercise **1.6.12** to expand $U_i \cap e_1{}^2e_1{}^1(U)$ in U_i to contain $f(\Delta_i)$ without moving points outside of U_i, in such a way that the image of $U_i \cap e_1{}^2e_1{}^1(U)$ contains $P \cup Q \cup f(\Delta_i)$. Let $e_t{}^3$ be the isotopy which is the composition of these moves, except do not include those corresponding to integers i for which diam $U_i > \epsilon/4$. Let $e_t = e_t{}^3e_t{}^2e_t{}^1$. It is easily verified that e_t is the desired isotopy.

Now consider the case $p = n - 3$. If we construct Σ_i as before, then the best estimate for dim Σ_i by general position is dim $\Sigma_i \leqslant n - 3$ so that the inductive hypothesis does not apply. However, the following lemma is exactly what is needed.

Infinite Piping Lemma 4.14.1. *For each i, there exists a map $\bar{f}_i\colon \Delta_i \to M - \widetilde{\infty}$, homotopic to f_i keeping $\sigma_i \cup v_i * \text{Bd}\,\sigma_i$ fixed and a subpolyhedron $J_i \subset \Delta_i$, such that*

(1) *If $\bar{f} = \bigcup_i \bar{f}_i$ then $S_i(\bar{f}) \subset J_i$,*

(2) dim $J_i \leqslant n - 4$,

(3) *Δ_i collapses to arbitrarily small regular neighborhoods of $v_i * \text{Bd}\,\sigma_i \cup J_i$ in Δ_i, and*

(4) diam$(\bar{f}_i(J_i)) \to 0$ *as $i \to \infty$.*

It is easy to complete the proof of Theorem **4.14.1** using piping Lemma **4.14.1** by essentially following along the proof of the previous case $p \leqslant n - 4$.

Remark 4.14.2. Recall that an intuitive discussion of piping was given in the proof of Lemma **4.6.1**. A rigorous development of finite piping is given in [Zeeman, 1]. The only difference in proving piping Lemma **4.14.1** and the finite piping lemma occurs when one attempts to construct all of the pipes. In the finite case all of the pipes can be constructed simultaneously, however, in the infinite case we have to construct the pipes in stages, corresponding to the sets Δ_i. One should note that the singularities of $\bar{f}$ which are formed by an intersection of $\bar{f}(\Delta_i - \sigma_i)$ with $P \cup Q$ lie in the $(n-5)$-skeleton of some triangulation of $S_i(\bar{f})$, so that there is no need to "pipe" around singularities of this type. A more general infinite piping Lemma than the above is given in [Rushing, 10].

Before stating the next infinite engulfing theorem which is proved in [Rushing, 10] and which is mentioned at the beginning of this section, let us make a definition.

Let M^n be a PL n-manifold, U an open subset of M, and P^k an infinite polyhedron in M. We say that **most of P can be pulled through M into U by a short homotopy** $H: (P - A) \times I \to M$, Cl A a compact subset of P, if

(1) $H(p, 0) = p$ for all $p \in P - A$.

(2) $H(p, 1) \in U$ for all $p \in P - A$, and

(3) given $\epsilon > 0$, there is a compact set $B \subset P$ such that $\operatorname{diam}(H(p \times I)) < \epsilon$ for all $p \in P - B$.

Strong Infinite Engulfing Theorem 4.14.2. *Suppose M, U, P, Q and $\tilde{\infty}$ are defined as in Theorem* **4.14.1** *except this time only assume that $M - \tilde{\infty}$ is ULC$^{\max(k,q)+k-n+2}$ and that $U - \tilde{\infty}$ is ULC$^{\max(k,q)+k-n+1}$, but also assume that most of P^k can be pulled through $M - \tilde{\infty}$ into $U - \tilde{\infty}$ by a short homotopy. Then, most of P^k can be engulfed by U in the sense of the conclusion of Theorem* **4.14.1**.

Remark 4.14.3. Bing noticed that Černavskiĭ's basic engulfing lemma can be improved by one codimension (see Remark 4 of [Bing, 11]). When we questioned him about this, he communicated a clever technique of proof which involves building a "more sophisticated shadow." It turns out that his technique will suffice to replace infinite piping in our proof of Theorem **4.12.2**.

Remark 4.14.4. The proof of Theorem **4.14.1** could be performed in a more classical way by constructing a short homotopy rather than the mapping of the Δ_i's given. However we preferred to follow the proof of Lemma 4.1 of [Cantrell and Lacher, 1] as closely as possible to show that our proof is a direct adaptation of their proof.

CHAPTER **5**

Taming and PL Approximating Embeddings

5.1. INTRODUCTION

Some important recent developments in the area of topological embeddings are presented in this last chapter. Also, many references are given to related work appearing in the literature. (See, for instance, the opening remarks of Sections **5.4–5.6**.) Section **5.2** is devoted to the development of, and applications of, a straightening technique due to Černavskiĭ. Our presentation of Černavskiĭ's technique incorporates modifications due to Cantrell and Lacher as well as modifications due to this author. In Section **5.3**, Černavskiĭ's technique is applied in the development of some work of this author on the taming of embeddings of PL manifolds around the boundary in all codimensions. Discussion and work on a topological embedding problem with many important implications is given in Section **5.4**, that is, the problem of PL approximating topological embeddings. The approximation technique expounded is due to Homma. We feel that Section **5.4** allows one to understand Homma's technique without undue agony. Results of the previous two sections are used in Section **5.5** to prove ϵ-taming theorems due to this author. Finally, Section **5.6** is devoted to work of Černavskiĭ, R. D. Edwards and Kirby on codimension zero taming and on local contractibility of the homeomorphism group of a manifold.

5.2. ČERNAVSKIĬ'S STRAIGHTENING TECHNIQUE APPLIED TO CELL PAIRS AND TO SINGULAR POINTS OF TOPOLOGICAL EMBEDDINGS

In this section we return to investigate a problem which was formulated as the β-statement in Sections **2.7** and **3.9**. Counterexamples to the β-statement were given for certain special cases in Section **2.7**. Most of the work in this section is based on [Černavskiĭ, 1, 5, 6], [Cantrell, 3], and [Cantrell and Lacher, 1]. For references to other work on the β-statement and for discussions of that statement see Sections **2.7** and **3.9**.

Before stating the main results of this section, let us make a few preliminary definitions. An embedding $h: M^{n-2} \to N^n$ of an $(n-2)$-manifold M into the n-manifold N is called **1-ALG** (Abelian local fundamental group) at $x \in M$ if for each neighborhood U of $h(x)$ in N there exists a neighborhood $V \subset U$ of $h(x)$ such that every mapping of S^1 into $V - h(M)$ which is null-homologous in $U - h(M)$ is null-homotopic in $U - h(M)$. (This property was defined in [Harrold, 1].) The next definition generalizes the notion of 1-LC given in Section **2.6**. If M^k is a k-submanifold of the n-manifold N^n, then $N - M$ is said to be **p-LC** (locally p-connected) at $x \in M$ if each neighborhood U of x contains a neighborhood V of x such that every mapping of S^p into $V - M$ is null-homotopic in $U - M$. An embedding $h: M^{n-2} \to N^n$ of an $(n-2)$-manifold M into an n-manifold N is said to be **locally homotopically unknotted** at $x \in M$ if h is 1-ALG at x and if $N - h(M)$ is p-LC at $h(x)$ for $2 \leqslant p \leqslant [n/2]$. We shall also say that a submanifold $M \subset N$ is **locally homotopically unknotted** at $x \in M$ if the inclusion embedding of M into N is locally homotopically unknotted at x.

We are now ready to state our first main result which has many of the β-statements as corollaries. (In this section, we let F be the k-dimensional hyperplane of E^n determined by $x_k = \cdots = x_{n-1} = 0$. We denote by B^{k-1} the closed unit ball in E^{n-1} and by B^k the closed unit ball in F. Finally $F_+ = F \cap E^n_+$, $F_- = F \cap E^n_-$, $B^k_+ = B^k \cap F_+$ and $B^k_- = B^k \cap F_-$; see Fig. **5.2.1**.)

Theorem 5.2.1. *Suppose that $h: B^k \to E^n$, $n \geqslant 5$, is a topological embedding such that $h \mid B^k_-$ and $h \mid (B^k_+ - B^{k-1})$ are locally flat. If $k = n-2$, suppose further that h is locally homotopically unknotted at the points of* Int B^{k-1}. *Then, h is locally flat.*

Corollary 5.2.1. *$\beta(n, k, k, k-1)$ is true for $n \geqslant 5$, $k \neq n-2$.*

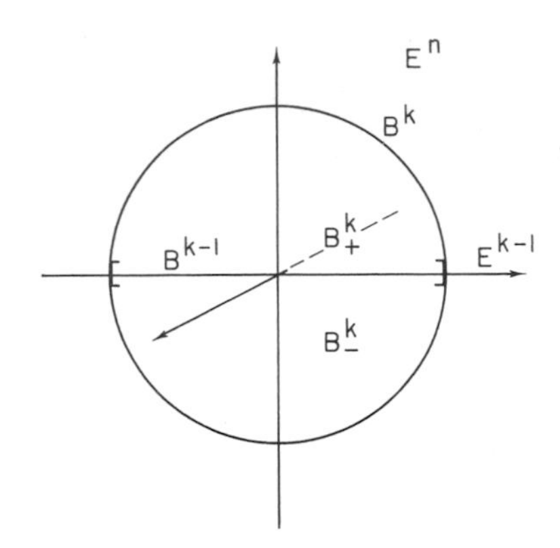

Figure 5.2.1

REMARK 5.2.1. Notice that for $2k + 2 \leqslant n$, Theorem **5.2.1** is a special case of Theorem **3.8.1**.

REMARK 5.2.2. Notice that Theorem **2.7.1** shows the necessity of the requirement that h be locally homotopically unknotted at points of Int B^{k-1} when $k = n - 2$.

Before beginning the proof of Theorem **5.2.1**, let us state and prove the second main result (Theorem **5.2.2**) of this section. The proof of Theorem **5.2.2** uses Theorem **5.2.1** in a fundamental manner and its origin is due to Cantrell [3]. If $h: M^k \to N^n$ is an embedding of a k-manifold M^k into an n-manifold N, then we say that the point $x \in M$ is a **singular point** of h if it has a neighborhood V in M such that h is locally flat at all points of $V - x$, and also possibly at x.

Theorem 5.2.2. *Suppose that $h: M^k \to E^n$ is an embedding of a k-dimensional manifold, possibly with boundary, into n-dimensional Euclidean space.*

(a) *If $n \geqslant 5$, $k \neq n - 2$, then h is locally flat at each singular point x of* Int M.

(b) *If $n \geqslant 4$, then h is locally flat at each singular point x of* Bd M.

(c) *If $n \geqslant 5$, $k = n - 2$, then h is locally flat at a singular point $x \in$* Int M *if and only if h is locally homotopically unknotted at x.*

PROOF. Part (b) is a special case of Corollary **3.4.3**. Hence, we may assume that our singular point x is in Int M. Then, let D be a locally flat k-cell neighborhood of x in Int M. We may express D as the union of two k-cells D_+ and D_- so that (D, D_+, D_-) is homeomorphic to (B^k, B^k_+, B^k_-) and so that $x \in D_+ \cap D_-$. Then, $h \mid D_+$ and $h \mid D_-$ are locally flat by Corollary **3.4.2** and so $h \mid D$ is locally flat at x by Theorem **5.2.1**.

REMARK 5.2.3. A proof of $\beta(n, n-1, n-1, n-2)$ is given in [Černavskiĭ, 4] which is different from the proof of this section. That proof is good for $n \geqslant 4$, and so it follows from the above proof that an embedding of an $(n-1)$-manifold into an n-manifold is locally flat at each singular point when $n \geqslant 4$. A proof of $\beta(n, n-1, n-1, n-2)$ is also given in [Kirby, 2].

We will postpone the proof of the following lemma until after the proof of Theorem **5.2.1**.

Engulfing Lemma 5.2.1. *Suppose that h: $E^n_+ \to E^n_+$ is a closed embedding which takes E^{k-1} into itself such that $h \mid N = 1$ for some neighborhood N of B^{k-1} in E^{k-1} and such that $h(E^n_+ - E^{k-1})$ is in* Int E^n_+. *Suppose that $n \geqslant 5$ and that $h(F_+) \cup F_-$ is locally homotopically unknotted at each point of N for $k = n-2$. Then, for any $\epsilon > 0$, there is an ϵ-homeomorphism f: $E^n \to E^n$ which is the identity outside the ϵ-neighborhood of B^{k-1} and on E^n_- such that $f(h(\text{Int } E^n_+) \cup B^{k-1})$ contains a neighborhood of B^{k-1} relative to B^k_+.*

The rest of this section will be organized as follows: First we will give several definitions. Next we will state and prove two lemmas preliminary to the proof of Theorem **5.2.1**. Then, we will prove Theorem **5.2.1**. Finally, we will prove Engulfing Lemma **5.2.1**.

The Planes π_α. Define a collection $\{\pi_\alpha\}$ of 2-dimensional hyperplanes filling up E^n as follows. For each point p in $E^n - F$, let p' be the (orthogonal) projection of p in F, and let L_p be the line in F through p' and orthogonal to E^{k-1}. The 2-dimensional hyperplane in E^n spanned by p and L_p for p in $E^n - F$ make up the collection $\{\pi_\alpha\}$. The collection $\{\pi_\alpha\}$ is indexed so that if $\alpha \neq \beta$, then $\pi_\alpha \neq \pi_\beta$.

It will be convenient to use a coordinate system for π_α in which the corresponding L_p is the abscissa and the origin 0_α is the point $L_p \cap E^{k-1}$. The ordinate axis in π_α is $\pi_\alpha \cap E^{n-1}$. The unit distance in π_α will be specified later.

Horns H_t. Let $\tilde{E}^{n-k+1}$ be the hyperplane of E^n determined by $x_1 = \cdots = x_{k-1} = 0$. For each real number t, let C_t be the "cone" in $\tilde{E}^{n-k+1}$ determined by the equation

$$x_n = t(x_k^2 + \cdots + x_{n-1}^2)^{1/2}$$

and let $H_t = E^{k-1} \times C_t \subset E^n$ (see Fig. **5.2.2**). The set H_t is called the **horn with coefficient** t. Notice that H_t is a closed copy of E^{n-1} in E^n.

For each t, the horn H_t intersects each plane π_α in a (topological) line containing 0_α, and 0_α divides this line into two (straight) rays each of which forms with the abscissa axis in π_α the angle arc cot t. These two

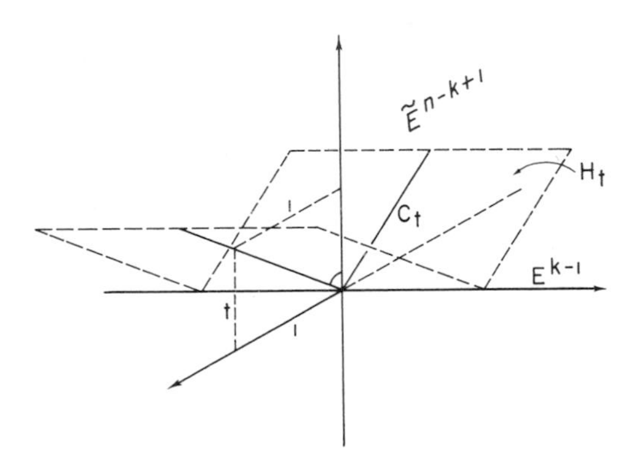

Figure 5.2.2

straight rays forming $H_t \cap \pi_\alpha$ lie in the first and fourth quadrants of π_α if $t > 0$ and in the second and third quadrants if $t < 0$. (For the case $t = 0$, $H_0 = E^{n-1}$, so that $H_0 \cap \pi_\alpha$ is the ordinate axis in π_α.)

Given t and t', then, by defining an appropriate homeomorphism on each plane π_α, one can construct a homeomorphism on E^n that is the identity on E^{k-1} and carries H_t onto $H_{t'}$. In particular, each H_t is a locally flat copy of E^{n-1} in E^n and divides E^n into two closed half-spaces; the one containing the "positive x_n-axis" is denoted by Q_t^+ and the other by Q_t^-. The closure of the region between H_t and $H_{t'}$ is denoted by $Q(t, t')$.

Tangentiality. A set X in E^n is said to be **tangential to** H_t if for each $\epsilon > 0$ there is a neighborhood N of Int B^{k-1} in E^n such that the intersection of X and N is nonempty and is contained in $Q(t - \epsilon, t + \epsilon)$. If further we always have $X \cap N \subset Q(t + \epsilon, t)$, we say that X is **tangential to** H_t **on the positive side**. Tangentiality on the negative side is defined similarly.

The Unit Length in π_α and the Pentagons P_α. We will define the unit length in π_α relative to a fixed point p on the positive x_n-axis. Let K_p be the join of p and B^{k-1}. For each point q in the join of p and Bd B^{k-1}, let I_q be the segment from q to the projection of q in B^{k-1}. For each plane

π_α that intersects Int B^{k-1}, there is a point q such that $I_q \subset \pi_\alpha$. We let I_q be the **unit length** in π_α. Notice that I_q lies on the abscissa of π_α, and that the "unit length in π_α" tends to zero as "π_α tends to Bd B^{k-1}." We will not need a unit length in those planes π_α which do not intersect Int B^{k-1}. If π_α intersects Int B^{k-1}, we define P_α to be the pentagon in π_α with vertices

$$(0, 1)_\alpha, \qquad (1, 1)_\alpha, \qquad (2, 0)_\alpha, \qquad (1, -1)_\alpha, \qquad \text{and} \qquad (0, -1)_\alpha.$$

Expansion Lemma 5.2.2. *Let h be an embedding of E^n_+ into E^n which takes E^{k-1} into itself such that $h \mid B^{k-1} = 1$ and such that for some p on the positive x_n-axis, $h(E^n_+) \supset K_p \cup (\bigcup_{\{\alpha \mid \pi_\alpha \cap \mathrm{Int}\, B^{k-1} \neq \emptyset\}} P_\alpha)$.* (See Fig. **5.2.3**.) *Then, there is a homeomorphism e of $E^n - B^{k-1}$ onto $E^n - K_p$ that is the identity on E^n_- and outside $h(E^n_+)$ which satisfies the following condition:*

A sequence of points $X \subset E^n - K_p$ converges to a point x in K_p if and only if $e^{-1}(X)$ converges to the projection of x in B^{k-1} and is tangential to $H_{t(x)}$, where $t(x)$ is the abscissa coordinate of x relative to the coordinate system in some π_α containing x.

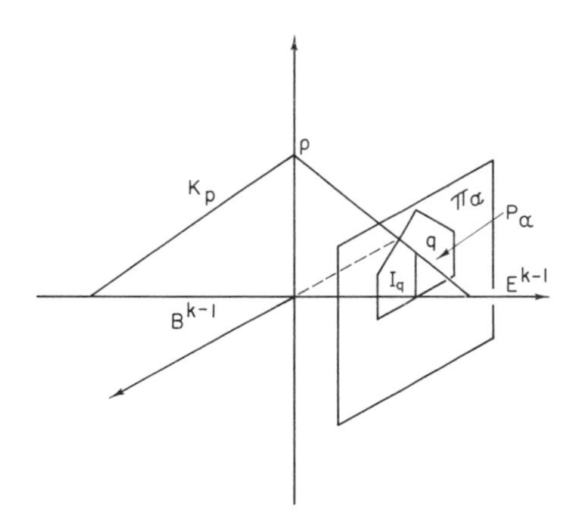

Figure 5.2.3

Proof. In each plane π_α (see Fig. **5.2.4**) we will define a homeomorphism e_α. If $\pi_\alpha \cap \mathrm{Int}\, B^{k-1} = \emptyset$, let $e_\alpha = 1$. If $\pi_\alpha \cap \mathrm{Int}\, B^{k-1} \neq \emptyset$, let e_α be the identity outside of the pentagon P_α, and define e_α on P_α as follows: For a point $r = (t, s)_\alpha$, $0 \leqslant t \leqslant 1$, on Bd P_α, we map the segment $[0_\alpha, r]$ from 0_α to r linearly onto the segment from $(t, 0)_\alpha$ to r and let e_α be the restriction of this map to the half-open interval $(0_\alpha, r]$. (In verifying that our map e has the desired properties, keep in mind that the segment $[0_\alpha, r]$ is contained in $H_t \cap \pi_\alpha$.) If $r = (t, s)_\alpha$, $1 \leqslant t \leqslant 2$,

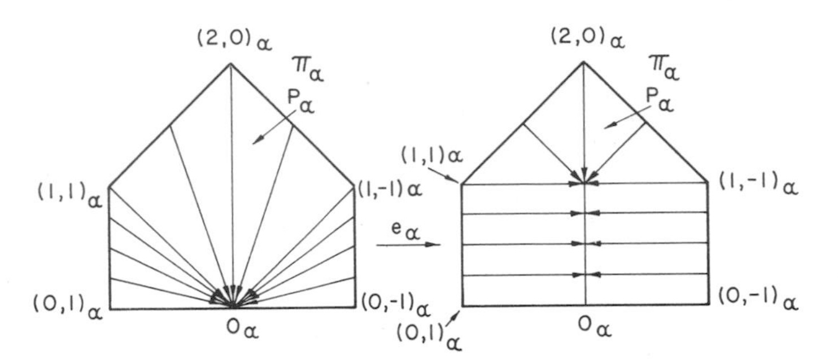

Figure 5.2.4

is on the boundary of P_α, map the segment $[0_\alpha, r]$ linearly onto $[(1, 0)_\alpha, r]$ and let e_α be the restriction of this map to $(0_\alpha, r]$.

Then let $e = e_\alpha$ on π_α. It is a routine matter to verify that e has the desired properties.

The next lemma, which is fundamental to the proof of Theorem **5.2.1**, employs the basic technique of "meshing a straight structure and a wiggly structure" which was probably developed by Brown when he showed that the monotone union of open n-cells is an open n-cell. A relative version of Brown's work was given in Theorem **3.4.2**.

Rearrangement Lemma 5.2.3. *Let h be an embedding of E^n_+ into E^n_+ which takes E^{k-1} into itself such that $h \mid N =$ identity for some closed neighborhood N of B^{k-1} in E^{k-1} and such that $h(E^{n-1} - E^{k-1})$ is in* Int E^n_+. *Suppose that $n \geqslant 5$ and that $h(F_+) \cup F_-$ is locally homotopically unknotted at each point of N for $k = n - 2$. Then, for any $\epsilon > 0$, there is an ϵ-homeomorphism r of E^n onto itself such that r is the identity on E^{k-1} and outside the ϵ-neighborhood of B^{k-1}, and such that for each t, $0 \leqslant t \leqslant 1$, $rh(H_t)$ is tangential to H_t. Also, for an appropriate choice of p we may assume that $rh(\text{Int } E^n_+) \cup B^{k-1} \supset K_p \cup (\bigcup_{\{\alpha \mid \pi_\alpha \cap \text{Int } B^{k-1} = \emptyset\}} P_\alpha)$.*

Proof. Let $t_0, t_1, t_2, \ldots$ be the dyadic rationals in I, $t_0 = 0$, $t_1 = 1$, $t_2 = 1/2$, $t_3 = 1/4$, $t_4 = 3/4$, $t_5 = 1/8$, ... For each i, we will construct a homeomorphism h_i on E^n such that, if $0 < j < 1$, then $(h_i h)\, H_{t_j}$ is tangential to H_{t_j} on the negative side and $(h_i h)^{-1}\, H_{t_j}$ is tangential to H_{t_j} on the positive side. The h_i will be constructed so that $r = \lim h_i$ will be a homeomorphism. It will be apparent that r is an ϵ-homeomorphism that is the identity on E^{k-1} and outside the ϵ-neighborhood of B^{k-1}. Also, it will be clear that $rh(H_{t_i})$ is tangential to H_{t_i} for $i = 0, 1, \ldots$. It is easy to see that if the tangentiality condition holds on a dense subset of I, then it holds on all of I.

Construction of h_0. We are given that $h(F_+) \cup F_-$ is locally homotopically unknotted at each point of N. Let $B_3 \subset \operatorname{Int} B_2 \subset \operatorname{Int} B_1$ be $(k-1)$-cells in N containing B^{k-1} which are concentric with B^{k-1}. (In the case $k-1=0$, $B_i^{k-1} = B^{k-1}$.) Let $K_p{}^i$, $P_\alpha{}^i$, and so forth, be defined in terms of B_i^{n-1}, $i = 1, 2, 3$, analogous to the way K_p, P_α, and so forth, were defined in terms of B^{n-1}.

By Lemma **5.2.1**, there is a homeomorphism e_1 of E^n which is the identity on E_-^n and such that $e_1(h(\operatorname{Int} E_+^n) \cup B_1^{k-1})$ contains some neighbothood of B_1^{k-1} in B_{1+}^k. Then, by properly defining a homeomorphism on each plane $\pi_\alpha{}^1$, one can easily construct a homeomorphism e_2 of E^n such that $e_2 \mid E^{k-1} =$ identity and such that $e_2 e_1(H_0)$ is tangential to H_0 on the negative side relative to B_1. Moreover, e_2 and e_1 can be chosen to be the identity outside of any preassigned neighborhood of B_1^{k-1} (hence, by picking B_1^{k-1} close to B^{k-1}, outside any preassigned neighborhood of B^{k-1}) and for an appropriate p we may construct e_2 such that

$$e_2 e_1(\operatorname{Int} E_+^n) \cup B_1^{k-1} \supset K_p{}^1 \cup \left(\bigcup_{\{\alpha \mid \pi_\alpha{}^1 \cap \operatorname{Int} B_1^{k-1} \neq \emptyset\}} P_\alpha{}^1 \right).$$

(See Fig. **5.2.5**.)

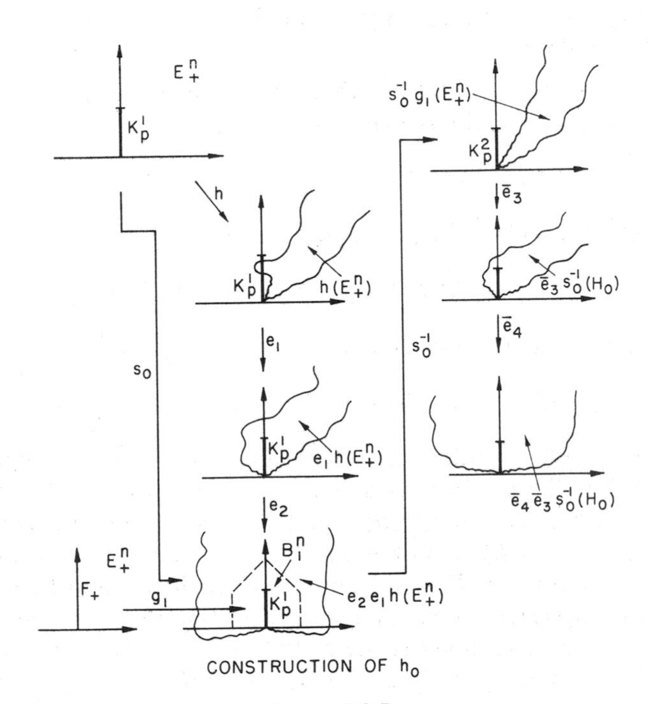

Figure 5.2.5

Let us denote the n-cell $K_p{}^i \cup (\bigcup_{\{\alpha \mid \pi_\alpha{}^i \cap \operatorname{Int} B_i^{k-1} \neq \emptyset\}} P_\alpha{}^i)$ by $A_i{}^n$, $i = 1, 2, 3$, and let $B_1{}^n$ denote Int $A_1{}^n \cup \operatorname{Int}(A_1{}^n \cap E^{n-1})$. Now construct a homeomorphism $g_1: (E_+^n, F_+, E^{k-1}) \twoheadrightarrow (B_1{}^n, B_1{}^n \cap F_+, B_1{}^n \cap E^{k-1})$ such that $g_1 \mid B_2^{k-1} =$ identity. Let $s_0 = e_2 e_1 h$ and notice that $s_0^{-1} g_1$ satisfies the hypothesis of Lemma **5.2.1**. Thus, there is a homeomorphism $\bar{e}_3$ on E^n which is the identity on E_-^n and such that $\bar{e}_3(s_0^{-1} g_1(\operatorname{Int} E_+^n) \cup B_2^{k-1})$ contains some neighborhood of B_2^{k-1} in B_{2+}^k. There is then a homeomorphism $\bar{e}_4$ on E^n such that $\bar{e}_4 \mid E_-^n =$ identity and $\bar{e}_4 \bar{e}_3 s_0^{-1}(H_0)$ is tangential to H_0 on the positive side relative to B_2^{k-1}.

Now we can define h_0. First, we let e_3 be defined on E^n by letting $e_3 =$ identity on $E^n - s_0(E_+^n)$ and $e_3 = s_0 \bar{e}_3^{-1} \bar{e}_4^{-1} s_0^{-1}$ on $s_0(E_+^n)$. Then let $h_0 = e_3 e_2 e_1$. It is clear that $(h_0 h) H_0$ is tangential to H_0 on the negative side and

$$(h_0 h)^{-1}(H_0) = h^{-1} e_1^{-1} e_2^{-1} e_3^{-1}(H_0) = h^{-1} e_1^{-1} e_2^{-1} s_0 \bar{e}_4 \bar{e}_3 s_0^{-1}(H_0)$$

$$= h^{-1} e_1^{-1} e_2^{-1} e_2 e_1 h \bar{e}_4 \bar{e}_3 s_0^{-1}(H_0) = \bar{e}_4 \bar{e}_3 s_0^{-1}(H_0)$$

is tangential to H_0 on the positive side. Moreover, if we are given $\epsilon_0 > 0$, we can construct h_0 such that it is an ϵ_0-homeomorphism of E^n which is the identity outside the ϵ_0-neighborhood of B^{k-1}. Finally, we note that $h_0 \mid B^{k-1} =$ identity.

Construction of h_1. Notice that in the construction of $\bar{e}_4$ in the preceding step, we could assure that

$$A_2{}^n \cap Q_1^+ \subset \bar{e}_4 \bar{e}_3 s_0^{-1}(E_+^n) = (h_0 h)^{-1}(E_+^n).$$

Hence, $h_0 h(A_2{}^n \cap Q_1{}^+) \subset E_+^n$. Let $B_2{}^n$ denote

$$(\operatorname{Int} A_2{}^n \cap \operatorname{Int} Q_1{}^+) \cup \operatorname{Int}(A_2{}^n \cap H_1).$$

Now construct a homeomorphism

$$g_2: (E_+^n, F_+, E^{k-1}) \twoheadrightarrow (B_2{}^n, B_2{}^n \cap F_+, B_2{}^n \cap H_1)$$

such that $g_2 \mid B_3^{k-1} =$ identity. Then, $h_0 h g_2$ satisfies the hypothesis of Lemma **5.2.1**. Thus, there is a homeomorphism e_5 on E^n that is the identity on E_-^n such that $e_5(h_0 h g_2(\operatorname{Int} E_+^n) \cup B_3^{k-1})$ contains some neighborhood of B_3^{k-1} in B_{3+}^k. As before there is a homeomorphism e_6 on E^n that is the identity on E_-^n (and outside a small neighborhood of B_3^{k-1}) such that $e_6 e_5 h_0 h(H_1)$ is tangential to H_1 on the negative side relative to B_3^{k-1}. We may also assume that $e_6 e_5 h_0 h(Q_1{}^+) \supset A_3{}^n \cap Q_1{}^+$. See Fig. **5.2.6**.

Let $B_3{}^n$ denote $(\operatorname{Int} A_3{}^n \cap \operatorname{Int} Q_1{}^+) \cup \operatorname{Int}(A_3{}^n \cap H_1)$. Construct a homeomorphism $g_3: (E_+^n, F_+, E^{k-1}) \twoheadrightarrow (B_3{}^n, B_3{}^n \cap F_+, B_3{}^n \cap H_1)$ such

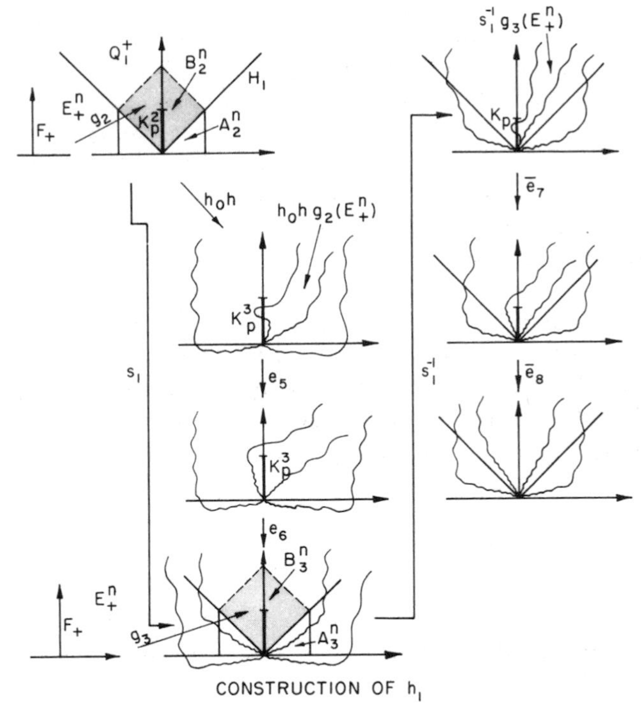

Figure 5.2.6

that $g_3 \mid B^{k-1} =$ identity. Let $s_1 = e_6e_5h_0h$ and notice that $s_1^{-1}g_3$ satisfies the hypothesis of Lemma **5.2.1**. Thus, there is a homeomorphism $\bar{e}_7$ on E^n which is the identity on Q_1^- (in applying Lemma **5.2.1** here, let Q_1^- play the role of E_-^n) such that $\bar{e}_7(s_1^{-1}g_3(\text{Int } E_+^n) \cup B^{k-1})$ contains some neighborhood of B^{k-1} in B_+^k. There is then a homeomorphism $\bar{e}_8$ on E^n that is the identity on Q_1^- and is such that $\bar{e}_8\bar{e}_7s_1^{-1}(H_1)$ is tangential to H_1 on the positive side.

Let e_7 be defined on E^n by $e_7 =$ identity on $E^n - s_1(E_+^n)$ and $e_7 = s_1\bar{e}_7^{-1}\bar{e}_8^{-1}s_1^{-1}$ on $s_1(E_+^n)$. Define $g_1 = e_7e_6e_5$, and let $h_1 = g_1h_0$. Again the tangentiality condition holds and given $\epsilon_1 > 0$, g_1 can be constructed to be an ϵ_1-homeomorphism of E^n onto itself, the identity outside the ϵ_1-neighborhood of B^{k-1}, and the identity on B^{k-1}. (In fact, g_1 is the identity on E_-^n.)

Construction of h_i for $i \geqslant 2$. (Engulfing Lemma **5.2.1** was used in the construction of e_1, $\bar{e}_3$, e_5, and $\bar{e}_7$. We will not need to apply Lemma **5.2.1** again. The construction of h_i for $i \geqslant 2$ is simpler than that of h_0 and h_1.) Suppose that h_{i-1}, $i \geqslant 2$, has been defined and let t_l, t_m be the numbers in $\{t_0, ..., t_{i-1}\}$ that are closest to t_i, with $t_l < t_i < t_m$

(see Fig. **5.2.7**). Consider $(h_{i-1}h)^{-1}$ and let $\bar{e}_9$ be a homeomorphism of E^n such that $\bar{e}_9 =$ identity on $(h_{i-1}h)^{-1}Q^-_{t_i} \cup Q^-_{t_i}$, and such that in a small neighborhood of Int B^{k-1}, $\bar{e}_9(H_{t_m}) \subset (h_{i-1}h)^{-1}Q^+_{t_m}$. Let e_9 be the identity on $Q^-_{t_i}$ and $(h_{i-1}h)\,\bar{e}_9(h_{i-1}h)^{-1}$ on $Q^+_{t_i}$. Then $e_9h_{i-1}h(H_{t_l} \cup H_{t_m})$ is "outside" while $e_9h_{i-1}h(H_{t_i})$ is inside $Q(t_l, t_m)$. Thus one easily constructs a homeomorphism e_{10} that is the identity outside $Q(t_l, t_m)$ and is such that $e_{10}e_9h_{i-1}(H_{t_i})$ is tangential to H_{t_i} on the negative side. To re-establish the tangentiality condition for t_m, we follow $e_{10}e_9h_{i-1}h$ by e_9^{-1}.

If $s_i = e_9^{-1}e_{10}e_9h_{i-1}h$, then, in a small neighborhood of Int B^{k-1}, $s_i^{-1}(H_{t_i})$ is in $Q(t_i, t_m)$. Hence, there exists a homeomorphism $\bar{e}_{11}$ on E^n that is the identity outside $Q(t_i, t_m)$ and is such that $\bar{e}_{11}e_i^{-1}(H_{t_i})$ is tangential to H_{t_i} on the positive side. We then let e_{11} be the identity outside $s_iQ(t_i, t_m)$ and $e_{11} = s_i\bar{e}_{11}^{-1}s_i^{-1}$ on $s_iQ(t_i, t_m)$.

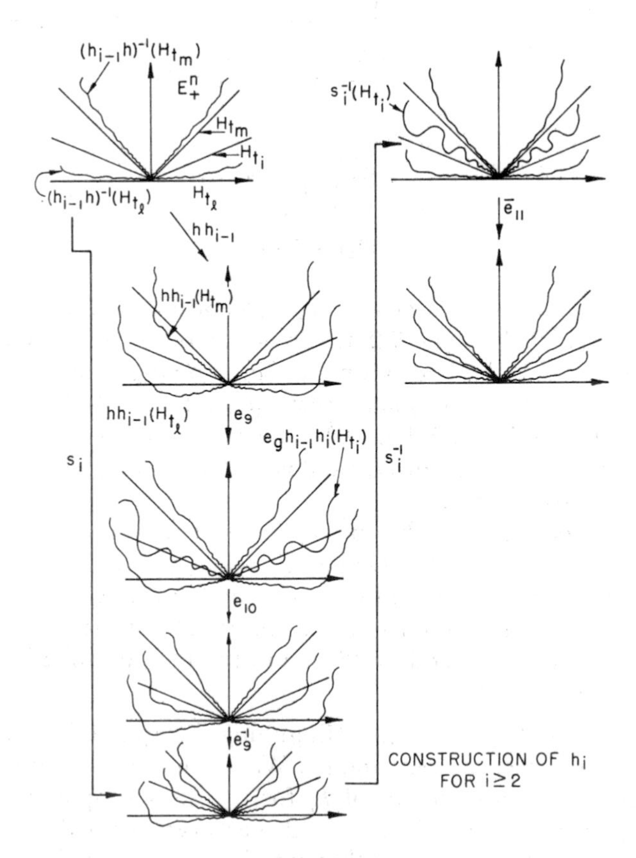

Figure 5.2.7

Finally, let $g_i = e_{11}e_9^{-1}e_{10}e_9$ and $h_i = g_ih_{i-1}$. Again, we observe that, given h_{i-1} and $\epsilon_i > 0$, g_i can be constructed so that it is an ϵ_i-homeomorphism of E^n, and is the identity on B^{k-1} and outside the ϵ_i-neighborhood of B^{k-1}.

Construction of r. If h and ϵ are as in the hypothesis, let $\epsilon_0 = \epsilon/2$, and construct $g_0 = h_0$ as above. When h_{i-1} has been constructed, let V be a δ-neighborhood of B^{k-1}, $\delta \leqslant \epsilon/2^i$, such that $\operatorname{dist}(x, h_{i-1}(x)) < \epsilon/2^{i+1}$ holds for all x in V. Then, choose $\epsilon_i > 0$ to be smaller than $\epsilon/2^{i+1}$ and to satisfy the following property: the closure of the ϵ_i-neighborhood U_i of B^{k-1} is contained in $h_{i-1}(V)$. Then, let g_i be constructed from h_{i-1} and ϵ_i as above, and let $h_i = g_ih_{i-1}$. Let $r = \lim_i h_i$. It is easily checked that r is an ϵ-homeomorphism of E^n onto itself which is the identity on B^{k-1} and outside of the ϵ-neighborhood of B^{k-1}. It is clear from the construction of r that $rh(H_{t_i})$ is tangential to H_{t_i} for $i = 0, 1, 2, \ldots$.

Proof of Theorem 5.2.1. We are given $h\colon B^k \to E^n$ such that $h \mid B^k_-$ is locally flat. It follows from Corollary **3.4.1**, that there is a homeomorphism $h_1\colon E^n \to E^n$ such that h_1h is the identity on B^k_-.

In a standard way construct a neighborhood of $B^k_- - B^{k-1}$ in E^n whose closure is an n-cell D whose boundary is locally flat in E^n and does not intersect $h_1h(B^n_+ - B^{k-1})$, and such that $(D, B^k_-) \approx (S^n_-, B^k_-)$. [Although it is not difficult to construct D in this simple situation, in the next section we will prove a general lemma (Lemma **5.3.3**) which will imply the existence of D as a corollary.] Now it is easy to construct a homeomorphism of the compactified E^n, $h_2\colon E^n \cup \infty \twoheadrightarrow E^n \cup \infty$ which takes the boundary of D onto $H_0 \cup \infty$ and which is the identity on B^k_-. Replacing the embedding h by the embedding h_2h_1h, we see that it is sufficient to consider the case where h is the identity on B^k_- and $h(B^k_+ - B^{k-1})$ lies inside E^n_+.

By using Theorem **3.4.3**, we can construct an n-cell $D_1 \supset h(B^k_+)$ such that $D_1 \subset \operatorname{Int} E^n_+ \cup B^{k-1}$ for which there is an extension of $h \mid B^k_+$ to an embedding of $2B^n_+$ onto D_1, where $2B^n$ is the ball of radius 2 in E^n.

We shall first prove that the embedding h is locally flat at each point of $\operatorname{Int} B^{k-1}$, for example at 0. Therefore we can replace B^k by a smaller concentric ball $\tilde{B}^k$ and thus we can assume that the embedding $h \mid \tilde{B}^k_+$ can be extended to a neighborhood U of $\tilde{B}^k_+$ in E^n_+ such that $h(U - E^{k-1}) \subset \operatorname{Int} E^n_+$. It is now easy to construct a homeomorphic mapping of E^n_+ into this neighborhood which takes E^{n-1} into itself and is the identity on a neighborhood of $\tilde{B}^k_+$ in $2\tilde{B}^k_+$. Then the composition of this mapping and the previous extension of h gives a new extension of h to a homeomorphic mapping h of the whole of E^n_+ into itself which takes E^{k-1} into itself such that $h \mid N =$ identity for some closed neighborhood N of $\tilde{B}^{k-1}$ in E^{k-1} and such that $h(E^{n-1} - E^{k-1})$ is in Int E^n_+. Also, in case $k = n - 2$, $h(F_+) \cup F_-$ is locally homotopically unknotted at each point of some neighborhood of N in E^{k-1}. (Let us denote $\tilde{B}^k$ by B^k from now on.)

From Lemma **5.2.3**, we obtain a homeomorphism r on E^n such that r is the identity on B^{k-1} and, for $0 \leqslant t \leqslant 1$, $rh(H_t)$ is tangential to H_t. It is easy to see that we may assume that r is the identity on B^k_-. Also by Lemma **5.2.3**, for some p on the positive x_n-axis, we may assume that

$$rh(\text{Int } E^n_+) \cup B^{k-1} \supset K_p \cup \Bigl(\bigcup_{\{\alpha | \pi_\alpha \cap \text{Int } B^{k-1} \neq \emptyset\}} P_\alpha \Bigr).$$

Let e be given by Lemma **5.2.2** for the homeomorphism rh. Define $\bar{f}: E^n \twoheadrightarrow E^n$ as follows (see Fig. **5.2.8**):

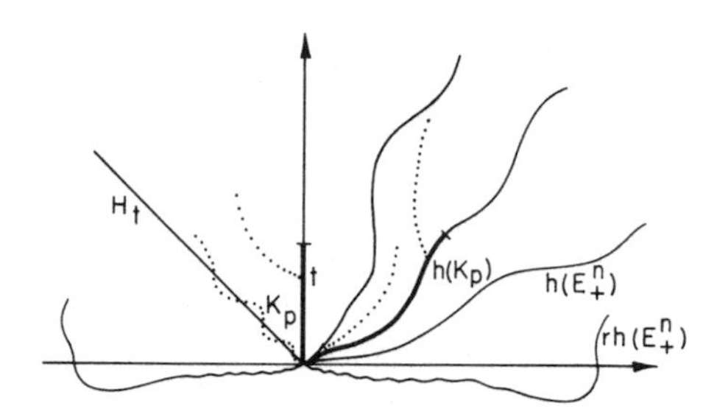

Figure 5.2.8

$$\bar{f} = \begin{cases} h^{-1} \mid h(K_p) & \text{on } h(K_p) \\ er(he^{-1}h^{-1}) & \text{on } h(E^n_+) - h(K_p) \\ r & \text{on } \text{Cl}(E^n - h(E^n_+)). \end{cases}$$

Then, $\bar{f}$ is the identity on $B^k_- = h(B^k_-)$ and $\bar{f}h(K) = h^{-1}h(K) = K$. We have therefore shown that h is locally flat at each point of Int B^{k-1}.

In order to complete the proof, we need to show that h is locally flat at points of Bd B^{k-1}. To do so, consider the n-cell G^n obtained from B^n by expanding E^n with coefficient 2 from E^{n-1}. (See Fig. **5.2.9**.) By a technique of construction similar to that employed in the proof of Theorem **3.4.1**, one can obtain an extension $\tilde{h}$ of the embedding h to G^n. Furthermore, there exists a homeomorphism $t: G^n \twoheadrightarrow G^n$ which is the identity on Bd G^n and which takes B^k onto B^k_-. The homeomorphism $\tilde{h}t\tilde{h}^{-1}$ can be extended identically onto the whole of E^n, and it takes $h(B^k)$ onto $h(B^k_-)$, that is, $h(B^k)$ is embedded in E^n, so as to be topologically equivalent to B^k_-. This proves the theorem.

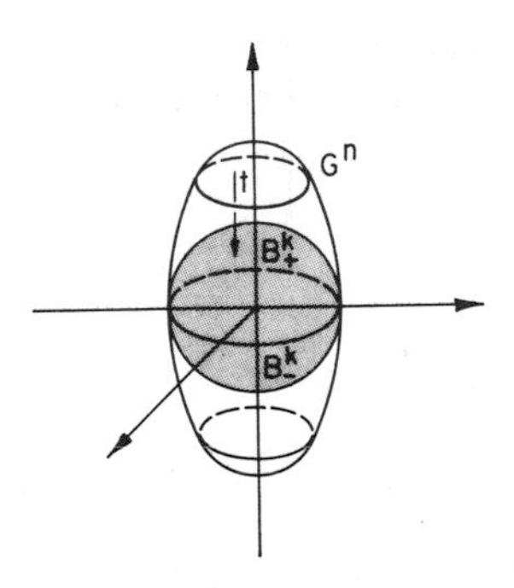

Figure 5.2.9

Proof of Engulfing Lemma 5.2.1 for $k \leqslant n - 3$. This proof consists of showing that for $k \leqslant n - 3$, Engulfing Lemma **5.2.1** is actually a corollary of Infinite Engulfing Theorem **4.14.1**. In applying Theorem **4.14.1**, let the role of $(M^n, U, P^k, Q, C, \epsilon)$ in Theorem **4.14.1** be played by

$$(\operatorname{Int} E^n_+, h(\operatorname{Int} B^n_+), B^k_+ - B^{k-1}, \emptyset, C, \epsilon/2)$$

of Lemma **5.2.1**, where C is a compact subpolyhedron of $B^k_+ - B^k$ such that $B^k_+ - C \subset N_{\epsilon/2}(B^{k-1}, E^n_+)$. It is clear that $B^k_+ - B^{k-1}$ tends to $h(\operatorname{Int}(B^n_+))$ and that $\operatorname{Int} E^n_+$ is ULC^k. Hence, in order to apply Theorem **4.14.1** it only remains to show that $h(\operatorname{Int} B^n_+)$ is ULC^{k-1}. Since B^n_+ is compact we can cover it with a finite number of open spherical neighborhoods $U_1, U_2, \ldots, U_s$ such that $\operatorname{diam}(h(U_i)) < \epsilon$ for $i = 1, \ldots, s$. Then, $h(U_1), \ldots, h(U_s)$ is an open covering of $h(B^n_+)$. Let δ be the Lebesgue number of that covering. Then any map f of a sphere S^p into $h(\operatorname{Int} B^n_+)$ such that $\operatorname{diam} f(S^p) < \delta$ lies in some $h(U_i)$ and so is null-homotopic through a homotopy whose track has diameter less than ϵ. Now we can apply Theorem **4.14.1** and obtain the isotopy $e_t\colon \operatorname{Int} E^n_+ \to \operatorname{Int} E^n_+$ of the conclusion. Extend e_1 over E^n_- by the identity and notice that this extension satisfies the requirements for f in the conclusion of Lemma **5.2.1**.

Before proving Engulfing Lemma **5.2.1** for the cases $k = n - 1$ and $k = n - 2$, let us state and prove the following preliminary lemma.

Lemma 5.2.4. *Let $k = n - 1$ or $n - 2$. Suppose that $h\colon E^n_+ \to E^n_+$ is a closed embedding which takes E^{k-1} into itself such that $h \mid N = 1$ for some neighborhood N of B^{k-1} in E^{k-1} and such that $h(E^n_+ - E^{k-1}) \subset \operatorname{Int} Q^+_{1/2}$. If $k = n - 2$, suppose that $h(F_+) \cup F_-$ is locally homotopically unknotted at each point of N. Then, for any $\epsilon > 0$ and $x \in N$ there exists a $\delta > 0$*

such that if $r\colon a * \dot{\Delta}^i \to N_\delta(x, E^n) - (h(Q_1{}^+) \cup Q_0{}^-)$ *is a mapping of the lateral surface of the* $(i+1)$*-dimensional simplex* $\Delta^{i+1} = a * \Delta^i$, $i \leqslant [n/2]$, *and either*

(a) $r(\dot{\Delta}^i) \subset h(Q_0{}^+ - Q_1{}^+)$ *or*
(b) $r(\dot{\Delta}^i) \subset Q_{1/2}^- - Q_0{}^-$,

then there exists an extension of r *to* Δ^{i+1} *such that*

$$r(\Delta^{i+1}) \subset N_\epsilon(x, E^n) - (h(Q_1{}^+) \cup Q_0{}^-)$$

and in Case (a), $r(\Delta^i) \subset h(Q_0{}^+ - Q_1{}^+)$, *and in Case* (b), $r(\Delta^i) \subset Q_{1/2}^- - Q_0{}^-$. (See Fig. **5.2.10**.)

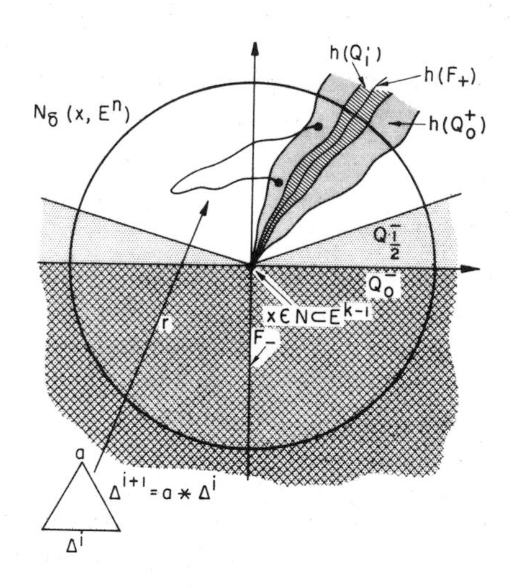

Figure 5.2.10

We shall consider Case (a). The proof of Case (b) is the same.

Proof of Lemma 5.2.4 for $k = n - 1$. Notice that for any t such that $0 \leqslant t \leqslant 1$, there is a natural way to define a map

$$g_t\colon E^n - h(F_+) \twoheadrightarrow E^n - h(\operatorname{Int}(Q_t{}^+)),$$

which is fixed on $E^n - h(Q_t{}^+)$. Also, there is a $\delta' > 0$ such that for any such t, $g_t(N_{\delta'}(x, E^n)) \subset N_\epsilon(x, E^n)$. Let $\delta'' > 0$ be small enough that if $D_{\delta''}^{n-1} = \{y \in F_+ \mid \operatorname{dist}(y, x) \leqslant \delta''\}$, then $h(D_{\delta''}^{n-1}) \subset N_{\delta'}(x, E^n)$. Now, let δ''' be small enough that $N_{\delta'''}(x, E^n) \cap h(F_+) \subset h(D_{\delta''}^{n-1})$. Finally, it is easy to see that there is a $\delta > 0$ such that any map of $\dot{\Delta}^i$ into

$N_\delta(x, E^n) \cap h(Q_0^+ - Q_1^+)$ can be extended to a map of Δ^i into $N_{\delta'''}(x, E^n) \cap h(Q_0^+ - Q_1^-)$.

We will show that the above constructed δ satisfies Lemma **5.2.4** for the chosen ϵ. Let $r: a * \dot{\Delta}^i \to N_\delta(x, E^n) - (h(Q_1^+) \cup Q_0^-)$ be given. Then, by choice of σ, r can be extended to take $(\dot{\Delta}^{i+1}, \Delta^i)$ into $(N_{\delta'''}(x, E^n) - (h(Q_1^+) \cup Q_0^-), h(Q_0^+ - Q_1^-))$. Now $N_{\delta'''}(x, E^n) - Q_0^-$ is contractible and so r can be extended to take Δ^{i+1} into $N_{\delta'''}(x, E^n) - Q_0^-$. Then, by choice of δ''', $r(\Delta^{i+1}) \cap h(F_+) \subset h(D_{\delta''}^{n-1})$ and by using Tietze's extension theorem as in the proof of Example **2.6.1**, we can "cut r off" on Int $h(D_{\delta''}^{n-1})$ and push it to the side of $h(F_+)$ containing $r(\dot{\Delta}^{i+1})$ by using a collar. We now have altered r so that it takes (Δ^{i+1}, Δ^i) into $(N_{\delta'}(x, E^n) - (h(Q_1^+) \cup Q_0^-), h(Q_0^+ - Q_1^-))$. There is a t such that $0 \leqslant t \leqslant 1$ and $r(\dot{\Delta}^{i+1}) \subset E^n - h(Q_t^+)$. Let g_t be defined as above. Then, $g_t r$, which we rename r, is the extension we were seeking.

Proof of Lemma 5.2.4 for $k = n - 2$

Case 1 ($i = 0$). In this case Δ^{i+1} is a segment with ends a and Δ^i. There is a natural way to define a homeomorphism

$$g: E^n - (F_- \cup h(F_+)) \twoheadrightarrow E^n - (Q_0^- \cup h(Q_1^+))$$

which is fixed on $E^n - (Q_{1/2}^- \cup h(Q_0^+))$. Let $\delta < \epsilon/2$ be a positive number which is small enough that points in $N_\delta(x, E^n)$ are moved less than $\epsilon/2$ under h. Since classically no n-manifold can be separated by a subset of dimension $\leqslant n - 2$ (see Corollary 1 of Theorem 14.4 of [Hurewicz and Wallman, 1]), it follows that $N_\delta(x, E^n)$ is not separated by $N_\delta(x, E^n) \cap (F_- \cup h(F_+))$. Thus, there is a path

$$\pi: a * \Delta^0 \to N_\delta(x, E^n) - (F_- \cup h(F_+)),$$

which connects $r(a)$ to an arbitrary point of $h(Q_0^+ - F_+)$. It follows from our construction that $r = g\pi$ is the desired extension.

Case 2 ($3 \leqslant i \leqslant [n/2]$). Notice that the sequence of open round balls $N_{1/j}(x, E^n)$, $j = 1, 2, \ldots$ has the property that

$$(N_{1/j}(x, E^n), N_{1/j}(x, E^n) \cap F) \approx (E^n, E^{n-2})$$

and so $N_{1/j}(x, E^n) - F$ has the homotopy type of S^1. Therefore, $N_{1/j}(x, E^n) \cap (Q_0^+ - Q_1^+)$ has the homotopy type of S^1 as does

$$D_j = h(N_{1/j}(x, E^n) \cap (Q_0^+ - Q_1^+)), \qquad j = 1, 2, \ldots.$$

The existence of D_j, $j = 1, 2, \ldots$ ensures:

Fact. Given $\delta' > 0$ there exists a $\delta > 0$ such that any map of $\dot{\Delta}_i$, $i \geqslant 2$, into $N_\delta(x, E^n) \cap h(Q_0^+ - Q_1^+)$ extends to a mapping of Δ^i into $N_{\delta'}(x, E^n) \cap h(Q_0^+ - Q_1^+)$. (We will use this fact soon.)

Since $h(F_+) \cup F_-$ is locally homotopically unknotted at x, it follows, by an argument similar to the one which used g_t in the proof above of Lemma **5.2.4** for $k = n - 1$, that there is a $\delta' > 0$ such that any map of $(\dot{\Delta}^{i+1}, \Delta^i)$ into $(N_{\delta'}(x, E^n) - (h(Q_1^+) \cup Q_0^-), h(Q_0^+ - Q_1^+))$ extends to a mapping of (Δ^{i+1}, Δ^i) into $(N_\epsilon(x, E^n) - (h(Q_1^+) \cup Q_0^-), h(Q_0^+ - Q_1^+))$.

Let the δ in the Fact given correspond to the δ' just mentioned. Then, given $r: (a * \dot{\Delta}^i, \dot{\Delta}^i) \to N_\delta(x, E^n) - (h(Q_1^+) \cup Q_0^-), h(Q_0^+ - Q_1^+))$ the Fact gives $r: (\dot{\Delta}^{i+1}, \Delta^i) \to (N_{\delta'}(x, E^n) - (h(Q_1^+) \cup Q_0^-), h(Q_0^+ - Q_1^+))$. Finally the last paragraph gives

$$r: (\Delta^{i+1}, \Delta^i) \to (N_\epsilon(x, E^n) - (h(Q_1^+) \cup Q_0^-), h(Q_0^+ - Q_1^+))$$

as desired.

Case 3 ($i = 2$). Let $B_i^+ = B_i \cap Q_0^+$, $i = 1, 2, \ldots$, where B_i is the closed round ball in E^n about x of radius $1/i$. Then, $(B_i^+, B_i^+ \cap F_+) \approx (I^n, I^{n-2})$. Let $F_i^n = h(B_i)$ and $F_i^{n-2} = h(B_i \cap F_+)$ (see Fig. **5.2.11**). Then, $(F_i^n, F_i^{n-2}) \approx (I^n, I^{n-2})$. Let l be a loop in $F_i^n - F_i^{n-2}$ which is null-homotopic in $E^n - (F_- \cup h(F_+))$. By pushing radially away from a point

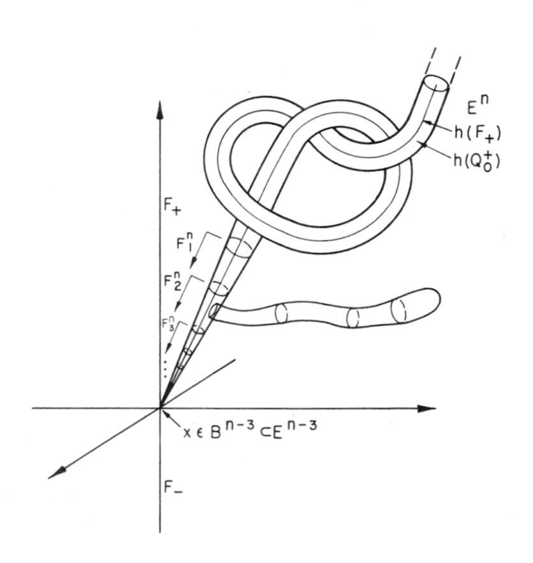

Figure 5.2.11

$p \in \operatorname{Int} F_i^{n-2}$, we see that l is homotopic in $F_i^n - F_i^{n-2}$ to a loop l' in $\operatorname{Bd} F_i^n - \operatorname{Bd} F_i^{n-2}$ which is null-homotopic in

$$E^n - ((\operatorname{Int} F_i^n) \cup F_- \cup h(F_+)).$$

We now wish to establish the following

Claim. l' is null-homotopic in $\operatorname{Bd} F_i^n - \operatorname{Bd} F^{n-2}$.

Since we know that l' is null-homotopic in $E^n - ((\operatorname{Int} F_i^n) \cup F_- \cup h(F_+))$, it will suffice to show that the injection

$$\pi_1(\operatorname{Bd} F_i^n - \operatorname{Bd} F_i^{n-2}) \to \pi_1(E^n - ((\operatorname{Int} F_i^n) \cup F_- \cup h(F_+))$$

is a monomorphism. In order to do so, consider the following Mayer-Vietoris sequence ([Spanier, 1, pp. 186–190]) where $X^{n-2} = F_- \cup h(F_+)$:

$$\cdots \to H_2(E^n - (X^{n-2} - \operatorname{Int} F_i^n)) \to H_1(\operatorname{Bd} F_i^n - \operatorname{Bd} F_i^{n-2})$$

$$\to H_1(E^n - (\operatorname{Int} F_i^n \cup X^{n-2})) \oplus H_1(F_i^n - (\operatorname{Bd} F_i^{n-2}))$$

$$\to H_1(E^n - (X^{n-2} - \operatorname{Int} F_i^{n-2})) \to \cdots .$$

By using Alexander duality [Spanier, 1, p. 296], this sequence becomes

$$\cdots \to 0 \to Z \to Z \oplus 0 \to 0 \to \cdots$$

Hence, the inclusion of $\operatorname{Bd} F_i^n - \operatorname{Bd} F_i^{n-2}$ into $E^n - (\operatorname{Int} F_i^n \cup X^{n-2})$ induces an isomorphism on first homology. But now any loop l in $\operatorname{Bd} F_i^n - \operatorname{Bd} F_i^{n-2}$ which is null-homotopic in $E^n - (\operatorname{Int} F_i^n \cup X^{n-2})$ is also null-homologous in $E^n - (\operatorname{Int} F_i^n \cup X_i^{n-2})$, consequently null-homologous in $\operatorname{Bd} F_i^n - \operatorname{Bd} F_i^{n-2}$. Since $\pi_1(\operatorname{Bd} F_i^n - \operatorname{Bd} F_i^{n-2})$ is Abelian it follows [Hu, 1, pp. 44–47] that l is null-homotopic in $\operatorname{Bd} F_i^n - \operatorname{Bd} F_i^{n-2}$ and so the injection

$$\pi_1(\operatorname{Bd} F_i^n - \operatorname{Bd} F_i^{n-2}) \to \pi_1(E^n - (\operatorname{Int} F_i^n \cup X^k))$$

is a monomorphism as desired.

We have just established that any loop in $F_i^n - F_i^{n-2}$ which is null-homotopic in $E^n - (F_- \cup h(F_+))$ is also null-homotopic in $F_i^n - F_i^{n-2}$. From this it follows easily that any loop l in $F_i^n - h(Q_1^+)$ which is null-homotopic in $E^n - (Q_0^- \cup h(Q_1^+))$ is also null-homotopic in $F_i^n - h(Q_1^+)$. From this follows

Fact. Given $\delta' > 0$, there exists a $\delta > 0$ such that any mapping

$r: (a * \dot{\Delta}^2, \dot{\Delta}^2) \to (N_\delta(x, E^n) - (Q_0^+ \cup h(Q_1^+)), h(Q_0^+ - Q_1^+))$ extends to

$$r: (\Delta^3, \dot{\Delta}^2) \to (N_{\delta'}(x, E^n) - (Q_0^+ \cup h(Q_1^+)), h(Q_0^+ - Q_1^+)).$$

The proof is now completed as in Case 2.

Case 4 $(i = 1)$. In order to prove this case we will need the following fact.

Fact. Suppose that we are given a closed $(n - 2)$-string X^{n-2} in E^n, a point $x \in X^{n-2}$ and $\epsilon > 0$. Then, there exists a $\delta > 0$ such that if F^n is an n-cell in $N_\delta(x, E^n)$ and $F^{n-2} = F^n \cap X^n$ is an $(n - 2)$-cell having the property that (F^n, F^{n-2}) is a trivial cell pair, then for any path l in $N_\delta(x, E^n) - X^{n-2}$ whose end points lie in F^n, there is another path l' in $F^n - F^{n-2}$ having the same end points such that the loop $l \cup l'$ is null-homologous in $N_\epsilon(x, E^n) - X^{n-2}$.

For examples of how l' would look in a couple of situations see Fig. **5.2.12**.

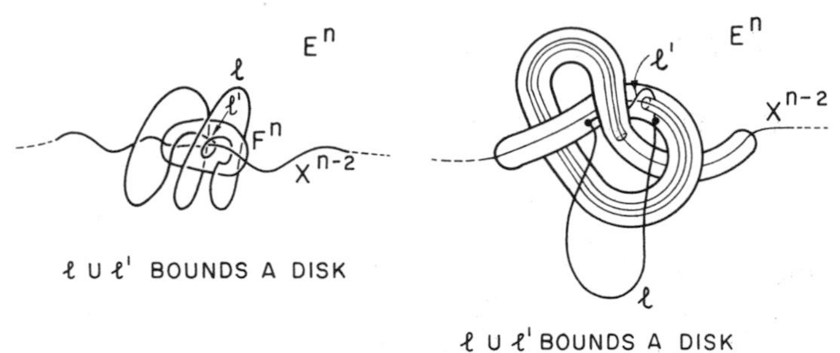

Figure 5.2.12

Since $F^n - F^{n-2}$ has the homotopy type of S^1, it should be fairly clear, intuitively, that one can obtain the desired l' by winding a certain number of circuits around $h(Q_1^+)$ in one direction or the other. A rigorous proof uses the notion of linking numbers and will not be presented here. The interested reader is referred to Volume 3 of [Alexandroff, 1]. We will now assume the fact and complete the proof.

For any t such that $0 \leqslant t \leqslant 1$, there is a natural way to define a map $g_t: E^n - (h(F_+) \cup F_-) \twoheadrightarrow E^n - (h(\mathrm{Int}\, Q_t^+) \cup \mathrm{Int}\, Q_0^-)$ which is fixed on $E^n - (h(\mathrm{Int}\, Q_t^+) \cup \mathrm{Int}\, Q_0^-)$. Also, there is an $\epsilon' > 0$ such that for any such t, $g_t(N_{\epsilon'}(x, E^n)) \subset N_\epsilon(x, E^n)$. Now since $h(F_+) \cup F_-$ is locally homotopically unknotted at x, there is a $\delta' > 0$ such that every loop in $N_{\delta'}(x, E^n) - (h(F_+) \cup F_-)$ which is homologous to zero in $N_{\epsilon'}(x, E^n) - (h(F_+) \cup F_-)$ is homotopic to zero in $N_{\epsilon'}(x, E^n) - (h(F_+) \cup F_-)$ If we take the ϵ of the fact to be δ', then we get a δ'' such that if F^n is an n-cell in $N_{\delta''}(x, E^n)$ and $F^{n-2} = F^n \cap X^{n-2}$ is an $(n-2)$-cell having the property that (F^n, F^{n-2}) is a trivial cell pair, then for any path l in $N_{\delta''}(x, E^n) - (h(F_+) \cup F_-)$ whose end points lie in F^n, there is another path l' in $F^n - F^{n-2}$ having the same end points such that the loop $l \cup l'$ is null-homologous in $N_{\delta'}(x, E^n) - (h(F_+) \cup F_-)$. We take F^n of the last sentence to be a sufficiently small F_i^n constructed in Case 3. Finally let δ be small enough that $N_\delta(x, E^n) \cap h(Q_0^+) \subset F^n$.

Suppose we have

$$r\colon (a * \dot{\Delta}^1, \dot{\Delta}^1) \to (N_\delta(x, E^n) - (h(Q_1^+) \cup Q_0^-), h(Q_0^+ - Q_1^+)).$$

Then, by choice of δ, $r(\dot{\Delta}^1) \subset F^n \subset N_{\delta''}(x, E^n)$ and so by choice of δ'' the fact assures that r can be extended to $r\colon \dot{\Delta}^2 \to N_{\delta''}(x, E^n) - (h(F_+ \cup F_-))$ where $r(\dot{\Delta}^2)$ is null-homologous in $N_{\delta'}(x, E^n) - (h(F_+) \cup F_-)$. Hence, by our choice of δ'', $r(\dot{\Delta}^2)$ is also null-homotopic in $N_{\delta'}(x, E^n) - (h(F_+) \cup F_-)$ and we obtain an extension $r\colon \Delta^2 \to N_{\delta'}(x, E^n) - (h(F_+) \cup F_-)$. There is a t such that $0 \leqslant t \leqslant 1$ and $r(a * \dot{\Delta}^1) \subset E^n - h(Q_t^+)$. Let g_t be defined as above. Then, $g_t r$, which we rename r, is the extension we were seeking. This completes the proof of Lemma **5.2.4**.

Proof of Engulfing Lemma 5.2.1 for $k = n - 1$ and $n - 2$ Throughout this proof let us abbreviate $N_\epsilon(B^{k-1}, E^n)$ to N_ϵ. Notice that we may assume, without loss of generality, that $h(E_+^n - E^{k-1}) \subset \operatorname{Int} Q_{1/2}^+$ so that Lemma **5.2.4** applies. Also, assume that ϵ is small enough that $N_\epsilon \cap E^{k-1} \subset N$. We take a triangulation T of $E^n - E^{k-1}$ refined to E^{k-1} and such that no closed simplex of T lying in N_ϵ simultaneously intersects two of the sets $h(H_1)$, $h(H_{1/2})$, $h(H_0) \cup H_{1/2}$, $H_{1/4}$, H_0.

We denote by P the polyhedron consisting of all closed simplexes of T which lie in N_ϵ and intersect $h(Q_1^+)$ and all of those closed simplexes of the $[n/2]$-skeleton of T which lie in N_ϵ and do not intersect Q_0^-. Suppose that P_* is the dual polyhedron consisting of those simplexes of the barycentric subdivision of T which lie in N_ϵ and have no vertices in P. It includes firstly all simplexes of the barycentric subdivision lying in N_ϵ which intersect Q_0^-, and secondly those simplexes of the $(n - [n/2] - 1)$-skeleton of that subdivision which do not intersect P nor $h(Q_1^+)$. (This is not correct with respect to the simplexes which intersect the boundary of N_ϵ, but we are not interested in those as we shall carry out all our constructions sufficiently far from the frontier of N_ϵ.)

Notice that the following fact can be established by using Lemma **5.2.4** and the technique of proof of Infinite Engulfing Theorem **4.14.1**. (Lemma **5.2.4** serves the purpose of the ULC assumptions in that technique.)

Fact. Suppose that Π is a polyhedron lying either (a) in $(E^n - Q_0^-) \cap N_\epsilon$ or (b) in $E^n - h(Q_1^+) \cap N_\epsilon$ and finite outside any neighborhood of E^{k-1}. We suppose that in Case (a)

$$\dim(\Pi \cap (E^n - h(Q_{1/2}^+))) \leqslant [n/2]$$

and in Case (b)

$$\dim(\Pi \cap (E^n - Q_{1/4}^-)) \leqslant [n/2].$$

Then, there exists an $\epsilon/3$-homeomorphism $f\colon E^n \to E^n$ which is the identity outside N_ϵ and on $h(Q_1^+) \cup Q_0^-$ and such that Π, with the exception of some finite part of it, lies in Case (a) in $fh(E_+^n)$ and Case (b) in $f(Q_{1/2}^-)$.

By the above fact, we can construct $\epsilon/3$-homeomorphisms $\bar{f}$ and $\bar{f}_*\colon E^n \to E^n$ which are the identity outside N_ϵ and on $h(Q_1^+) \cup Q_0^-$ so that $\bar{f}h(Q_0^+) \supset P \cap N_{\epsilon'}$ and $\bar{f}_*(Q_{1/2}^-) \supset P_* \cap N_{\epsilon'}$ for some $\epsilon' > 0$.

The next fact follows from the trick of Stallings which was used five times in Chapter **4**. The first time the trick was used was in the proof of Theorem **4.4.1**. (See the proof of Statement A in Section **4.13**.)

Fact. If U and U_* are two regions in $E^n - E^{k-1}$ which contain $(P \cap N_{\epsilon'}) \cup h(Q_1^+)$ and $(P_* \cap N_{\epsilon'}) \cup Q_0^-$ respectively, then there exists an $\epsilon/3$-homeomorphism $\check{f}\colon E^n \to E^n$ which is the identity outside N_ϵ and on $(P \cap N_{\epsilon'}) \cup h(Q_1^+) \cup (P_* \cap N_{\epsilon'}) \cup Q_0^-$ and is such that $\check{f}(U) \cup U_*$ contains some neighborhood of B^{k-1} in E^n.

In accordance with the above fact we can construct an $\epsilon/3$-homeomorphism $\check{f}\colon E^n \to E^n$ which is the identity outside N_ϵ and on $(P \cap N_{\epsilon'}) \cup h(Q_1^+) \cup (P_* \cap N_{\epsilon'}) \cup Q_0^-$ by taking $\bar{f}h(\operatorname{Int} Q_0^+)$ for U and $\bar{f}_*(\operatorname{Int} Q_{1/2}^-)$ for U_*. Then, $\check{f}\bar{f}h(Q_0^+) \cup \bar{f}_*(Q_{1/2}^-)$ contains some neighborhood of B^{k-1} in E^n and so $\bar{f}_*^{-1}\check{f}\bar{f}h(Q_0^+) \cup Q_{1/2}^-$ also contains some neighborhood of B^{k-1} in E^n. Thus, $\bar{f}_*^{-1}\check{f}\bar{f}h(Q_0^+)$ contains some neighborhood of B^{k-1} in $Q_{1/2}^+$. Since, in addition, $\bar{f}_*^{-1}\check{f}\bar{f}$ is an ϵ-homeomorphism and is the identity outside N_ϵ, we can take it as the required f and the proof of Engulfing Lemma **5.2.1** is complete.

REMARK 5.2.4. Closely related to the β-statements proved in this section are the so-called γ-statements. See [Lacher, 5] for a succinct discussion of the γ-statements. Also see Exercise **5.5.2** for the γ-statement and a discussion. The purpose of this remark is to point out that the following proposition, which includes many γ-statements, is a consequence of the techniques of this section.

Proposition 5.2.1. *Suppose that* $h: B_+^k \to E^n$, $k \leqslant 2/3n - 1$, *is a topological embedding such that* $h \mid (B_+^k - B^{k-1})$ *and* $h \mid B^{k-1}$ *are locally flat. Then, h is locally flat.*

The idea of the proof of Proposition **5.2.1** is basically the same as that described for Theorem **5.2.1**. First, one carries $h(B^{k-1})$ back to B^{k-1}. However, this time we cannot push the image of B_+^k up into E_+^n staying fixed on B^{k-1}. Even so, the same collapsing-and-expanding technique and the same meshing technique will suffice to move the image of B_+^k back to B_+^k. The only difference is that in this case the meshing technique requires a better engulfing lemma. The following lemma will suffice.

Engulfing Lemma 5.2.5. *Suppose that* $h: E_+^n \to E_+^n$ *is a closed embedding which takes* E^{k-1}, $k \leqslant 2/3n - 1$, *into itself such that* $h \mid N = 1$ *for some neighborhood* N *of* B^{k-1} *in* E^{k-1}. *Then for each* $\epsilon > 0$ *there is an* ϵ*-homeomorphism* $f: E^n \to E^n$ *which is the identity outside the* ϵ*-neighborhood of* B^{k-1} *and on* E^{k-1} *such that* $f(h(\text{Int } E_+^n) \cup B^{k-1})$ *contains a neighborhood of* B^{k-1} *relative to* B_+^k.

Lemma **5.2.5** will follow from Infinite Engulfing Theorem **4.14.2** by letting $(E^n, E^{k-1}, h(\text{Int } E_+^n), B_+^k - B^{k-1})$ correspond to $(M, \underset{\infty}{\sim}, U, P)$. Certainly, $h(\text{Int } E_+^n)$ is ULC^{2k-n+1}. Since $E^n - E^{k-1}$ has the homotopy type of S^{n-k}, it follows that $E^n - E^{k-1}$ is ULC^{n-k-1}. Thus, Theorem **4.14.2** applies whenever $2k - n + 2 \leqslant n - k - 1$ or $k \leqslant 2/3n - 1$ as hypothesized. One can easily construct the necessary short homotopy which pulls $B_+^k - B^{k-1}$ through $E^n - E^{k-1}$ into $h(\text{Int } E_+^n)$. (See [Černavskiĭ, 6].)

EXERCISE 5.2.1. (The author has recently seen that the following fact can be established by using the technique of this section along with Exercise **3.3.3**. The proof can be accomplished for $n = 3$ and $n = 4$ because the hypothesis allows one to avoid the engulfing part of the technique of this section.) Let Δ^n denote the standard n-simplex spanned by the origin and the unit vectors in E^n and let $f: \Delta^n \to E^n$, n arbitrary, be an embedding such that $f \mid \Delta^n - \Delta^{n-3}$ is locally flat. Then, there is a continuous function $\epsilon(x): \Delta^n \to E_+^1$, $\epsilon(x) > 0$ for $x \in \Delta^n - \Delta^{n-3}$, for which the existence of a locally flat $\epsilon(x)$-approximation $g: \Delta^n \to E^n$ of f implies that f is locally flat.

5.3. TAMING EMBEDDINGS OF PL MANIFOLDS AROUND THE BOUNDARY IN ALL CODIMENSIONS

As the title advertises, in this section we are not going to tame embeddings everywhere, but are going to concentrate on taming around the boundary of the ambient manifold. We will return to tame everywhere in Section **5.5**. Most of the main ideas of this section were developed in [Rushing, 2, 4], although the present form of the work of this section has not appeared in print elsewhere.

A topological embedding $f: M^k \to N^n$ of the PL k-manifold M into the PL n-manifold N is said to be **allowable** if $f^{-1}(\partial N)$ is a PL $(k-1)$-submanifold (possibly empty) of ∂M. A manifold pair (N, M) is said to be **allowable** if the inclusion of M into N is allowable. Notice that under the definition of "locally homotopically unknotted" given in the last section, it makes perfectly good sense to speak of an allowable embedding $f: M^{n-2} \to N^n$ as being locally homotopicially unknotted at points of $\operatorname{Int}(f^{-1}(\partial N))$.

The following is the main theorem of this section.

Taming around Boundary Theorem 5.3.1. *Let*

$$f: M^k \to N^n, \qquad n \geqslant 5,$$

be an allowable embedding (proper embedding if $n - k = 0, 1,$ or 2) of the PL *k-manifold M into the* PL *n-manifold N such that $f \mid M - f^{-1}(\partial N)$ is locally flat and $f \mid f^{-1}(\partial N)$ is* PL. *In the case that $k = n - 2$, suppose further that $f(f^{-1}(\partial N))$ is locally flat in ∂N and that f is locally homotopically unknotted at points of* $\operatorname{Int}(f^{-1}(\partial N))$. *Then, given $\epsilon > 0$, there exists a neighborhood U of $f^{-1}(\partial N)$ in M and an ϵ-push e_t of $(N, f(f^{-1}(\partial N)))$ which is fixed on ∂N such that $e_1 f \mid U: U \to N$ is* PL.

The rest of this section will be organized as follows: First we will state and prove a lemma, and then prove Theorem **5.3.1** for the codimension zero case. After that, we will prove several other lemmas and conclude by proving Theorem **5.3.1** for the other codimensions.

Uniqueness of Collars Lemma 5.3.1. *Let M be a topological manifold and let C_1:* Bd $M \times [0, 1] \to M$ *and C_2:* Bd $M \times [0, 1] \to M$ *be collars of* Bd M *in M, that is, C_i, $i = 1, 2$, is a homeomorphism of* Bd $M \times [0, 1]$ *onto a neighborhood of* Bd M *in M such that $C_i(x, 0) = x$ for all $x \in$* Bd M (see Fig. **5.3.1**). *Then, given $\epsilon > 0$, for some $s > 0$ there is an ϵ-isotopy e_t, $0 \leqslant t \leqslant s$, of M such that*

(1) $e_0 = 1$,

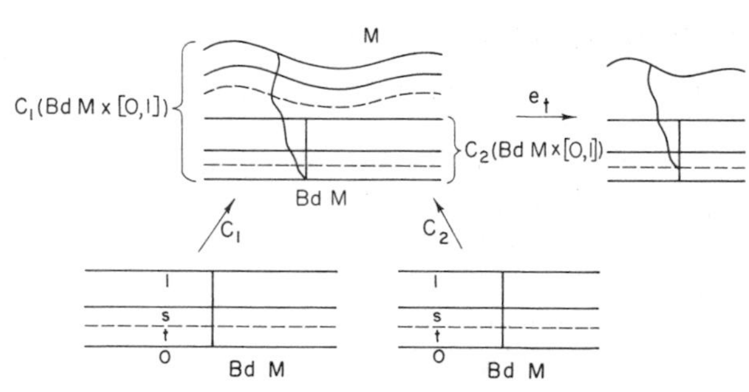

Figure 5.3.1.

(2) $e_t \mid \mathrm{Bd}\, M \cup (M - (C_1(\mathrm{Bd}\, M \times [0, 1]) \cup C_2(\mathrm{Bd}\, M \times [0, 1]))) = 1$,

(3) $e_s C_1 \mid \mathrm{Bd}\, M \times [0, s] = C_2 \mid \mathrm{Bd}\, M \times [0, s]$.

PROOF. Let e_t, $0 \leqslant t \leqslant s$, be defined to be the identity on $M - (C_1(\mathrm{Bd}\, M \times [0, 1]) \cup C_2(\mathrm{Bd}\, M \times [0, 1]))$, and let e_t be defined to be $C_2(C_1)^{-1}$ on $C_1(\mathrm{Bd}\, M \times [0, t])$. It remains to define e_t on $C_1(\mathrm{Bd}\, M \times [t, 1]) \cup C_2(\mathrm{Bd}\, M \times [0, 1]) - C_1(\mathrm{Bd}\, M \times [0, t])$. Define a homeomorphism f_1: $\mathrm{Bd}\, M \times [t, 1] \twoheadrightarrow \mathrm{Bd}\, M \times [0, 1]$ on each fiber $x \times [t, 1]$ by taking (x, t) to $(x, 0)$ and $(x, 1)$ to itself and extending linearly. Let $\bar{f}_1$: $C_1(\mathrm{Bd}\, M \times [t, 1]) \twoheadrightarrow C_1(\mathrm{Bd}\, M \times [0, 1])$ be defined by $\bar{f}_1 = C_1 f_1 (C_1)^{-1}$. Now extend $\bar{f}_1$ over

$$C_2(\mathrm{Bd}\, M \times [0, 1]) - C_1(\mathrm{Bd}\, M \times [0, 1])$$

by the identity. Next, denote $(f_1)^{-1}$: $\mathrm{Bd}\, M \times [0, 1] \twoheadrightarrow \mathrm{Bd}\, M \times [t, 1]$ by f_2. Let $\bar{f}_2$: $C_2(\mathrm{Bd}\, M \times [0, 1]) \twoheadrightarrow C_2(\mathrm{Bd}\, M \times [t, 1])$ be defined by $\bar{f}_2 = C_2 f_2 (C_2)^{-1}$. Extend $\bar{f}_2$ over $C_1(\mathrm{Bd}\, M \times [0, 1]) - C_2(\mathrm{Bd}\, M \times [0, 1])$ by the identity. Finally, define e_t on

$$C_1(\mathrm{Bd}\, M \times [t, 1]) \cup (C_2(\mathrm{Bd}\, M \times [0, 1]) - C_1(\mathrm{Bd}\, M \times [0, t])$$

to be $\bar{f}_2 \bar{f}_1$. It is easy to check that e_t is the desired isotopy.

Proof of Theorem 5.3.1 for Codimension Zero Case. (It is not necessary to assume $n \geqslant 5$ in this case.) We are given a proper embedding f: $M^n \to N^n$ of the PL n-manifold M^n into the PL n-manifold N^n such that $f \mid \partial M$ is PL. By Corollary **4.4.4** or by Remark **4.4.1**, there is a PL Collar C_1 of ∂M, that is, there is a PL homeomorphism C_1: $\partial M \times I \to M$ such that $C_1(x, 0) = x$ for all $x \in \partial M$. Also, there is a PL collar

$$\begin{array}{ccc} \partial M \times I & \xrightarrow{\bar{f}} & \partial N \times I \\ \uparrow{\scriptstyle C_1} & & {\scriptstyle C_2}\uparrow \\ M & \xrightarrow{f} & N \end{array}$$

Diagram 5.3.1

C_2: $\partial N \times I \to N$ of ∂N. There is a natural PL homeomorphism $\bar{f}$: $\partial M \times I \twoheadrightarrow \partial N \times I$ defined by $\bar{f}(x, t) = (f(x), t)$. Then

$$f C_1 \bar{f}^{-1}: \partial N \times I \to N$$

is also a collar of ∂N. Now given $\epsilon > 0$, Lemma **5.3.1** gives for some $s > 0$ an ϵ-push e_t, $0 \leqslant t \leqslant s$, of $(N, \partial N)$ such that $e_t \mid \mathrm{Bd}\, N = 1$ and $e_s f C_1 \bar{f}^{-1} \mid \mathrm{Bd}\, N \times [0, s] = C_2 \mid \mathrm{Bd}\, N \times [0, s]$. Consequently, $e_s f \mid C_1 \bar{f}^{-1}(\mathrm{Bd}\, N \times [0, 1]) = C_2 \bar{f} C_1^{-1} \mid C_1 \bar{f}^{-1}(\mathrm{Bd}\, N \times [0, 1])$ which is

PL. Hence, $U = C_1\bar{f}^{-1}(\text{Bd}\, N \times [0, 1])$ is the desired neighborhood and e_t is the desired ϵ-push.

REMARK 5.3.1. It is quite natural to wonder whether a similar argument to that given for codimension zero will work in other codimensions. The obvious attempt involves using Theorem **1.7.7** to get certain relative collars and then applying Lemma **5.3.1**. A difficulty results, however, from the fact that one of these collars is only topological and not PL. Attempts by this author at resolving this difficulty (for instance, by trying to prove a certain stronger relative collaring theorem) have not panned out.

The rest of this section is devoted to handling the other codimensional cases in a different manner. We will again in Section **5.5**, in the process of proving general taming theorems, employ the technique we are about to develop.

Before proving Theorem **5.3.1** in codimensions other than zero, we will give some definitions and establish four preliminary lemmas. If S is a set (which may or may not be contained in the polyhedron $|J|$), then

$$N(S, J) = \{\sigma \in J \mid \sigma \text{ is a face of a simplex of } J \text{ which meets } S\},$$
$$C(S, J) = \{\sigma \in J \mid \sigma \cap S = \emptyset\}, \text{ and } \partial N(S, J) = N(S, J) \cap C(S, J).$$

If K is a subcomplex of J, we say that K is a **full** subcomplex of J if $\sigma \cap K$ is a simplex for each $\sigma \in J$. (This definition is obviously equivalent to the one given in Section **1.6, B.**) If K is a complex, then $\sigma \in K$ is called a **principle simplex** if σ is not a proper face of any simplex of K. If X and Y are subpolyhedra of some larger polyhedron, recall that in Section **1.6, C**, we defined

$$X_R = \text{Cl}(X - Y) \quad \text{and} \quad Y_R = \text{Cl}(X - Y) \cap Y.$$

The following lemma was first proved in [Cohen, 1].

Natural Parameterization Lemma 5.3.2 (Cohen). *If K and L are full subcomplexes of the complex J, then every simplex σ of $N(K - L, J)$ is uniquely expressible as $\alpha * \beta * \gamma$, where $\alpha \in L_R$, $\beta \in C(L_R, K_R) = C(L, K)$ and $\gamma \in \partial N(K_R, N(K - L, J))$. Furthermore, if σ is a principle simplex, then $\beta \neq \emptyset$.*

PROOF. Notice that if we can show that

(1) L_R is full in K_R and that K_R is full in $N(K - L, J)$ (see Fig. **5.3.2**), then it follows immediately that

(1∗) every simplex $\sigma \in N(K_R, N(K - L, J))$ is uniquely expressible as $\sigma = \alpha * \beta * \gamma$ where $\alpha \in L_R$, $\beta \in C(L_R, K_R)$ and $\gamma \in \partial N(K_R, N(K - L, J))$.

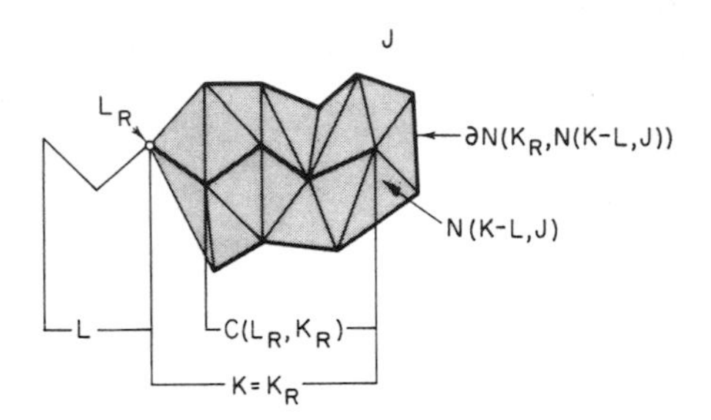

Figure 5.3.2

Moreover, it follows that if σ is a principle simplex of $N(K_R, N(K-L, J))$, then $\beta \neq \emptyset$.

If furthermore we can show that

(2) $N(K-L, J) = N(K_R, N(K-L, J))$,

then the proof will be complete.

Since L is full in J, it follows immediately that L_R is full in K_R because the only place a simplex of K_R can hit L is in L_R, since $L_R = K_R \cap L$ by definition. Also, since K is full in J, it will follow immediately that K_R is full in $N(K-L, J)$ if we can show that

(3) $K_R = K \cap N(K-L, J)$.

In fact, notice that we will be through if we establish (3), because (2) also follows immediately from (3). Since $K_R = N(K-L, K) = N(K-(L \cap K), K)$ it will suffice to show that

$$K \cap N(K-(L \cap K), J) = N(K-(L \cap K), K).$$

If σ is a principle simplex of $N(K-(L \cap K), K)$, then σ is a simplex of K meeting $(K-(L \cap K))$. So $\sigma \in K \cap N(K-(L \cap K), J)$. If on the other hand, $\sigma \in K \cap N(K-(L \cap K), J)$, then $\sigma \in K$ and there exists a simplex τ such that $\sigma < \tau \in J$ and $\tau \cap (K-(L \cap K)) \neq \emptyset$. Then, since K is full, $\tau \cap K = \tau_0$ is a simplex which clearly contains $\sigma \cup (\tau \cap K)$. Thus, τ_0 is a simplex of K such that $\sigma < \tau_0$ and $\tau_0 \cap (K-(L \cap K)) \neq \emptyset$. Hence, $\sigma \in N(K-(L \cap K), K)$.

The next definition generalizes very nicely an idea originally due to Whitehead. Let $a = (0, 0)$, $b_0 = (1, 0)$, and $b_1 = (1, 1) \in E^2$. Let $\Delta = a * b_0 * b_1 \subset E^2$. If K and L are full subcomplexes of J, then the **natural parameterization** of $N(K-L, J)$ is the unique simplicial mapping $\eta\colon N(K-L, J) \to \Delta$ such that $\eta(\alpha) = a$, $\eta(\beta) = b_0$, and $\eta(\gamma) = b_1$ for every simplex $\alpha * \beta * \gamma \in N(K-L, J)$.

The next lemma first appeared in [Rushing, 2].

Crushing Lemma 5.3.3. *Let K and L be full subcomplexes of the complex J, let $N = N(K - L, J')$ and let $C \subset |J|$ be a compact set such that $C \cap K_R \subset L_R$ (see Fig. **5.3.3**). Then, given $\epsilon > 0$, there is a C-regular neighborhood N_* of K mod L in J and an ϵ-push e_t of $(|J|, L_R)$ keeping $K_R \cup \mathrm{Cl}(|J| - |N|)$ fixed such that $(e_1(C) \cap N_*) \subset L_R$. (Thus, e_t crushes $C \cap N$ against $|J| - N$.)*

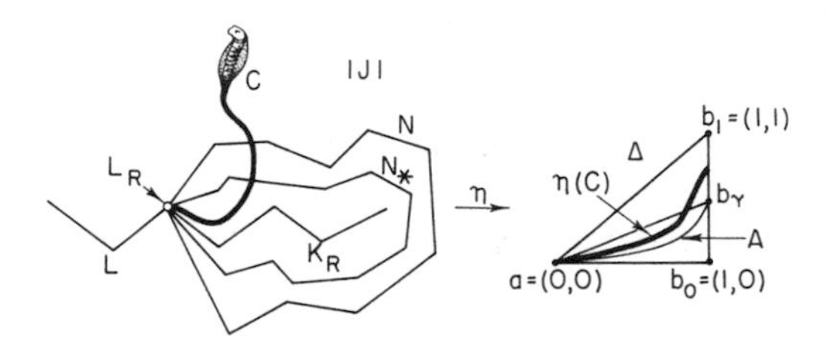

Figure 5.3.3

Proof. (We will just get an isotopy; however, it will be clear from the proof that we can get an ϵ-push.) Let η: $N(K - L, J') \to \Delta$ be the natural parameterization of $N(K - L, J')$. Consider the set $\eta(C \cap N) \subset \Delta$. Since $\eta(C) \cap ((a * b_0) - a) = \emptyset$, if $b_\gamma = (1, \gamma)$, then there is a $\gamma > 0$ such that $\eta(C) \cap b_0 * b_\gamma = \emptyset$. Let $\delta = a * b_0 * b_\gamma$. Then, $N_* = \eta^{-1}(\delta)$ is a regular neighborhood of K mod L in J; however, C may intersect N_* off of L_R. We denote by S_ϵ the segment $(\epsilon, 0) * b_1$, $0 \leqslant \epsilon \leqslant 1$. Let ψ be a continuous function defined on $[0, 1]$ such that $\psi(0) = 0$, $\psi(\epsilon) > 0$ for $\epsilon > 0$, and if $x_\epsilon \in S_\epsilon$ is the point such that $\mathrm{dist}((\epsilon, 0), x_\epsilon)/\mathrm{dist}((\epsilon, 0), b_1) = \psi(\epsilon)$, then $(\epsilon, 0) * x_\epsilon \cap \eta(C) = \emptyset$ for $\epsilon > 0$. (We can assume that $b_\gamma = x_1$.) Let A denote the arc consisting of the points x_ϵ, $0 \leqslant \epsilon \leqslant 1$. Then, there is an obvious way to define an isotopy h_t of Δ which is the identity on $\partial\Delta$ such that $h_0 = 1$ and which when restricted to a segment S_ϵ slides the point x_ϵ "linearly" to the point $S_\epsilon \cap a * b_\gamma$. Thus, if W is the region bounded by the simple closed curve $a * b_0 \cup b_0 * b_\gamma \cup A$, we see that the isotopy h_t takes W onto δ.

We are now ready to define an isotopy e_t of $|J|$ that takes $\eta^{-1}(W)$ onto $N_* = \eta^{-1}(\delta)$. We define e_t to be the identity on $K_R \cup \mathrm{Cl}(|J| - |N|)$ and so it remains only to define e_t on the rest of $N = N(K - L, J')$. Let σ be an arbitrary simplex of $N(K - L, J')$. We will show how to define e_t on σ. Recall that $\sigma = \alpha * \beta * \gamma$ as defined above. Let l be a segment in the join structure between $\alpha * \beta$ and γ. Then, l is mapped isomorphically onto a unique S_ϵ under η. Thus, we define e_t on l by $e_t \mid l = \eta^{-1} h_t \eta | l$. This completes the proof.

Let F be the k-dimensional hyperplane of E^n determined by

$x_k = \cdots = x_{n-1} = 0$. We let $\tilde{I}^k = I^{k-1} \times 0 \times \cdots \times 0 \times I \subset F$. Also, we let $F_+ = F \cap E_+^n$, $F_- = F \cap E_-^n$, $\tilde{I}_+^k = \tilde{I}^k \cap F_+$ and $\tilde{I}_-^k = \tilde{I}^k \cap F_-$.

Straightening Lemma 5.3.4. *Let $f: \tilde{I}_+^k \to E_+^n$, $n \geqslant 5$, be an embedding such that $f \mid I^{k-1} = 1$ and $f(\tilde{I}_+^k - I^{k-1})$ is locally flat in $E_+^n - E^{n-1}$. In the case that $k = n - 2$, suppose further that f is locally homotopically unknotted at points of* Int I^{k-1}. *Then, there exists a homeomorphism $g: E_+^n \twoheadrightarrow E_+^n$ which is the identity on E^{n-1} and outside some compact set and is such that $gf \mid \hat{I}_+^k = 1$ where $\hat{I}_+^k = \tilde{I}_+^k$ if $k \leqslant n - 3$ and where $\hat{I}_+^k$ is the upper half of a k-cell in the interior of, and concentric with I^k, if $k = n - 2$ or $n - 1$.*

Proof. Recall that we defined H_t, Q_t^+, Q_t^-, and $Q(t, t')$ in the last section. We may assume that f, in fact, embeds $\tilde{I}_+^k - I^{k-1}$ in Int Q_1^+. Let $j: E_+^n \to E^n$ be an embedding such that $j \mid Q_1^+ = 1$ and

$$((jQ(\tfrac{1}{2}, 1)), j(Q(0, \tfrac{1}{2}))) = (Q(0, 1), Q(-1, 0)).$$

Now the technique of proof of Theorem **5.2.1**, in the last section, gives us a homeomorphism $\bar{g}: Q_{-1}^+ \twoheadrightarrow Q_{-1}^+$ which is the identity on H_{-1} and outside some compact set and is such that $\bar{g}f \mid \hat{I}_+^k = 1$. The reason that we have that $\bar{g}$ is the identity on H_{-1} is that the homeomorphism r in the proof of Theorem **5.2.1** may be assumed to be the identity on Q_{-1}^-. We are able to get that $\bar{g}f \mid \hat{I}_+^k = 1$ for $\hat{I}_+^k = \tilde{I}_+^k$ when $k \leqslant n - 3$, because in that range we are able to apply Engulfing Lemma **5.2.1** and do not have to employ the Stallings trick. (If Section **5.2** had been developed exactly as Černavskiĭ presented it in [5], then we would not have had this stronger result in codimension three and so would only have been able to prove Theorem **5.3.1** for proper embeddings rather than allowable embeddings.) Now it is easy to see that $g = j^{-1}\bar{g}j$ is the desired homeomorphism.

Collaring Lemma 5.3.5. *Let R^{k-1} be a* PL-*manifold of dimension $k - 1$ which is contained in the boundary of a k-dimensional* PL-*manifold M^k and let $\epsilon > 0$ be given. Then, there is a* PL-*homeomorphism $\lambda: R \times [0, 1] \to M$ such that*

(a) $\lambda(r, 0) = r$,

(b) $\operatorname{diam}(\lambda(r \times [0, 1])) < \epsilon$, *and*

(c) $\lambda(R \times [0, 1])$ *is a neighborhood of R in M.*

Proof. Let N be a regular neighborhood of R in M. Let $\eta: \partial M \times I \to M$ be a PL collaring of ∂M in M; that is, $\eta(x, 0) = x$ for all $x \in \partial M$. It is easy to see that N and $\eta(R \times I)$ are both HZ-regular neighborhoods of R modulo ∂R such that there is a triangulation J of M for which N and $\eta(R \times I)$ satisfy condition β given in Section **1.6.5,C**. Hence, by Part 2 of Theorem **1.6.5**, there is a PL homeomorphism $h: \eta(R \times I) \twoheadrightarrow N$ such that $h(\eta(x, 0)) = x$ for all $x \in R$. Thus, $\lambda: R \times I \to M$ defined by $\lambda(x, t) = h\eta(x, t)$ is the desired PL homeomorphism. (Condition 2 of the conclusion follows by choosing the regular neighborhoods small and by uniform continuity.)

Proof of Theorem 5.3.1. First let us prove the theorem for $n - k \geqslant 3$. We will leave the easy modification of this proof necessary to handle the cases $n - k = 1$ and 2 as an exercise. We will not worry about getting an ϵ-push, but will simply get a push. An ϵ-push can be obtained by making small choices for our triangulations, neighborhoods, collars, and so forth. First, we will get a push e_t which satisfies the conclusion except for being the identity on ∂N, although it will be the identity on $f(f^{-1}(\partial N))$ and outside a small neighborhood of $f(f^{-1}(\partial N))$. Then, we will show how to modify the proof so as to have e_t be the identity on ∂N.

Let $R = f^{-1}(\partial N)$ and suppose that $R = \bigcup_{i=0}^{p} H^{k-1}$ is a decomposition of R as a "handlebody" assured by Theorem **1.6.12**. By Lemma **5.3.5**, there is a PL homeomorphism $\lambda: R \times I \to M$ ($I = [0, 1]$ here) such that $\lambda(r, 0) = r$, diam$(\lambda(r \times I))$ is small and $\lambda(R \times I)$ is a neighborhood of R in M. Now consider the collection $\lambda(H_0 \times I), \lambda(H_1 \times I), ..., \lambda(H_p \times I)$. This is a covering of $U = \lambda(R \times I)$ with k-balls. Note that it follows from Theorem **1.6.12** that $\lambda(H_i \times I)$ meets $\partial M \cup (\bigcup_{j=0}^{i-1} \lambda(H_j \times I))$ in a $(k - 1)$-ball.

We will get a sequence of isotopies e_t^i, $i = 0, 1, ..., p$, of N onto itself such that $e_0^i = 1$, $e_t^i \mid f(R) = 1$, and $e_1^i e_1^{i-1} \cdots e_1^0 f \mid \bigcup_{j=0}^{i} \lambda(H_j \times 1)$ is PL. We will construct the e_t^i so that $e_t^i \mid e_1^{i-1} \cdots e_1^0 f(\bigcup_{j=0}^{i-1} \lambda(H_j \times I))$ is the identity. Let $\lambda_*: \partial N \times I \to N$ be a PL collaring of ∂N. Then, $D_0 = \lambda_*(f(H_0) \times I)$ is a PL k-ball in N such that $D_0 \cap \partial N = f(H_0)$. Let V be the interior of a regular neighborhood of $f(H_0)$ in ∂N. Then, $(V, f(H_0)) \approx (E^{n-1}, I^{k-1})$ and $(\lambda_*(V \times [0, 1]), f(H_0)) \approx (E_+^n, I^{k-1})$. Let $g: \lambda(H_0 \times I) \twoheadrightarrow D_0$ be a PL homeomorphism such that $g \mid \lambda(H_0 \times 0) = f \mid \lambda(H_0 \times 0)$. Then, by applying Lemma **5.3.4**, we can get an isotopy $e_t^0: N \twoheadrightarrow N$ such that $e_0^0 = 1$, $e_t^0 \mid \partial N = 1$ and $e_1^0 f \mid \lambda(H_0 \times I) = g$. Thus, $e_1^0 f \mid \lambda(H_0 \times I)$ is PL.

Now, we will show how to construct e_t^1 and then it will be clear how to construct e_t^i, $i = 2, 3, ..., p$. Since N is a PL manifold and $e_1^0 f(\lambda(H_0 \times I))$ is link-collapsible on $\mathrm{Cl}(\lambda_*(f(R) \times I) - \lambda_*(f(H_0 \times I))$, it follows from Part 1 of Theorem **1.6.5** that we can choose a regular neighborhood of $e_1{}^0f(\lambda(H_0 \times I))$ mod $\mathrm{Cl}(\lambda_*(f(R) \times I) - \lambda_*(f(H_0) \times I))$. Hence, K^0 is a PL manifold which collapses to $e_1{}^0f(\lambda(H_0 \times I))$ and so is an n-ball. Furthermore, it follows from Theorem **1.6.5** that K^0 meets the boundary of N regularly, that is, in an $(n - 1)$-ball. Now apply Lemma **5.3.3** where the compact set C of that lemma is $e_0{}^1f(\bigcup_{j=1}^p \lambda(H_j \times I))$ and get a regular neighborhood $K_*{}^0$ of

$$e_1{}^0f(\lambda(H_0 \times I)) \bmod \mathrm{Cl}(\lambda_*(f(R) \times I) - \lambda_*(f(H_0) \times I))$$

and an isotopy $\tilde{e}_t{}^1$ which is the identity on $\lambda_*(f(R) \times I)$ and such that $\tilde{e}_1{}^1e_1{}^0f(\bigcup_{j=1}^p \lambda(H_j \times I)) \cap K_*{}^0 = \lambda_*(f(H_0) \times I) \cap (\bigcup_{j=1}^p \lambda_*(f(H_j) \times I))$. We now form a new PL manifold $N_1 = \mathrm{Cl}(N - K_*{}^0)$. Then, we have the embedding $\tilde{e}_1{}^1e_1{}^0f \mid \lambda(H_1 \times I)\colon \lambda(H_1 \times I) \to N_1$ and we have the PL k-ball $D_1 = \lambda_*(f(H_1) \times I) \subset N_1$ such that

$$\tilde{e}_1{}^1e_1{}^0f(\lambda(H_1 \times I)) \cap \partial N_1 = D_1 \cap \partial N_1$$

is a $(k - 1)$-ball. Thus, by applying Lemma **5.3.4**, similarly to the way we did in the construction $e_t{}^0$, we get an isotopy $\hat{e}_t{}^1$ of N_1 which is the identity on ∂N_1 and is such that $\hat{e}_1{}^1\tilde{e}_1{}^1e_1{}^0f \mid \lambda(H_1 \times I)$ is PL. Now, extend $\hat{e}_t{}^1$ to N by way the identity. Then, $e_t{}^1 = \hat{e}_t{}^1\tilde{e}_t{}^1$ is the desired isotopy and we see our way clear.

To finish the proof for $n - k \geqslant 3$, it remains only to show why we can assume that e_t is the identity on ∂N. Let K be a regular neighborhood of ∂N mod $f(R)$. Then by Lemma **5.3.3**, there is a regular neighborhood K_* of ∂N mod $f(R)$ and a small isotopy h_t of N keeping ∂N fixed such that $h_1f(M) \cap K_* = f(R)$. [The compact set C of Lemma **5.3.3** is $f(M)$.] Now, $N_* = \mathrm{Cl}(N - K_*)$ is a PL manifold and $h_1f\colon M \to N_*$ is an allowable embedding. Thus, we can go through the preceding proof and get an isotopy $\bar{e}_t$ of N_* such that $\bar{e}_1h_1f \mid U\colon U \to N_*$ is PL and $\bar{e}_t \mid f(R) = 1$. But, it follows from Remark **1.6.5** that K_* is a collar of ∂N in N pinched at $f(R)$ and so we can use K_* to extend $\bar{e}_t$ to N so that $\bar{e}_t \mid \partial N = 1$. Then, $e_t = \bar{e}_th_t$ is the desired isotopy.

Exercise 5.3.1. Formulate the necessary modification of the above proof to establish Theorem **5.3.1** for $n - k = 1$ and 2.

5.4. PL APPROXIMATING TOPOLOGICAL EMBEDDINGS

The problem of PL approximating topological embeddings has become an important part of the theory of topological embeddings. In this section, we will discuss the work which has been done in that area as well as develop an approximation technique fundamentally due to Homma. The approximation theorem proved in this section will be used in the next section. The problem of PL approximating topological embeddings is simply this:

When can a topological embedding of one polyhedron into another be approximated with a PL *embedding?*

We have already seen that a PL approximation theorem can play an important role in the proof of a taming theorem. Specifically, recall that the key two parts of the proof of taming Theorem **3.6.1** involved showing "solvability" and "denseness." The "denseness" part simply required showing that, in the trivial range, any topological embedding of a polyhedron into a PL manifold can be approximated with a PL embedding. This followed in that range by a straightforward application of general position.

The PL approximation problem becomes much more involved above the trivial range. Homma [2] attempted to show that a topological embedding $h: M \to N$ of a closed PL m-manifold M into the interior of a PL n-manifold N can be approximated by a PL embedding $g: M \to N$ whenever $m \leqslant n - 3$. Berkowitz [1] showed that Homma's proof contains certain difficulties. A portion of [Homma, 2], the most difficult part, was restated and published as [Homma, 3]. However, the main difficulties of [Homma, 2] are also present there. Berkowitz [1] was able to modify Homma's technique so as to establish the result whenever $m \leqslant \frac{3}{4}n - \frac{5}{4}$. (His proof appears in abbreviated form in [Berkowitz and Dancis, 1].) Later Berkowitz [2] adapted Homma's techniques to prove that a topological embedding of a (possibly noncompact) polyhedron into a PL manifold can be approximated by a PL embedding in the metastable range. (That proof of Berkowitz's appears in abbreviated form in [Berkowitz and Dancis, 2].) Weber [1, 2] had previously established, by different techniques, that result for compact polyhedra. It should be remarked that [Homma, 3] was followed up by [Homma, 4] which establishes a different sort of approximation theorem.

Recently proofs of codimension three PL approximation theorems have been announced. Černavskiĭ [7] announced that he could approximate topological embeddings into E^n of cells and spheres with PL embeddings in codimensions greater than two. A fairly detailed proof of this result appeared in [Černavskiĭ, 8]. However, there seems to be a gap in Černavskiĭ's proof resulting basically from the fact that he tried to take an HZ-relative regular neighborhood without having the link-collapsibility condition. Bryant [3] assumed Černavskiĭ's approximation theorem and proved PL approximation theorems for allowable embeddings of PL manifolds and for embeddings of polyhedra into the interior of PL manifolds in codimensions greater than three. Černavskiĭ [9], in a supplement to his article, sketched a proof that embeddings of manifolds can be PL approximated in codimensions greater than three. Miller [1] filled in Černavskiĭ's gap and gave a proof of the approximation theorem for cells in codimensions greater than three. Miller also obtained the approximation theorem for manifolds.

A couple of different types of approximation theorems have been proved recently. Price and Seebeck [1, 2] have obtained a codimension one approximation theorem for locally nice embeddings of manifolds with a flat spot, and Štan'ko [1] has proved that codimension three embeddings of compacta can be approximated by locally nice embeddings. The rest of this section will be devoted to proving the following metastable range approximation theorem for manifolds. The proof will be carried out by restricting [Homma, 2] to the metastable range where everything goes through nicely.

Metastable PL Approximation Theorem 5.4.1 (Homma). *If M is a (possibly noncompact)* PL *m-manifold without boundary, Q is a* PL *q-manifold where $m \leqslant \frac{2}{3}q - 1$, $h\colon M \to \operatorname{Int} Q$ is a topological embedding of M into Q and $\epsilon\colon M \to (0, \infty)$ is a continuous mapping of M into the positive real numbers, then there exists a* PL *homeomorphism $g\colon M \to Q$ such that $\operatorname{dist}(g(x), h(x)) < \epsilon(x)$ for each $x \in M$.*

Furthermore, if M_* is a PL *m-submanifold of M and if $h \mid U$ is* PL *where U is a neighborhood of* $\operatorname{Cl}(M - M_*)$, *then $g \mid \operatorname{Cl}(M - M_*)$ may be taken to be $h \mid \operatorname{Cl}(M - M_*)$.*

Although Homma does not say so in his papers, it is quite evident that he was familiar with the Penrose–Whitehead–Zeeman technique presented in Section **4.7**, because Homma's ingenious technique is a modification of that technique. It seems that the best way to convey Homma's technique painlessly is to exhibit it as such a modification. Hence, before beginning the proof of Theorem **5.4.1**, let us reconsider Metastable Range Embedding Theorem **4.7.2**.

Suppose that M is a closed PL m-manifold and that $h\colon M \to \operatorname{Int} Q$ is a map taking M into the interior of the PL q-manifold Q where $m \leqslant \frac{2}{3}q - 1$. The Penrose–Whitehead–Zeeman Theorem says that if M is $(2m - q)$-connected and Q is $(2m - q + 1)$-connected, then h can be homotoped to an embedding g.

Recall that the proof of the Penrose–Whitehead–Zeeman Theorem went basically as follows (see Fig. **5.4.1**). First homotop h to a map f in general position. Let $S(f)$ denote the singular set of f. Then since M is

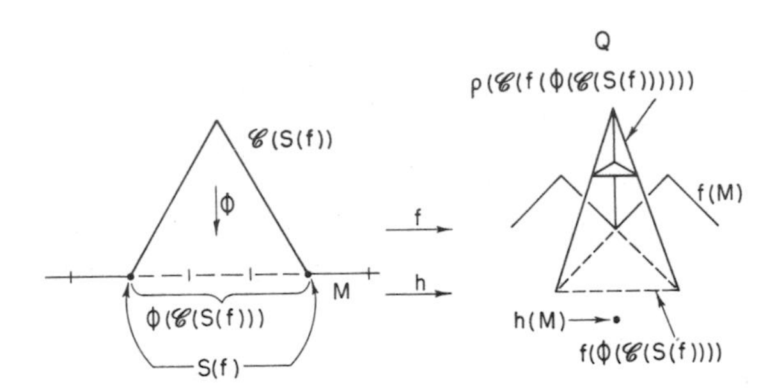

Figure 5.4.1

$(2m - q)$-connected and the cone $\mathscr{C}(S(f))$ is of dimension at most $2m - q$, $\mathscr{C}(S(f))$ can be mapped into M keeping the base fixed. Hence, by general position there is an embedding ϕ taking $\mathscr{C}(S(f))$ into M which keeps $S(f)$ fixed. Now consider the cone $\mathscr{C}(f(\phi(\mathscr{C}(S(f)))))$. Since it is at most $(2m - q + 1)$-dimensional and since Q is $(2m - q + 1)$-connected, we may map $\mathscr{C}(f(\phi(\mathscr{C}(S(f)))))$ into Q keeping the base fixed. By general position we can obtain an embedding ρ of $\mathscr{C}(f(\phi(\mathscr{C}(S(f)))))$ into Q which keeps $f(\phi(\mathscr{C}(S(f))))$ fixed such that

$$\rho(\mathscr{C}(f(\phi(\mathscr{C}(S(f)))))) \cap f(M) = f(\phi(\mathscr{C}(S(f)))).$$

Take a second derived neighborhood B_M of $\phi(\mathscr{C}(S(f)))$ and a second derived neighborhood B_Q of $\rho(\mathscr{C}(f(\phi(\mathscr{C}(S(f))))))$. Of course, B_M and B_Q are both balls. The embedding g is then taken to be f outside of B_M and to be a "conewise" extension of $f \mid \partial B_M$ on B_M which properly embeds B_M in B_Q.

In order to view Homma's technique as a modification of the above Penrose–Whitehead–Zeeman technique, one should visualize the last part of the above technique in a slightly different manner. In particular, suppose that ϕ and ρ have been constructed just as above. There is a natural way to construct a PL manifold $\tilde{M}$ which is homeomorphic to $M/\phi(\mathscr{C}(S(f)))$. Specifically, $\tilde{M} = (M - \text{Int } B_M)) \cup \tilde{B}_M$ where $\tilde{B}_M$ is the cone $\mathscr{C}(\text{Bd } B_M)$. There is a natural PL map $\tilde{\pi}\colon M \twoheadrightarrow \tilde{M}$ which is the linear extension of the map which takes vertices of $M - (\text{Int } B_M)$ to themselves and which takes vertices of $\phi(\mathscr{C}(S(f)))$ to the cone-point of $\tilde{B}_M$ (see Fig. **5.4.2**). Also, there is a PL homeomorphism $\lambda\colon \tilde{M} \twoheadrightarrow M$ which is the identity on $M - \text{Int } B_M$ and which takes $\tilde{B}_M$ onto B_M. Analogously, define

(1) $\tilde{Q} = (Q - \text{Int } B_Q) \cup \tilde{B}_Q$ where $\tilde{B}_Q = \mathscr{C}(\text{Bd } B_Q)$,

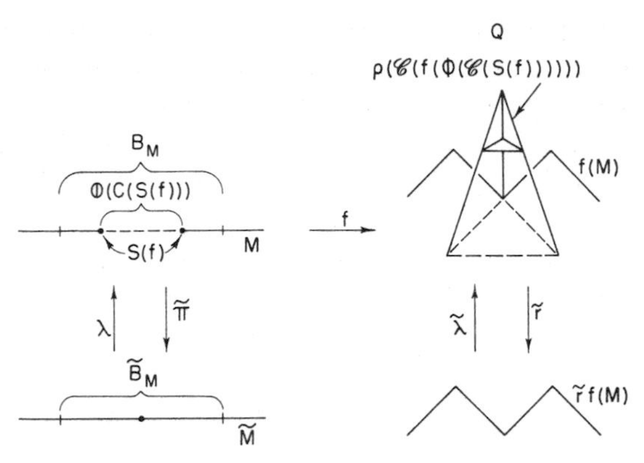

Figure 5.4.2

(2) a PL map $\tilde{r}: Q \twoheadrightarrow \tilde{Q}$, and

(3) a PL homeomorphism $\tilde{\lambda}: \tilde{Q} \twoheadrightarrow Q$.

It is easy to see that $\psi: \tilde{M} \to \tilde{Q}$ defined by $\psi = \tilde{r}f\tilde{\pi}^{-1}$ is a PL embedding of $\tilde{M}$ into $\tilde{Q}$. Thus, the g required for the conclusion of the Penrose–Whitehead–Zeeman Theorem may be taken to be $\tilde{\lambda}\psi\lambda^{-1}$.

We now give a couple of definitions before beginning the proof of Theorem **5.4.1**. First, let us recall the definition of i-LC given in Section **5.2**. A space X is **i-locally connected at a point** $x \in X$ (i-LC at x) if every open set U containing x contains an open set V containing x such that every mapping of S^i into V is null-homotopic in U. A space is i-LC if it is i-LC at every point. A space is **locally contractible at a point** x if every open set U containing x contains an open set V containing x such that V is contractible in U to a point. The space is **locally contractible** if it has the property at every point.

Exercise 5.4.1. (a) Every manifold is locally contractible. (b) Locally contractible implies i-LC for all i.

Proof of Theorem 5.4.1 (Homma's approximation technique). (We will not worry about showing that the approximation g which we will obtain can be taken to agree with h on M_*, because it will be obvious how to prove this added condition once one understands the proof of the absolute case.) The hypothesis of Theorem **5.4.1** gives us a topological embedding $h: M \to Q$. (Recall that the hypothesis of the Penrose–Whitehead–Zeeman Theorem only gave us a map $h: M \to Q$; however, we shall see that by requiring h to be an embedding, one eliminates the necessity of any connectivity conditions on M and Q.) It is easy to find an open set U in Q such that $h(M)$ is a closed subset of U. Of course U

inherits a PL structure from Q. From now on assume that $Q = U$ so that we have that $h: M \to Q$ is a closed embedding. Again homotop h a small amount to a closed general position PL map f and consider the singular set $S(f)$. This time, instead of taking the cone over $S(f)$, we take the mapping cylinder C_f of $f|S(f)$. It is instructive to think of C_f as being a bunch of cones which are stuck together; that is, $\gamma_f: C_f \to f(S(f))$ is the projection, the point inverses under γ_f are cones. Furthermore, since f approximates the embedding h, the base of each cone $\gamma_f^{-1}(x)$ is small in diameter. Hence, one can use the local contractibility (more particularly, the local i-connectivity) of M to map C_f into M keeping the base fixed. (Thus, because f approximates an embedding closely, the local contractibility of M serves the purpose here of the connectivity condition on M in the Penrose–Whitehead–Zeeman technique.) The way one uses the i-LC property to map C_f into M keeping the base fixed goes as follows: First, get a fine triangulation T of C_f so that $\gamma_f: C_f \to f(S(f))$ is simplicial. Map each vertex v of T to some point in $\gamma_f^{-1}(v)$. Now, extend the map to the 1-skeleton of T using 0-LC, then extend to the 2-skeleton by using 1-LC, and so forth. Notice that the image under this map of $\gamma_f^{-1}(x)$ is small in diameter for each $x \in f(S(f))$. As before, general position gives a PL embedding $\phi: C_f \to M$ which keeps the base fixed. Furthermore, $\phi\gamma_f^{-1}(x)$ is small in diameter for each $x \in f(S(f))$.

Now consider $f(\phi(C_f))$ and consider the mapping cylinder C_r where $r: f(\phi(C_f)) \twoheadrightarrow \phi(f(S(f)))$ is defined by $r = \phi\gamma_f\phi^{-1}f^{-1}$. (The mapping cylinder over a mapping cylinder, C_r, plays a role here analogous to the role played by the cone over a cone in the Penrose–Whitehead–Zeeman technique.) Let $\gamma_r: C_r \to \phi(f(S(f)))$ be the projection. This time local contractibility of Q gives a map of C_r into Q which keeps the base $f(\phi(C_f))$ of C_r fixed and such that the image under this map of $\gamma_r^{-1}(x)$ is small in diameter for each $x \in \phi(f(S(f)))$. General position gives an embedding $\rho: C_r \to Q$ which is the identity on $f(\phi(C_f))$ such that $\rho(C_r) \cap f(M) = f(\phi(C_f))$ and such that $\rho\gamma_r^{-1}(x)$ is small in diameter for each $x \in \phi(f(S(f)))$.

Define $\pi: \phi(C_f) \twoheadrightarrow \phi(f(S(f)))$ by $\pi = \phi\gamma_f\phi^{-1}$ (see Fig. **5.4.3**). Suppose that we can find a PL manifold $\tilde{M}$ which contains a copy of $\phi(f(S(f)))$ as a subpolyhedron and a PL map $\tilde{\pi}: M \twoheadrightarrow \tilde{M}$ which extends π and which is a PL homeomorphism off of $\phi(C_f)$. Also suppose we can find a PL homeomorphism $\lambda: \tilde{M} \twoheadrightarrow M$ such that $\lambda\tilde{\pi}$ approximates the identity on M.

We have the PL map $\gamma_r\rho^{-1}: \rho(C_r) \twoheadrightarrow \phi(f(S(f)))$. Suppose that we can find a PL manifold $\tilde{Q}$ which contains a copy of $\phi(f(S(f)))$ as a subpolyhedron and a PL map $\tilde{r}: Q \twoheadrightarrow \tilde{Q}$ which extends $\gamma_r\rho^{-1}$ and which is a PL homeomorphism off of $\rho(C_r)$. In addition, suppose that we can

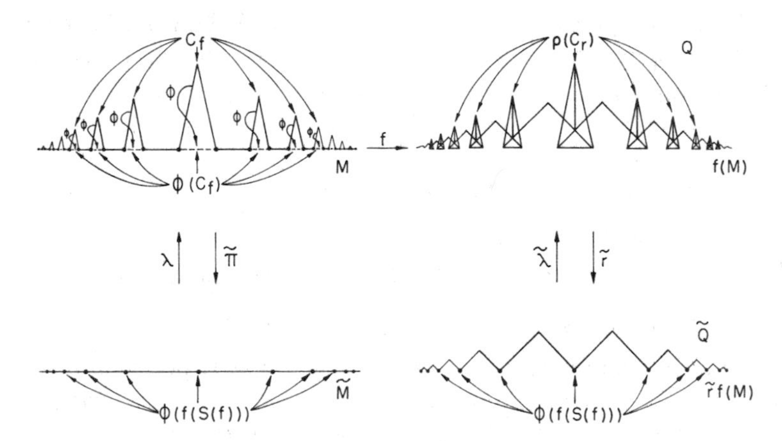

Figure 5.4.3

find a PL homeomorphism $\tilde{\lambda}: \tilde{Q} \twoheadrightarrow Q$ such that $\tilde{\lambda}\tilde{r}$ approximates the identity on Q.

Given $\tilde{M}$, $\tilde{Q}$, λ, $\tilde{\pi}$, $\tilde{\lambda}$, and $\tilde{r}$, the proof of Theorem **5.4.1** is completed just as the conclusion of our formulation of the Penrose–Whitehead–Zeeman technique above. That is, we define the PL embedding $\psi: \tilde{M} \to \tilde{Q}$ by $\psi = \tilde{r}f\tilde{\pi}^{-1}$ and we take $g = \tilde{\lambda}\psi\lambda^{-1}$ as the desired approximation.

One will understand Homma's technique from the above presentation once he sees how to construct $\tilde{M}$, $\tilde{Q}$, λ, $\tilde{\pi}$, $\tilde{\lambda}$, and $\tilde{r}$. Their existence will follow from Lemma **5.4.1**. Before stating and proving Lemma **5.4.1**, it is necessary to make some definitions.

A subcomplex L of a complex K is **locally collapsible** if for any simplex σ of K, the set $\cup\{\tau \in K \mid \sigma < \tau\} \cap |L|$ is collapsible. Let K be a complex and $\{L_i\}_J$ be collection of finite subcomplexes of K such that

$$N(L_i, K) \cap N(L_j, K) \subset \partial N(L_i, K) \cap \partial N(L_j, K) \qquad \text{for} \quad i \neq j.$$

Let $\{v_i\}_J$ be a collection of points not in $|K|$. The complex

$$\left(K - \bigcup_J \mathrm{St}(L_i, K)\right) \bigcup_J (v_i * \mathrm{Lk}(L_i, K))$$

is denoted by $K/\{L_i\}_J$ and is called the **quotient complex of** K **with respect to** $\{L_i\}_J$. We define the **projection mapping** $p: K \to K/\{L_i\}_J$ simplicially by defining p on the vertices of K as follows: (i) if v is a vertex of L_i, then $p(v) = v_i$; (ii) if v is a vertex of $K - \bigcup_J\{L_i\}$, then $p(v) = v$.

Let K be a complex and L a subcomplex of K. **A derived subdivision of K with respect to L**, denoted $K'(L)$, is defined inductively as follows:

(i) For each simplex $\sigma \in K - L$, let $a(\sigma)$ be a point of Int σ.

(ii) The set of vertices $K'(L)^0$ of $K'(L)$ is $\{v \mid v \text{ is a vertex of } L\} \cup \{a(\sigma) \mid \sigma \in K - L\}$.

(iii) Assume that $K'(L)^{i-1}$ has been defined. Let

$$K'(L)^i = K'(L)^{i-1} \cup \{\sigma \in L \mid \dim \sigma = i\} \cup \{a(\sigma) * \tau \mid \sigma \in K - L,\ \tau \in K'(L)^{i-1} \text{ and } \tau \subset \text{Bd}\, \sigma\}.$$

(iv) Finally, if $\dim K = k$, define $K'(L) = K'(L)^k$.

If π is a PL mapping of a (possibly noncompact) polyhedron F onto a possibly noncompact polyhedron G such that $\pi^{-1}(x)$ is collapsible for each $x \in G$, then the triple $\{F, G, \pi\}$ is called a **semi-forest**.

Let $\{F, G, \pi\}$ be a semi-forest and let M be a PL manifold such that F is contained in M as a closed subset. Let $K \subset H, L$ be triangulations of $F \subset M$, G, respectively with $\pi: K \to L$ simplicial. Let $H'(K)$ be a derived subdivision of H with respect to K. Let $H(L)$ be the complex containing L as a subcomplex such that σ is a simplex of $H(L)$ if and only if (a) $\sigma \in L$ or (b) $\sigma \in (H'(K) - N(K, H'(K))) \cup \partial N(K, H'(K))$ or (c) $\sigma = \langle v_0, \ldots, v_r, w_0, \ldots, w_t \rangle$, where $\langle v_0, \ldots, v_r \rangle \in L$, $\langle w_0, \ldots, w_t \rangle \in H'(K) - K$ and there exists $\tau \in K$ such that $\pi(\tau) = \langle v_0, \ldots, v_r \rangle$ and $\tau * \langle w_0, \ldots, w_t \rangle \in H'(K)$.

It is easy to show that (a) $\tilde{\pi}^{-1}(x) = \pi^{-1}(x)$ for $x \in |L|$ and (b) $\tilde{\pi}^{-1}(x)$ is collapsible for $x \in |H(L)| - |L|$. Thus, if $\{F, G, \pi\}$ is a semi-forest, then $\{M, |H(L)|, \tilde{\pi}\}$ is a semi-forest.

Forms of the following lemma were proved independently by Cohen [2] and Homma [2]. The proof given here is essentially Berkowitz's [2] unpublished formulation of Homma's proof.

Lemma 5.4.1. *Let $\{F, G, \pi\}$ be a semi-forest such that for all $p \in G$, $\text{diam}\, \pi^{-1}(p) < \delta$ for some fixed $\delta > 0$. Let M be a* PL *m-manifold without boundary with F contained in M as a closed subset. Then, there is a standard extension $\tilde{\pi}: M \to \tilde{M}$ and a* PL *homeomorphism $\lambda: \tilde{M} \to M$ such that* $\text{dist}(\lambda\tilde{\pi}, \text{identity}) < \delta$ (see Fig. **5.4.4**).

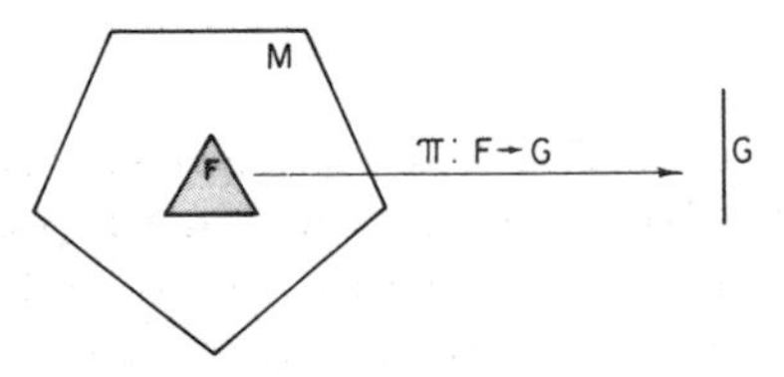

Figure 5.4.4

Proof. (We will first construct a PL homeomorphism $\lambda: \tilde{M} \to M$ and not worry about the condition that $\text{dist}(\lambda\tilde{\pi}, \text{identity}) < \delta$. Then, at the end of the proof, we will indicate how we can obtain that condition.) Let $K \subset H, L$ be triangulations of $F \subset M$, G respectively with $\pi: K \to L$ simplicial. Let $H(L)_b$ denote the barycentric subdivision of $H(L)$. Let $H'(K)_{\tilde{\pi}}$ denote a first derived subdivision of $H'(K)$ such that $\tilde{\pi}: H'(K)_{\tilde{\pi}} \to H(L)_b$ is simplicial. For each i, $0 \leqslant i \leqslant m$, order the i-simplices of $H(L)$: $\sigma_{i,1}, \sigma_{i,2}, \ldots$. Let $a_{i,j}$ be the barycenter of $\sigma_{i,j}$. Let J_i be the index set for $\sigma_{i,j}$. Consider the following sequence of quotient complexes:

$$\begin{aligned} H'(K)_{\tilde{\pi}} &= H_{-1} \xrightarrow{\pi_0} H_{-1}/\{\tilde{\pi}^{-1}(a_{0,j})\}_{J_0} = H_0 \xrightarrow{\pi_1} H_0/\{\pi_0\tilde{\pi}^{-1}(a_{1,j})\}_{J_1} \\ &= H_1 \xrightarrow{\pi_2} \cdots \xrightarrow{\pi_m} H_{m-1}/\{\pi_{m-1} \cdots \pi_0\tilde{\pi}^{-1}(a_{m,j})\}_{J_m} = H_m \end{aligned}$$

(see Fig. **5.4.5**).

Step 1. Since $\tilde{\pi}^{-1}(a_{0,j})$ is collapsible, $Lk(\tilde{\pi}^{-1}(a_{0,j}), H_{-1})$ is a PL-sphere. Hence, $v_{0,j} * \text{Lk}(\tilde{\pi}^{-1}(a_{0,j}), H_{-1})$ is a PL ball in H_0 which is PL homeomorphic to $\text{St}(\tilde{\pi}^{-1}(a_{0,j}), H_{-1})$ under a homeomorphism which is the identity on $\text{Lk}(\tilde{\pi}^{-1}(a_{0,j}), H_{-1})$. Let $M_0 = | H_0 |$. Then, there is a PL homeomorphism $\lambda_0: M_0 \to M$ such that

(a) $\lambda_0\pi_0(x) = x$, for $x \in \text{Cl}(M - \bigcup_{J_0} (\text{Int St}(\tilde{\pi}^{-1}(a_{0,j}), H_{-1})))$, and
(b) $\lambda_0\pi_0(x) \in \text{St}(\tilde{\pi}^{-1}(a_{0,j}), H_{-1})$ iff $x \in \text{St}(\tilde{\pi}^{-1}(a_{0,j}), H_{-1})$.

Step 2. Let us assume the following PL fact.

Fact 1. *$H_i \mid \pi_i \cdots \pi_0\tilde{\pi}^{-1}(a_{i+1,j})$ is full and locally collapsible in H_i for each $j \in J_{i+1}$.*

Let $M_i = | H_i |$ for $0 \leqslant i \leqslant m$. By induction we can assume that M_i is a PL manifold which is PL homeomorphic to M. Thus, it follows from Fact 1 and the following PL fact that $\text{St}(\pi_i \cdots \pi_0\tilde{\pi}^{-1}(a_{i+1,j}), H_i)$ is a PL ball.

Fact 2. *Let $K \supset L$ be a triangulation of a* PL *manifold and its subpolyhedron P such that*

(a) *L is full in K, and*
(b) *L is locally collapsible in K.*

Then, $\text{St}(L, K)$ *is a regular neighborhood of P.*

An argument similar to that in Step 1 applies to show the existence of a PL homeomorphism $\lambda_{i+1}: M_{i+1} \to M_i$ satisfying

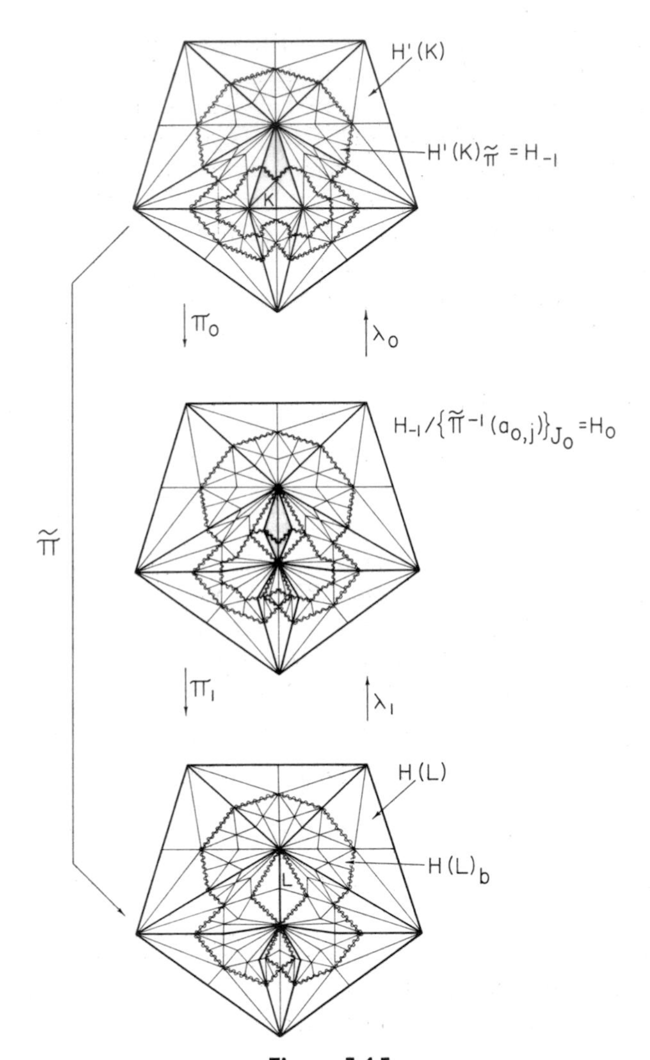

Figure 5.4.5

(a) $\lambda_{i+1}\pi_{i+1}(x) = x$, for

$$x \in \mathrm{Cl}\Big(M_i - \bigcup_{J_{i+1}} (\mathrm{Int}\ \mathrm{St}(\pi_i \cdots \pi_0\tilde{\pi}^{-1}(a_{i+1,j}), H_i))\Big),$$

(b) $\lambda_{i+1}\pi_{i+1}(x) \in \mathrm{St}(\pi_i \cdots \pi_0\tilde{\pi}^{-1}(a_{i+1,j}), H_i)$ if and only if

$$x \in \mathrm{St}(\pi_i \cdots \pi_0\tilde{\pi}^{-1}(a_{i+1,j}), H_i).$$

Step 3. Since $M_m = |H(L)|$, let $\lambda = \lambda_0 \cdots \lambda_m: M_m \to M$ and we have our PL homeomorphism.

Note that $\lambda\tilde{\pi}(x) = x$ for any $x \notin \operatorname{Int} N(|K|, H_{-1})$. Also note that because of Condition (b) of Step 2 and the fact that the interiors of $\operatorname{St}(\pi_i \cdots \pi_0\tilde{\pi}^{-1}(a_{i+1,j}), H_i)$ are disjoint for $j \in J_{i+1}$, we can insure that $\operatorname{dist}(\lambda\tilde{\pi}, \text{identity}) < \delta$ by choosing the mesh of H small enough in the preceding proof.

5.5. ε-TAMING ALLOWABLE EMBEDDINGS OF PL MANIFOLDS

At the beginning of Section **3.6**, we discussed some of the first work which was done on the taming of embeddings of general objects. In particular, we discussed the taming in the trivial range of embeddings of polyhedra and PL manifolds. In Sections **3.6**, **3.7**, and **3.8**, proofs of certain trivial range taming theorems were given. In this section, we will first discuss generalizations to lower codimensions of the results just mentioned. Then, we will discuss and prove some such generalizations which were developed in [Rushing, 2–4].

The first significant work done on taming general objects in codimensions lower than the trivial range was by Černavskiĭ. This work was sketched in [Černavskiĭ, 10] and presented in detail in [Černavskiĭ, 2]. The main result of these papers was that an embedding of a polyhedron into a combinatorial manifold which is locally flat on open simplexes is ϵ-tame, in the metastable range. Unfortunately, in 1968 this author discovered a mistake in [Černavskiĭ, 2]. Specifically, the mistake occurred in obtaining the homotopy φ_t^i which was asserted to exist following Proposition (H) of that paper. The error in Černavskiĭ's method for obtaining φ_t^j resulted from the fact that he was not careful enough about choosing triangulations. However by changing around Černavskiĭ's paper and doing some additional work, one can patch up Černavskiĭ's proof.

Bryant and Seebeck have also done fairly recently some work on taming. In [Bryant and Seebeck, 1] the following taming theorem is proved: Each locally nice embedding f of a k-dimensional polyhedron P into E^n, $n \geqslant 5$, $2k + 2 \leqslant n$, is ϵ-tame. A key part of the proof of this taming theorem is contained in an engulfing theorem. That Bryant–Seebeck Engulfing Theorem is a very useful result. For instance, it was used in [Bryant and Seebeck, 1] to establish the result which is mentioned in Exercise **5.5.2** at the end of this section. A form of the engulfing theorem as well as some other machinery, was used [Bryant and Seebeck, 3] to establish some codimension three taming theorems.

Quite recently, two students of Kister at the University of Michigan, Miller and Connelly, proved piecewise linear unknotting theorems which lead to proofs of taming theorems. Miller's work [2] was done first and is concerned with unknotting close PL embeddings of PL manifolds. Connelly's work [1] concerns unknotting close PL embeddings of polyhedra. (See the Appendix.)

A number of long outstanding problems in topological embeddings have been answered via some powerful machinery developed recently by Kirby, Siebenmann, and Wall. In particular, some results on taming were announced to follow from that machinery in [Kirby and Siebenmann, 1]. In addition to giving a taming theorem for codimensions greater than two, Kirby and Siebenmann give an obstruction to "strongly" taming in low codimensions. (For related work, see Theorem 6.1 of [Rourke and Sanderson, 1].)

Some results on low codimensional taming appear in [Cantrell and Rushing, 1].

The main results of this section are a codimension three taming theorem for allowable embeddings of PL manifolds, Theorem **5.5.1**, and a codimension three taming theorem for embeddings of certain polyhedra, Theorem **5.5.2**, both of which were proved in [Rushing, 4]. In applying taming theorems, one quite often needs a taming theorem for an allowable embedding, rather than for a proper embedding or for an embedding into the interior of the ambient manifold. (For instance, see [Rushing, 8].) The idea of the proof of Taming Theorem **5.5.1** is quite natural. One just takes a handlebody decomposition of the embedded manifold and then tames one handle at a time, each time keeping all of the previously tamed handles fixed. The proof is extremely geometrical in nature. One can actually see the PL handle onto which the topological handle is to more, form, and then one can see the topological handle move onto that PL handle. The taming around the boundary of the ambient manifold has already been handled (no pun intended) in Section **5.3**, and there again one could visualize the movement.

The main results of this section follow.

Taming Theorem 5.5.1. *Let* $f\colon M^k \to Q^n$, $n - k \geqslant 3$, *be an allowable embedding of the* PL *manifold* M^k *into the* PL *manifold* Q^n *such that* $f \mid f^{-1}(\partial Q)$ *and* $f \mid (M - f^{-1}(\partial Q))$ *are locally flat. Then,* f *is* ϵ-*tame.*

Addendum 5.5.1. *If $f \mid f^{-1}(\partial Q)$ is* PL, *then the taming isotopy is fixed on ∂Q.*

Addendum **5.5.1** can be sharpened as follows.

Addendum 5.5.2. *If the inclusion $M' \subset M - f^{-1}(\partial Q)$ is allowable where M' is a (compact)* PL *k-submanifold of $M - f^{-1}(\partial Q)$ and the inclusion $M'' \subset f^{-1}(\partial Q)$ is allowable where M'' is a* PL *$(k-1)$-submanifold of $f^{-1}(\partial Q)$ and if $f \mid M' \cup M''$ is* PL, *then f can be ε-tamed by an isotopy e_t such that*

$$e_t \mid [f(M' \cup M'')] \cup [\partial Q - N_\epsilon(f(f^{-1}(\partial Q) - M''), \partial Q)] \cup [Q - N_\epsilon(f(M - M'), Q)] = 1$$

for all t.

If $V \subset P$ are polyhedra, then the pair (V, P) is said to be **admissible** if $P = V \cup (\bigcup_{i=0}^{r} M_i^{k_i})$, where $M_i^{k_i}$ is a k_i-dimensional PL manifold such that $M_j^{k_i} \cap (V \cup (\bigcup_{j=0}^{i-1} M_j^{k_j}))$ is either empty or a $(k_i - 1)$-dimensional PL submanifold of $\partial M_i^{k_i}$ such that $V \cup (\bigcup_{j=0}^{i-1} M_j^{k_i})$ is link-collapsible on $M_i^{k_i} \cap (V \cup (\bigcup_{j=0}^{i-1} M_j^{k_j}))$, $i = 0, 1, \ldots, r$.

Taming Theorem 5.5.2. *Let (V, P) be an admissible pair of polyhedra where* $\mathrm{Cl}(P - V)$ *is k-dimensional and let Q^n, $n - k \geqslant 3$, be a* PL *n-manifold. Suppose that $f\colon P \to \operatorname{Int} Q$ is an embedding which is locally flat on the open simplexes of some triangulation of P and is such that $f \mid V$ is* PL. *Then, f can be ε-tamed by an isotopy e_t such that*

$$e_t \mid f(V) \cup (Q - N_\epsilon(\mathrm{Cl}(f(P - V))), Q) = 1.$$

Remark 5.5.1. Theorem **5.5.2** can also be formulated for embeddings which hit the boundary of Q.

Corollary 5.5.1. *Let $f_i\colon I^{k_i} \to I^n$, $n - k_i \geqslant 3$, $i = 1, 2, \ldots, r$ be proper embeddings such that* (a) *$f_i \mid \partial I^{k_i}\colon I^{k_i} \to \partial I^n$ is locally flat,* (b) *$f_i \mid \operatorname{Int} I^{k_i}$ is locally flat and* (c) *$f_i(I^{k_i}) \cap f_j(I^{k_j}) = \emptyset$ when $i \neq j$. Then, $f\colon \bigcup_{i=1}^{r} I^{k_i} \to I^n$ ($\bigcup_{i=1}^{r} I^{k_i}$ = disjoint union) defined by $f \mid I^{k_i} = f_i$ is ε-tame.*

Remark 5.5.2. The next two corollaries illustrate how results in the topological category can be obtained from results in the PL category via a taming theorem. The first corollary follows from Corollary **5.5.1** and the following result which was established in [Lickorish, 1]: *Piecewise linear spheres Σ^p and Σ^q contained in S^n, $n - p \geqslant 3$, $n - q \geqslant 3$, are unlinked if and only if regarding*

S^n *as* ∂I^{n+1}, Σ^p *and* Σ^q *bound disjoint* PL *balls* B^{p+1}, B^{q+1}, *whose interiors are contained in the interior of* B^{n+1}. (A collection of spheres of S^n is said to be **unlinked** if there is a collection of disjoint n-cells in S^n such that each sphere is contained in one of the n-cells and each of the n-cells contains exactly one of the spheres.)

The second corollary below follows from either of the above taming theorems combined with the Alexander isotopy (Exercise **4.11.2**) and the following unknotting theorem which was proved in [Zeeman, 6]: *Any proper* PL *ball pair* (B^n, B^k), *where* $n - k \geqslant 3$, *is* PL *homeomorphic to the standard pair* (I^n, I^k).

Corollary 5.5.2. *Spheres* $S_i^{k_i}$, $i = 1, 2, \ldots, r$, *contained locally flatly in* S^n, $n - k_i \geqslant 3$, *are unlinked if and only if regarding* S^n *as* ∂I^{n+1}, $S_i^{k_i}$, $i = 1, 2, \ldots, r$ *bound disjoint cells* $D_i^{k_i+1}$, $i = 1, 2, \ldots, r$, *respectively, whose interiors are contained in the interior of* I^{n+1} *and are locally flat there.*

Corollary 5.5.3. *Let* $f\colon I^k \to I^n$, $n - k \geqslant 3$, *be a proper embedding such that* $f \mid \partial I^k\colon \partial I^k \to \partial I^n$ *and* $f \mid \operatorname{Int} I^k$ *are locally flat. Then, there is an isotopy* $e_t\colon I^n \twoheadrightarrow I^n$ *such that* $e_0 = 1$ *and* $e_1 f = 1$. *Furthermore, if* $f \mid \partial I^k = 1$, *then* $e_t \mid \partial I^n = 1$.

Taming Theorem 5.5.3. *Let* M^k *and* Q^n, $n - k \geqslant 3$, *be* PL *manifolds and suppose that* $P^p \subset M^k$ *is a polyhedron. Let* $f\colon M \to \operatorname{Int} Q$ *be an embedding such that* $f \mid P$ *and* $f \mid (M - P)$ *are* PL. *Then,* f *can be* ϵ*-tamed by an isotopy which is the identity outside the* ϵ*-neighborhood of* $f(P)$.

REMARK 5.5.3. Theorem **5.5.3** can be formulated for allowable embeddings.

Before beginning the proofs of the above three theorems, we will formulate a few "statements" and prove a couple of preliminary lemmas.

Let P be a polyhedron contained in the PL ball H. Then an embedding $f\colon P \cup \operatorname{Int} H \to \operatorname{Int} Q$ of $P \cup \operatorname{Int} H$ into the interior of the PL manifold Q is said to be **piecewise linear** (PL) if for every C-regular neighborhood N of ∂H mod P in H, it is true that $f \mid \operatorname{Cl}(H - N)$ is PL is the usual sense.

STATEMENT $\eta^1(n, k, p)$. *Let* $D^n \subset E^n$ *be a locally flat n-cell, let* $P^p \subset I^k$ *be a polyhedron and let* $f\colon I^k \to D$ *be a proper, locally flat embedding such that* $f \mid P$ *is* PL. *Then, there is a proper embedding* $g\colon I^k \to D$ *such that* (1) $g \mid P \cup \partial I^k = f \mid P \cup \partial I^k$ *and* (2) $g \mid P \cup \operatorname{Int} I^k$ *is* PL.

REMARK 5.5.4. $\eta^1(n, k, p)$ is true for $n \geqslant 2k + 1$ simply by general position.

STATEMENT $\eta^2(n, k, m)$. *Let* $D^n \subset E^n$ *be a locally flat n-cell and let R* *and R' be regular neighborhoods of $\partial I^m \times I^{k-m}$ in I^k such that R' is contained in the point-set interior of R. Suppose that $f\colon I^k \to D$ is a proper, locally flat embedding such that $f \mid R$ is* PL. *Then, there is a proper embedding $g : I^k \to D$ such that*

(a) $g \mid R' \cup \partial I^k = f \mid R' \cup \partial I^k$, and

(b) $g \mid R' \cup \operatorname{Int} I^k$ *is* PL.

REMARK 5.5.5. (a) $\eta^2(n, k, m)$ is true for $n \geqslant 2k + 1$ by general position. (b) $\eta^2(n, k, m)$ follows directly from Theorem **5.4.1** whenever $k \leqslant \frac{2}{3}n - 1$. (This fact, let us emphasize, makes the proof of Theorem **5.5.1** complete in this book for the metastable range.) (c) For $k < \frac{3}{4}n - 1$, $\eta^2(n, k, m)$ follows from Berkowitz's modification [1] of Homma's technique. (Berkowitz's modification was mentioned in the last section.) (d) $\eta^2(n, k, k)$ follows from Exercise **4.7.1** for $n - k \geqslant 3$. (e) For $n - k \geqslant 3$, $\eta^2(n, k, m)$ follows from the various codimension three approximation theorems mentioned in the last section. Two other methods for obtaining $\eta^2(n, k, m)$ are given in [Rushing, 4].

STATEMENT $\omega(n, k)$. *Let $M^k \subset E^n$, $n - k \geqslant 3$, be a (possibly infinite) k-dimensional* PL *manifold. Suppose that $f\colon E^n \twoheadrightarrow E^n$ is a (topological) homeomorphism such that $f \mid M^k$ is* PL. *If $\epsilon(x) > 0$ is a continuous real-valued function on E^n, then there is a* PL *homeomorphism $g\colon E^n \twoheadrightarrow E^n$ which is an $\epsilon(x)$-approximation of f such that $g \mid M = f \mid M$.*

REMARK 5.5.6. (a) Theorem **4.11.1** shows that $\omega(n, k)$, $n \geqslant 5$, is true whenever $M = \emptyset$; that is, $\omega(n, -1)$ is true. It is possible that one could prove the ω-statement for other cases by appropriately modifying the technique of proof of that theorem. (b) It should be pointed out that in [Siebenmann and Sondow, 1] it is shown that for $n \geqslant 5$ there is a PL $(n-2)$-sphere K in S^n and a (topological) homeomorphism $h : S^n \to S^n$ such that $h \mid K$ is PL and for which there is no PL homeomorphism $g\colon S^n \to S^n$ such that $g \mid K = h \mid K$. However, K is not locally flat in S^n in these examples.

Proposition 5.5.1. $\omega(n, k) \Rightarrow \eta^2(n, k, m)$, $m = 0, 1, \ldots, k$ for $n - k \neq 2$.

PROOF. Extend f of η^2 to take I^n onto D^n. Let $h\colon \operatorname{Int} D^n \twoheadrightarrow \operatorname{Int} I^n$ be a PL homeomorphism. Consider $\operatorname{Int} I^n$ to be E^n. Then, $hf\colon E^n \twoheadrightarrow E^n$ is a topological homeomorphism which is PL when restricted to the infinite PL manifold $R \cap \operatorname{Int} I^n$. Let $\epsilon(x) > 0$ be a continuous function on E^n such that $\epsilon(x) \to 0$ as $x \to \infty$. By applying $\omega(n, k)$, we can get a PL homeomorphism $g\colon E^n \twoheadrightarrow E^n$ which is an $\epsilon(x)$-approximation of hf such that $g \mid R \cap \operatorname{Int} I^n = hf \mid R \cap \operatorname{Int} I^n$. Then, $h^{-1}g\colon \operatorname{Int} I^n \twoheadrightarrow \operatorname{Int} D^n$ can be extended to take I^n onto D^n by means of $f \mid \partial I^n$ and this extension when restricted to I^k satisfies the conclusion of η^2.

STATEMENT $\tau^1(n, k, p)$. *Let $P^p \subset K^k$ be polyhedra and let $f: K \to \operatorname{Int} Q^n$ be an embedding of K into the interior of the* PL *n-manifold Q such that $f \mid P$ and $f \mid K - P$ are* PL. *Then, f can be ϵ-tamed by an isotopy which is the identity outside the ϵ-neighborhood of $f(P)$.*

REMARK 5.5.7. (a) Notice that Theorem **5.5.3** establishes τ^1 for the case $n - k \geqslant 3$ and K is a PL manifold. (b) The following special case of τ^1 is proved in [Rushing, 4]: *Let M^k and Q^n, $n - k \geqslant 3$, be* PL *manifolds and suppose that $P_i^{p_i}$, $i = 1, 2, ..., r$, are disjoint collapsible polyhedra (for instance balls) contained in M such that either $P_i \cap \partial M = \emptyset$ or $P_i \cap \partial M$ is collapsible. Let $\epsilon > 0$ be given. If $f: M \to \operatorname{Int} Q$ is an embedding such that $f \mid \bigcup_{i=1}^r P_i$ and $f \mid M - \bigcup_{i=1}^r P_i$ are* PL, *then f is tame, and the taming isotopy is the identity outside the ϵ-neighborhood of $f(\bigcup_{i=1}^r P_i)$.*

The final statement is weaker than τ^1; however, it does not follow from Remark **5.5.7**(b).

STATEMENT $\tau^2(n, k, m)$. *Let M^k be a k-dimensional* PL *manifold. Suppose $M^k = M_*{}^k \cup H^k$ where $M_*{}^k$ is a k-dimensional* PL *manifold and $(H^k, H^k \cap M_*{}^k) \overset{\text{PL}}{\approx} (I^k, \partial I^m \times I^{k-m})$ for some m such that $0 \leqslant m \leqslant k$. Furthermore, suppose that $f: M \to \operatorname{Int} Q^n$ is an embedding such that $f \mid (M^k - (H^k \cap M_*{}^k))$ and $f \mid M_*{}^k$ are* PL. *Then, f can be ϵ-tamed by an isotopy which is the identity outside the ϵ-neighborhood of $f(H^k \cap M_*{}^k)$.*

Let us now consider two lemmas. The proof of the first lemma is straightforward and will be left as an exercise (see Theorem **1.6.12**).

Handlebody Lemma 5.5.1. *Let M^k be a k-dimensional* PL *manifold, let V^k be a k-dimensional* PL*-submanifold of M such that the inclusion $V \subset M$ is allowable, and let $\epsilon > 0$ be given. Then, $M = V \cup (\bigcup_{i=0}^p H_i)$, where*

$$\left(H_j, H_j \cap \left(V \cup \left(\bigcup_{i=0}^{j-1} H_i\right)\right)\right) \overset{\text{PL}}{\approx} (I^k, \partial I^{m_j} \times I^{k-m_j})$$

for some $m_j \leqslant k$, $j = 0, 1, ..., p$. Furthermore, diam $H_j < \epsilon$ *for $j = 0, 1, ..., p$. (H_j is called a handle of index m_j.)*

Question 5.5.1. Does Lemma **5.5.1** hold if we drop the requirement that the inclusion $V \subset M$ be allowable and let V be any codimension zero submanifold.

Lemma 5.5.2. *Let P be a polyhedron in the* PL *ball H and let C be a compact set such that $C \cap H \subset \partial H \cap P$ and let $X = C \cup H$. Suppose that $f: X \to \operatorname{Int} Q$ is an embedding of X into the interior of the* PL *manifold*

Q such that $f \mid H$ is locally flat and $f \mid P \cup \operatorname{Int} H$ is PL. *Then, given $\epsilon > 0$, $f \mid H$ can be ϵ-tamed by an isotopy e_t such that*

$$e_t \mid f(C) \cup f(P) \cup (Q - N_\epsilon(f(H), Q)) = 1.$$

PROOF. By using Theorem **3.4.1** and Crushing Lemma **5.3.3**, we can get a locally flat n-cell $D \subset \operatorname{Int} Q$ in a small neighborhood of $f(H)$ such that $(D, D \cap f(X)) = (D, f(H))$ is a trivial cell pair and D is contained in an open n-cell $U \subset \operatorname{Int} Q$ (see Fig. **5.5.1**). Let h: $H \twoheadrightarrow I^k$ be a PL homeo-

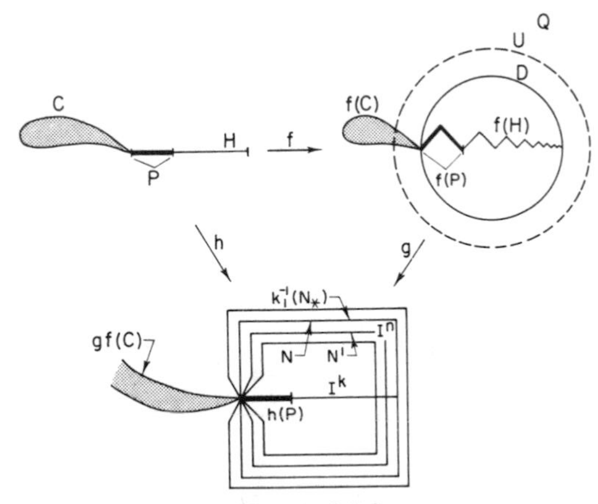

Figure 5.5.1

morphism. Then, we can get a homeomorphism g: $(D, f(H)) \twoheadrightarrow (I^n, I^k)$ such that $gfh^{-1} = 1$. Hence, we have the following commutative diagram,

$$\begin{array}{ccc} (-, H, P) & \xrightarrow{f|H} & (D, f(H)), f(P)) \\ & {\scriptstyle h}\searrow \quad \swarrow{\scriptstyle g} & \\ & (I^n, I^k, h(P)). & \end{array}$$

Let N be a small C-regular neighborhood of ∂I^n mod $h(P)$ in I^n. Then, by Remark **1.6.5**, N is a collar of ∂I^n pinched at $h(P) \cap \partial I^k$, that is,

$$N \approx (\partial I^n \times I)/((y, t) = (y, 0) \quad \text{if} \quad y \in h(P) \cap \partial I^k, \quad 0 \leqslant t \leqslant 1).$$

Also $\operatorname{Cl}(I^n - N)$ is a PL n-ball and $\operatorname{Cl}(I^k - (N \cap I^k))$ is a PL k-ball. Thus, $fh^{-1}(\operatorname{Cl}(I^k - (N \cap I^k))) = g^{-1}(\operatorname{Cl}(I^k - (N \cap I^k)))$ is a PL k-ball. Now let N' be a C-regular neighborhood of $\partial(\operatorname{Cl}(I^n - N))$ mod $h(P)$. Similar remarks hold for N' that held for N.

The homeomorphism g: $D \twoheadrightarrow I^n$ can be extended to take U onto E^n

by using Generalized Schoenflies Theorem **1.8.2**. Let N'' be a small C-regular neighborhood of $\partial I^n \bmod h(P) \cap \partial I^k$ in the PL manifold $E^n - \operatorname{Int} I^n$. Crushing Lemma **5.3.3** says that there is a small C-regular neighborhood N_* of $\partial I^n \bmod h(P) \cap \partial I^k$ in $E^n - \operatorname{Int} I^n$ and a small push k_t of $(E^n - \operatorname{Int} I^n, h(P) \cap \partial I^k)$ keeping $\partial I^n \cup (E^n - N'')$ fixed such that $k_1 gf(C) \cap N_* \subset h(P) \cap \partial I^k$. Then N_* is a collar of ∂I^n in $E^n - \operatorname{Int} I^n$ which is pinched at $h(P) \cap \partial I^k$ and $I^n \cup N_*$ is an n-ball. Let B^n be $I^n \cup k_1^{-1}(N_*)$. Now we can construct a small isotopy $\bar{e}_t \colon B^n \twoheadrightarrow B^n$ such that

(1) $\bar{e}_t \mid (\partial B^n \cup \operatorname{Cl}(I^n - (N \cup N'))) = 1$,
(2) $\bar{e}_1(k^{-1}(N_*)) = k_1^{-1}(N_*) \cup N$,
(3) $\bar{e}_1(N \cup N') = N'$, and
(4) $\bar{e}_1 \mid I^k \colon I^k \twoheadrightarrow \operatorname{Cl}(I^k - (N \cap I^k))$ is PL.

Define $e_t \colon g^{-1}(B^n) \twoheadrightarrow g^{-1}(B^n)$ by $e_t = g^{-1}\bar{e}_t$ and extend e_t to all of Q by the identity. This is clearly the desired isotopy and the proof is complete.

Theorem 5.5.4. *Theorem* **5.5.1** *with Addendums* **5.5.1** *and* **5.5.2** *holds if* $\eta^2(n, k, m)$ *and* $\eta^2(n-1, k-1, m-1)$ *are true for every* m *such that* $0 \leqslant m \leqslant k$.

REMARK 5.5.8. Theorem **5.5.1** with Addendums **5.5.1** and **5.5.2** follows from Theorem **5.5.4** and Remark **5.5.5**. Remark **5.5.5** suggests several ways of establishing η^2 in various codimensions.

Proof of Theorem 5.5.4 *Case* 1 (*Assume that* $f(M) \subset \operatorname{Int} Q$ *and that* $n \geqslant 5$). Apply Lemma **5.5.1**, letting M' play the role of V and get the type of handlebody decomposition of M assured by that lemma where the diameters of the handles are small in comparison with the ϵ of Theorem **5.5.1**. Suppose that $f \mid M' \cup (\bigcup_{i=0}^{j-1} H_i)$ can be ϵ-tamed by an isotopy e_t^{j-1} such that $e_t^{j-1} \mid f(M') \cup (Q - N_\epsilon(f(\bigcup_{i=0}^{j-1} H_i), Q)) = 1$ for all t, where j is such that $0 \leqslant j \leqslant p$. (If $j = 0$, then we define $\bigcup_{i=0}^{j-1=-1} H_i = \emptyset$.) We will now show how to ϵ-tame $f \mid M' \cup (\bigcup_{i=0}^{j} H_i)$ keeping $f(M') \cup (Q - N_\epsilon(f(\bigcup_{i=0}^{j} H_i), Q))$ fixed (see Figure **5.5.2**).

Let M_{j-1} denote $M' \cup (\bigcup_{i=0}^{j-1} H_i)$. Certainly $e_1^{j-1}f(M_{j-1})$ is link-collapsible on $e_1^{j-1}f(H_j \cap M_{j-1})$. Let T be a triangulation of Q^n which contains $e_1^{j-1}f(M_{j-1})$ and $e_1^{j-1}f(H_j \cap M_{j-1})$ as subcomplexes and let N be the simplicial neighborhood of $e_1^{j-1}f(M_{j-1} - (H_j \cap M_{j-1}))$ in a second derived subdivision of T. Now by applying Crushing Lemma **5.3.3**, where the compact set C of that lemma is $e_1^{j-1}f(H_j)$, we get N_* (which is both a C-regular neighborhood and a HZ-regular neighborhood of $e_1^{j-1}f(M_{j-1}) \bmod e_1^{j-1}f(H_j \cap M_{j-1}))$ and a small push $\bar{e}_t^j$ of $(Q, e^{j-1}f(H_j \cap M_{j-1}))$ keeping $e^{j-1}f(M_{j-1}) \cup \mathrm{Cl}(Q - N)$ fixed such that $\bar{e}_1{}^j e_1^{j-1}f(H_j) \cap N_* = e_1^{j-1}f(H_j \cap M_{j-1})$.

By the Alexander–Newman Theorem (see Remark **1.8.3**),

$$Q_* = \mathrm{Cl}(Q - N_*)$$

is a PL manifold and $\bar{e}_1{}^j e_1^{j-1}f \mid H_j \colon H_j \to Q_*$ is an allowable embedding which satisfies the hypotheses of Taming around Boundary Theorem **5.3.1**. Hence by that theorem, we can get a regular neighborhood R of $H_j \cap M_{i-1}$ in H_j and a small push $\hat{e}_t^j$ of $(Q_*, \bar{e}_1{}^j e_1^{j-1}f(H_j \cap M_{j-1}))$ which is fixed on ∂Q_* such that $\hat{e}_1{}^j \bar{e}_1{}^j e_1^{j-1}f \mid R \colon R \to Q_*$ is PL. Extend $\hat{e}_t^j$ to Q by way of the identity. (Notice that if H_j is a handle of index 0, then this paragraph and the preceding one say nothing.)

By using Theorem **3.4.2** and Crushing Lemma **5.3.3**, we can get a locally flat n-cell $D^n \subset \mathrm{Int}\, Q^n$ of small diameter such that

$$(D^n, D^n \cap \hat{e}_1{}^j e_1{}^j e_1^{j-1}f(M)) = (D^n, \hat{e}_1{}^j e_1{}^j e_1^{j-1}f(H_j))$$

is a trivial cell pair (see Fig. **5.5.2**). Of course, D^n is contained in an

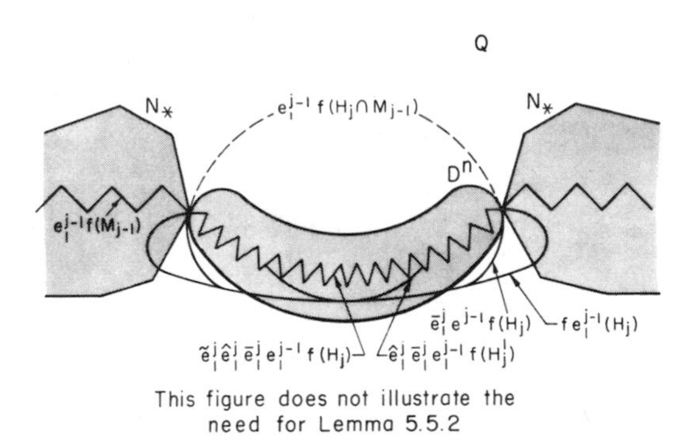

Figure 5.5.2

open n-cell $U \subset \mathrm{Int}\, Q$ and U inherits a PL structure from Q^n such that $U \approx E^n$ by Corollary **4.4.1**. Now apply $\eta^2(n, k, m)$, where (D^n, E^n, I^k, f, R) of η^2 corresponds to $(D^n, U, H_j, \hat{e}_1{}^j \bar{e}_1{}^j e_1^{j-1}f \mid H_j, R)$. Then we get a proper embedding $g \colon H_j \to D$ such that for a regular neighborhood of R' of $H_j \cap M_{j-1}$ in H_j, contained in the point-set interior of R,

(1) $g \mid R' \cup \partial H_j = f \mid R' \cup \partial H_j$, and

(2) $g \mid R' \cup \mathrm{Int}\, H_j$ is PL.

We would now like to get an isotopy $\tilde{e}_t^j\colon D \twoheadrightarrow D$ such that

(1) $\tilde{e}_0^j = 1$,

(2) $\tilde{e}_t^j \mid \partial D = 1$, and

(3) $\tilde{e}_1^j \hat{e}_1^j \bar{e}_1^j e_1^{j-1} f \mid H_j = g$.

By going to the standard situation and applying Taming around Boundary Theorem **5.3.1**, Unknotting Cell Pairs Theorem **4.5.2**, and the Alexander isotopy (Exercise **4.11.2**), we can get such an isotopy. Extend $\tilde{e}_t^j$ to Q by way of the identity.

It now follows that we can ϵ-tame $f \mid M' \cup (\bigcup_{i=0}^{j} H_i)$ keeping $f(M') \cup (Q - N_\epsilon(f(\bigcup_{i=0}^{j-1} H_i), Q))$ fixed by applying Lemma **5.5.2** where $(R', H_j, M_{j-1}, \tilde{e}_1^j \hat{e}_1^j e_1^{j-1} f, Q)$ correspond to (P, H, C, f, Q) of that lemma. This establishes the inductive step and the proof of Case 1.

Case 2 (*Assume that* $f(M) \subset \operatorname{Int} Q$ *and* $n = 4$). This case follows from Corollary **3.6.1**.

Case 3 (*Assume that* $f(M) \cap \partial Q \neq \emptyset$). First apply Case 1 or Case 2 to $f \mid f^{-1}(\partial Q)\colon f^{-1}(\partial Q) \to \partial Q$ and tame this embedding without moving far. Then use a small collar of ∂Q in Q to extend the taming isotopy to Q. Now we are in position to apply Taming around Boundary Theorem **5.3.1** to tame on a small regular neighborhood N of $f^{-1}(\partial Q)$ in M. Let N' denote a smaller regular neighborhood of $f^{-1}(\partial Q)$ in M. Then, Case 3 follows by applying Case 1 or Case 2 to $f \mid \operatorname{Cl}(M - N')$ after throwing $\operatorname{Cl}(N - N')$ in with M'. This completes the proof of Theorem **5.5.1**.

The next theorem follows from Theorem **5.5.4** and Proposition **5.5.1**.

Theorem 5.5.5. *Theorem* **5.5.1** *with Addendums* **5.5.1** *and* **5.5.2** *holds if* $\omega(n, k)$ *and* $\omega(n-1, k-1)$ *are true.*

Theorem 5.5.6. *Let* $f\colon M^k \to Q^n$, $n - k \geqslant 3$ *be an allowable embedding of the* PL *manifold* M^k *into the* PL *manifold* Q^n *such that* $f \mid f^{-1}(\partial Q)$ *and* $f \mid M - f^{-1}(\partial Q)$ *are locally flat. Also, let* $f \mid P$ *be* PL *where* P *is a subpolyhedron of* M *such that* $P \cap (M - f^{-1}(\partial Q))$ *is compact. If* $\eta^1(n, k, p)$ *and* $\eta^1(n-1, k-1, p-1)$ *are true for every* p *such that* $0 \leqslant p \leqslant k$, *then* f *can be* ϵ*-tamed by an isotopy* e_t *such that* $e_1 \mid P = f \mid P$.

REMARK 5.5.9. Remark **5.5.4** allows us to drop the η^1 conditions from the hypothesis of Theorem **5.5.6** whenever $n \geqslant 2k + 1$.

Proof of Theorem 5.5.5. (The proof of this theorem is quite similar to that of Theorem **5.5.4**; however, there is one major difference which occurs at the first of the proof. We will indicate this difference.) Again we start by assuming that $f(M) \subset \text{Int}\, Q$ and this time we get a handlebody decomposition $\{H_0, H_1, ..., H_p\}$ of all of M, where the diameters of the handles are small. Suppose that $f \mid \bigcup_{i=1}^{j-1} H_i$ can be tamed by an isotopy e_1^{j-1} such that $e_1^{j-1}f \mid P = 1$ for some j such that $0 \leqslant j \leqslant p$. Now we want to ϵ-tame $f \mid \bigcup_{i=0}^{j-1} H_j$ and in doing so bring $f(P)$ back where it starts.

Let $M_{j-1} = \bigcup_{i=0}^{j-1} H_i$ and let R be a small regular neighborhood of $H_j \cap M_{j-1}$ in M_{j-1}. Then $H_j \cup R$ is a PL k-ball. The proof now proceeds in much the same way as the proof of Theorem **5.5.4** where we let the k-ball $H_j \cup R$ play the role of H_j in that proof and $e_1^{j-1}f$ play the role of $\hat{e}_1{}^j\bar{e}_1{}^je_1^{j-1}f$ in that proof. When we come to the appropriate place in that proof we would of course apply $\eta^1(n, k, p)$ rather than $\eta^2(n, k, m)$, and the P of the η^1-statement would be $(P \cap H_j) \cup R$.

The proof of the following theorem is a simplification of the preceding two proofs.

Theorem 5.5.7. *Theorem* **5.5.1** *holds if*

$$\tau^2(n, k, m) \qquad \textit{and} \qquad \tau^2(n-1, k-1, m-1)$$

are true.

Theorem 5.5.8. *If $\tau^1(n, k, p)$ is true for $0 \leqslant p \leqslant k$, then any embedding $f: P^k \to Q^n$, $n - k \geqslant 3$, of a polyhedron P into the interior of a* PL-*manifold Q which is locally flat on the open simplexes of some triangulation is ϵ-tame.*

Exercise 5.5.1. Use Exercise **5.5.2**(a), below, Crushing Lemma **5.3.3**, Theorem **4.11.1**, and the stable Homeomorphism theorem (see Remark **4.11.1**) to prove Theorem **5.5.8**. (Although one should easily be able to do this exercise without assitance, we will point out that the key step of the exercise is stated as Lemma 1 of [Cantrell and Rushing, 1].)

Proof of Theorem 5.5.2. Express P as $V \cup (\bigcup_{i=0}^{r} M_i^{k_i})$ in the manner assured by the hypothesis that (V, P) is an admissible pair. Let N_j denote $V \cup (\bigcup_{i=0}^{j} M_j^{k_i})$ and inductively suppose that $f \mid N_{j-1}$ is PL. By Crushing Lemma **5.3.3**, there is a regular neighborhood N_* of $f(N_{j-1}) \bmod f(N_{j-1} \cap M_j^{k_j})$ and a small push $\bar{e}_t$ of $(Q, f(N_{j-1} \cap M_j^{k_j})$. By the Alexander–Newman Theorem (see Remark **1.8.3**), $Q_* = \text{Cl}(Q - N_*)$ is a PL manifold and $\bar{e}_1 f \mid M_j^{k_j}: M_j^{k_j} \to Q_*$ is an allowable embedding. The inductive step now follows by an application of Theorem **5.5.1** and Exercise **5.5.2** below.

REMARK 5.5.10. Various refinements of Theorem **5.5.2** can be made by applying the refinements of Theorem **5.5.1** given above.

EXERCISE 5.5.2. A form of the following γ-statement was first given in [Cantrell and Lacher, 1].

$\gamma(n, m, k)$: *Let D^m be an m-cell in E^n and let D^k, $k \leqslant m - 1$, be a k-cell in* Bd D^m *such that $D - E$ is locally flat in E^n and D^k is locally flat in* Bd D^m *and E^n. Then, D^m is locally flat in E^n (hence flat).*

In [Bryant and Seebeck, 1], $\gamma(n, m, k)$ is established for $0 \leqslant k < m - 3$ and $n \geqslant 5$. $\gamma(n, n, n - 2)$ and $\gamma(n, n - 1, n - 2)$ are proved for $n \geqslant 4$ in [Cantrell–Price–Rushing, 1].

This exercise consists of showing the following two facts.

(a) Use $\gamma(n, m, k)$ and $\beta(n, m, m - 1)$ to show: *If P^p is a polyhedron, $p \leqslant n - 3$, and if $q: P^p \to E^n$, $n \geqslant 5$, is a topological embedding which is locally flat on each open simplex of some triangulation T of P, then q is locally flat on each closed simplex of T.*

(b) Use the γ and β-statements to show: *If $q: M^m \to N^n$ is an embedding of a manifold M^m (possibly with boundary) into the topological n-manifold N where $m \leqslant n - 3$, $n \geqslant 5$, and if q is locally flat on the open simplexes of some combinatorial triangulation T of M, then q is a locally flat embedding.*

(A better result than part (b) is proved in [Cantrell and Lacher, 2].)

Proof of Theorem 5.5.3. Let $\epsilon > 0$ be given, let T be a triangulation of (M, P) of mesh less than ϵ, and let V be the simplicial neighborhood of P in a second derived subdivision. Then, $\mathrm{Cl}(M - V)$ is a PL n-submanifold of M such that the inclusion $\mathrm{Cl}(M - V) \subset M$ is allowable. [This follows from the easily proved fact that if M^k is a PL manifold and $V^k \subset M^k$ is a codimension zero PL submanifold, then the inclusion $\mathrm{Cl}(M - V) \subset M$ is allowable.] Notice that it follows from Exercise **5.5.2** that the embedding f of Theorem **5.5.3** is locally flat. Hence, Theorem **5.5.3** follows from Theorem **5.5.1** and Addendum **5.5.2** by letting $\mathrm{Cl}(M - V)$ play the role of M'.

5.6. LOCAL CONTRACTIBILITY OF THE HOMEOMORPHISM GROUP OF A MANIFOLD AND CODIMENSION ZERO TAMING

Let us begin this section by giving some history concerning the type of problem which we will be considering. Let $\mathscr{H}(M^m)$ denote the group of homeomorphisms of an m-manifold M onto itself. One topology, the compact-open topology (CO-topology) which can be put on $\mathscr{H}(M)$ was introduced in [Fox, 2]. It is well known that $\mathscr{H}(M)$ with the CO-topology is a topological group (see Exercise **5.6.2**). (We will define a couple of other topologies for $\mathscr{H}(M)$ in this section.) The general type question considered here is: What homeomorphisms of a manifold M can be connected by a small homotopy in $\mathscr{H}(M)$? Although questions of this type were first treated explicitly and for their own sake about twenty years ago, they had been handled implicitly in work done by Alexander [4] (see Exercise **4.11.3**) and Kneser [1] almost fifty years ago. Fort [2] proved that the space $\mathscr{H}(P)$ of homeomorphisms of the plane P onto itself is locally arcwise connected and Floyd and Fort [1] proved that the space of homeomorphisms of the 2-sphere onto itself is uniformly locally connected. (Roberts [1] had previously announced a proof that the space of homeomorphisms of the plane onto itself has exactly two components.) Hamstrom and Dyer [1] showed that $\mathscr{H}(M^2)$ is locally contractible for a general compact 2-manifold M^2. In [Hamstrom, 1], it was shown that the space of homeomorphisms of a compact 3-manifold with boundary onto itself is p-LC for all p. (Related results had been considered in [Roberts, 2], [Sanderson, 1], [Fisher, 1], and [Kister, 2].) Kister [1] proved the local contractibility of the group of homeomorphisms of E^n when given the uniform topology. (We will define the uniform topology later in this section. This result of Kister's was given as Exercise **4.11.4**.)

In 1968, Černavskiĭ announced remarkable results on the local contractibility of the homeomorphism groups of manifolds [11]. Complete proofs of these results appeared in 1969 in [Černavskiĭ, 12]. (A translation into English of a manuscript of Černavskiĭ's complete work was made by Walker at the University of Georgia prior to the appearance of [Černavskiĭ, 12].) In this work Černavskiĭ considered the group $\mathscr{H}(M)$ supplied with one of the three topologies: the CO, the uniform, or the majorant. The following are some of the results of that paper: For any metrizable manifold M, the group $\mathscr{H}(M)$ is locally contractible when supplied with the majorant topology (Fundamental Theorem). If the manifold M is compact, then the group $\mathscr{H}(M)$ is locally contractible with the CO-topology (Theorem 1). If the manifold M is the interior of a compact manifold N, then $\mathscr{H}(M)$ is locally contractible when supplied with the CO-topology or the uniform topology if the metric of M is induced by that of N (Theorem 2). (It follows immediately from this that $\mathscr{H}(E^n)$ is locally contractible.) A relative local contractibility modulo locally flat submanifolds theorem followed from that work (Section 5.1). Certain covering homotopy theorems also followed in the space of embeddings of manifolds (Section 5.3). The main key to obtaining the above results was

Černavskiĭ's Handle Straightening Theorem (Local Theorem **2.16**).

In 1971, another paper on local contractibility by Edwards and Kirby appeared [1]. The major contribution of that paper was to give two short and elegant proofs of Černavskiĭ's Handle Straightening Theorem mentioned above. (This is Lemma 4.1 of [Edwards and Kirby, 1].) Černavskiĭ's proof of that result is quite laborious to say the least, although the basic idea is simple enough. The basic idea is to use Morton Brown's "technique of meshing a straight structure and a wiggly structure" to pull a slighly perturbed handle back. (Brown's technique was employed in Section **3.4** and Section **5.2**.) The results and proofs (with the exception of the proofs of the handle straightening theorem) of Edwards and Kirby are basically the same as those of [Černavskiĭ, 12].

This section will be devoted to some of the work of Černavskiĭ and Edwards and Kirby mentioned above. Our proof of Černavskiĭ's Handle Straightening Theorem will be the elegant one of Edwards and Kirby. This proof is a generalization of results which first appeared in [Kirby, 4]. (An oversimplification of the generalization is "to cross the proof of [Kirby, 4] with B^k.")

Before stating and proving the results of this section, let us give some preliminary definitions and discussions. We will always consider our manifold M to have a fixed metric $\rho(x, y)$. Denote the group $\mathscr{H}(M)$ when endowed with the **CO**, **uniform**, or **majorant topologies** by $\mathscr{H}_c(M)$, $\mathscr{H}_u(M)$, or $\mathscr{H}_m(M)$, respectively, or by $\mathscr{H}_\tau(M)$ if the topology is unspecified. A basis of neighborhoods of the identity $e = e(M)$ (the identity mapping) is given in $\mathscr{H}_c(M)$ by the pairs (K, ϵ), where $\epsilon > 0$ and K is a compact subset of M. The neighborhood determined by the pair (K, ϵ) is denoted by $N_{K,\epsilon}(e)$ and consists of all homeomorphisms $h : M \to M$ such that $\rho(x, h(x)) < \epsilon$ for $x \in K$.

Exercise 5.6.1. Show that the above definition of the CO topology is equivalent to the usual one.

A basis of neighborhoods of e in $\mathscr{H}_u(M)$ is given by numbers $\epsilon > 0$. The neighborhood determined by ϵ is denoted by $N_\epsilon(e)$ and consists of all homeomorphisms h such that $\rho(x, h(x)) < \epsilon$ for all $x \in M$. In $\mathscr{H}_m(M)$ a basis of neighborhoods of the identity is given by the continuous strictly positive functions on M, which we shall call **majorants**. The neighborhood determined by the majorant $f: M \to (0, \infty)$ is denoted by $N_f(e)$ and consists of all h such that $\rho(x, h(x)) < f(x)$ for all $x \in M$.

Exercise 5.6.2. Show that for all three values of τ, the group $\mathscr{H}_\tau(M)$ is a topological group, and a topological group of transformations of M. To do this one must verify the continuity of the following three maps:

(a) $\mathscr{H}_\tau(M) \times \mathscr{H}_\tau(M) \to \mathscr{H}_\tau(M)$: $(h, h') \to hh'$,

(b) $\mathscr{H}_\tau(M) \to \mathscr{H}_\tau(M)$: $h \to h^{-1}$, and

(c) $\mathscr{H}_\tau(M) \times M \to M$: $(h, x) \to h(x)$. (See [Arens, 1].)

Exercise 5.6.3. Show that for compact manifolds M, $\mathscr{H}_\tau(M)$ has the same topology for $\tau = \text{c, u, or m}$. [Hint: Even though the topology $\tau = \text{u}$ depends in general on the metric in M, the exercise can be established by showing that the identity mappings

$$\mathscr{H}_{\text{m}}(M) \to \mathscr{H}_{\text{u}}(M) \to \mathscr{H}_{\text{c}}(M)$$

are continuous and by showing that open sets in $\mathscr{H}_{\text{m}}(M)$ are open in $\mathscr{H}_{\text{c}}(M)$.]

In this section, an **isotopy** of the manifold M means a layer homeomorphism of $M \times [0, 1]$ onto itself. Being a **layer homeomorphism** means that the isotopy $H: M \times [0, 1] \to M \times [0, 1]$ determines homeomorphisms $h_t: M \to M$ such that $H(x, t) = (h_t(x), t)$ for each point $(x, t) \in M \times [0, 1]$. We shall say that H joins the homeomorphisms h_0 and h_1, or that it takes h_0 into h_1.

Exercise 5.6.4. Our definition of isotopy does not depend on the topology in the group of homeomorphisms $\mathscr{H}(M)$. It might seem that for the study of homotopic properties of these groups, for example local contractibility, the definition of isotopies as paths in $\mathscr{H}_\tau(M)$ would be more natural. It is well known that the two definitions are equivalent for $\tau = \text{c}$. Show that this is not so in the other two cases. In fact, for $\mathscr{H}_{\text{m}}(M)$, show that a path can only join homeomorphisms which coincide outside some compact set, which implies that, in general, $\mathscr{H}_{\text{m}}(M)$ is not even locally arcwise connected when M is noncompact.

We consider **the set of all isotopies**, $\mathscr{I}(M)$, of M as a subgroup of the group $\mathscr{H}(M \times [0, 1])$, and we topologize it as a subspace of $\mathscr{H}_\tau(M \times [0, 1])$. The group $\mathscr{I}(M)$ with this topology is denoted by $\mathscr{I}_\tau(M)$. (The direct product metric is taken in $M \times [0, 1]$.) We denote the unit of the group $\mathscr{I}_\tau(M)$ (the identity isotopy) by E, or if necessary, by $E(M)$.

Exercise 5.6.5. Show that for any neighborhood $N(e)$ in the group $\mathscr{H}_\tau(M)$, there is a neighborhood $N(E)$ in the group $\mathscr{I}_\tau(M)$ such that for $h \in N(E)$ all the homeomorphisms h_t lie in $N(e)$.

We are now ready to give definitions concerning homotopies in $\mathscr{H}_\tau(M)$. Let A, B, and C be subsets of $\mathscr{H}_\tau(M)$ such that $A \cup C \subset B$.

Then, A **deforms on** B **into** C if there is a continuous mapping $\Gamma: A \to \mathscr{I}_\tau(M)$ such that for $h \in A$ we have

(a) $(\Gamma(h))_0 = h$,
(b) $(\Gamma(h))_t \in B$ for $t \in [0, 1]$, and
(c) $(\Gamma(h))_1 \in C$.

Now let us introduce the main concept of this section. The group $\mathscr{H}_\tau(M)$ is called **locally contractible** if there is a neighborhood of e in $\mathscr{H}_\tau(M)$ which deforms on $\mathscr{H}_\tau(M)$ into e.

REMARK 5.6.1. In the above definition of local contractibility we may always assume that $\Gamma(e) = E$ since any contraction $\Gamma(h)$ can be replaced by the contraction $\Gamma'(h) = (\Gamma(e))^{-1}\,\Gamma(h)$, having this property. Furthermore, an arbitrary neighborhood $N(e)$ may be taken as B, provided that a sufficiently small neighborhood $N'(e)$, of e is taken as A. For, by Exercise **5.6.5** there is a neighborhood $N(E) \subset \mathscr{I}_\tau(M)$ such that for each isotopy $\Phi \in N(E)$ all the homeomorphisms Φ_t lie in $N(e)$. Since Γ is continuous, there is a neighborhood $N'(e)$ such that $\Gamma(N'(e)) \subset N(E)$. Thus, if $\mathscr{H}_\tau(M)$ is locally contractible in the sense of the above definition, then for a given neighborhood $N(e)$ there is a neighborhood $N'(e)$ which deforms into e in $N(e)$. This agrees with the usual definition of locally contractible as given in Section **5.4**.

EXERCISE 5.6.6. In Exercise **5.6.4**, you showed that a path in $\mathscr{H}_{\mathrm{m}}(M)$ can only join homeomorphisms which coincide outside some compact set. This indicated that if we are to prove the local contractibility of $\mathscr{H}_{\mathrm{m}}(M)$, we need to be somewhat careful about our definitions. It is quite easy to visualize "nice" homeomorphisms of E^n, for example, which are not the identity outside any compact set and which one feels he can pull back to the identity. In that spirit, given an arbitrary neighborhood B of e in $\mathscr{H}_{\mathrm{m}}(E^n)$, define a homeomorphism $A \in \mathscr{H}_{\mathrm{m}}(E^n)$ which is not the identity outside any compact set such that A deforms on B to e. [This should help to convince one that our definition of locally contractible is quite natural for $\mathscr{H}_{\mathrm{m}}(M)$.]

Example 5.6.1. In order to motivate consideration of the majorant topology, it might be well at this point to consider the following instructive example. Remove a countable number of pairs of open disks from the open unit disk and then obtain N by attaching a countable number of handles whose diameters tend to zero as indicated in Fig. **5.6.1**. Now, it is intuitively clear (and quite easy to prove) that there are homeomorphisms in arbitrarily small neighborhoods of e which can not be deformed to e in the uniform topology much less the CO topology, that is, twist some of the small handles out toward infinity. Thus, it

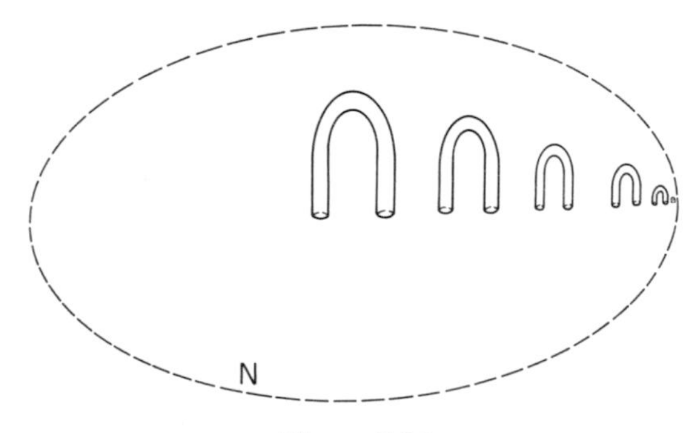

Figure 5.6.1

follows that neither $\mathscr{H}_c(N)$ or $\mathscr{H}_u(N)$ is locally contractible. [We will prove later (Theorem **5.6.3**) that for any manifold M which can be realized as the interior of a compact manifold both $\mathscr{H}_u(M)$ and $\mathscr{H}_c(M)$ are locally contractible, but this is not true of N.] However, if one defines $\epsilon: N \to (0, \infty)$ small enough, then any homeomorphism of N within ϵ of the identity cannot twist the handles and so one should be able to pull such homeomorphisms back to the identity. Indeed, we will prove that $\mathscr{H}_m(M)$ is locally contractible for any manifold M (Theorem **5.6.4**).

If U is a subset of a manifold M, a **proper** embedding of U into M is an embedding $h: U \to M$ such that $h^{-1}(\partial M) = U \cap \partial M$. An **isotopy of** U **into** M is a layer homeomorphism $H: U \times [0, 1] \to M \times [0, 1]$. As in the case $U = M$ defined previously, each isotopy $H: U \times [0, 1] \to M \times [0, 1]$ determines embeddings $h_t: U \to M$ such that $H(x, t) = (h_t(x), t)$ for each point $(x, t) \in U \times [0, 1]$. An isotopy H is **proper** if each such embedding h_t for the isotopy is proper.

If C and U are subsets of M with $C \subset U$, let $E(U, C; M)$ denote the set of proper embeddings of U into M which are the identity on C, and let $E(U, M)$ denote $E(U, \phi; M)$. Let $E(U, C; M)$ be provided with the CO topology. Thus, a typical basic neighborhood of $h \in E(U, C; M)$ is of the form $N_{K,\epsilon}(h) = \{g \in E(U, C; M) \mid \rho(g(x), h(x)) < \epsilon \text{ for all } x \in K\}$, where K is a compact subset of U, $\epsilon > 0$ and ρ is the metric on M.

If P and B are subsets of $E(U; M)$, then a **deformation of** P **into** B is a map $\Phi: P \times I \to E(U; M) \times I$ such that $\Phi_0 \mid P = 1_P$ and $\Phi_1(P) \subset B$. (This is compatible with our previous definition of deformation since we are dealing with the CO topology here.) We may equivalently regard Φ as a map $\Phi: P \times I \times U \to M$ such that for each $h \in P$ and $t \in I$, the map $\Phi(h, t, U): U \to M$ is a proper embedding. Thus, a deformation of P is simply a collection $\{h_t: U \to M, t \in I \mid h \in P\}$ of proper isotopies of U into M, continuously indexed by P, such that

$h_0 = h$. If W is a subset of U, a deformation $\Phi: P \times I \to E(U; M) \times I$ is **modulo** W if $\Phi_t(h) \mid W = h \mid W$ for all $h \in P$ and $t \in I$.

Suppose that $\Phi: P \times I \to E(U, M) \times I$ and $\psi: Q \times I \to E(U, M) \times I$ are deformations of subsets of $E(U, M)$, and suppose that $\Phi_1(P) \subset Q$. Then, the **composition of** ψ **with** Φ denoted by $\psi * \Phi: P \times I \to E(U, M) \times I$ is defined by

$$\psi * \Phi(h, t) = \begin{cases} (\Phi_{2t}(h), t) & \text{if} \quad t \in [0, \frac{1}{2}], \\ (\psi_{2t-1}\Phi_1(h), t) & \text{if} \quad t \in [\frac{1}{2}, 1]. \end{cases}$$

We shall denote the cube $\{x \in E^n \mid |x_i| \leqslant r, 1 \leqslant i \leqslant n\}$ by I_r^n. We regard S^1 as the space obtained by identifying the endpoints of $[-4, 4]$ and we let $p: E^1 \to S^1$ denote the natural covering projection, that is, $p(x) = (x + 4)_{(\text{mod } 8)} - 4$. Let T^n be the n-fold product of S^1. Then, I_r^n can be regarded as a subset of T^n for $r < 4$. Let $p^n: E^n \to T^n$ be the product covering projection and let $p^{k,n}: I_1^k \times E^n \to I_1^k \times T^n$ be the map $1_{I_1^k} \times p^n$. These maps will each be denoted by p when there is no possibility of confusion.

Let B^n be the unit n-ball in E^n and let S^{n-1} be its boundary as usual. We regard $S^{n-1} \times [-1, 1]$ as a subset of E^n by identifying (x, t) with $(1 + t/2) \cdot x$.

With the above discussions, definitions and notation out of the way, we are ready to start formulating some lemmas preliminary to the proofs of the main results of this section.

A discussion of, and a geometrical proof of, our first lemma will be postponed until the end of this section. (An **immersion** of one space into another is a continuous map which is locally an embedding.)

Immersion Lemma 5.6.1. *There is an immersion* $\alpha: T^n - B^n \to E^n$ *of the punctured torus into* E^n.

For a picture of α in the case $n = 2$, see Fig. **5.6.2**.

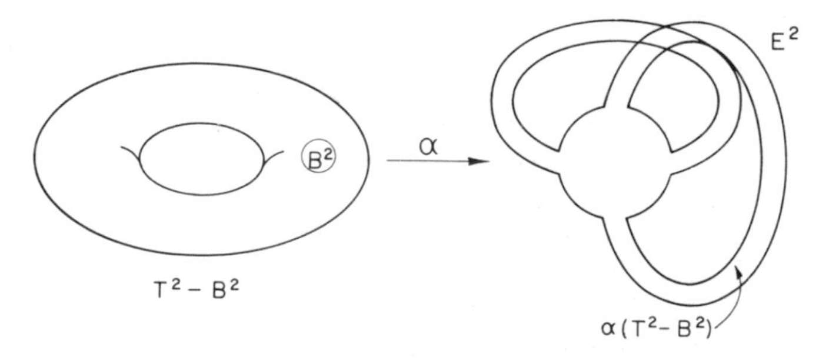

Figure 5.6.2

The next lemma can be proved by redoing the proof of Generalized Schoenflies Theorem **1.8.2** in a canonical way. We leave this as an exercise. Work related to this lemma can be found in Theorem 1.2 of [Huebsch and Morse, 1], [Gauld, 1], Theorem 1 of [Kister, 3], [Wright, 2], and [Smith, 1].

Weak Canonical Schoenflies Lemma 5.6.2. *There exists an $\epsilon > 0$ such that for any embedding $f: S^{n-1} \times [-1, 1] \to E^n$ within ϵ of the identity, $f \mid S^{n-1}$ extends canonically to an embedding $\bar{f}: B^n \to E^n$. The embedding $\bar{f}$ is canonical in the sense that $\bar{f}$ depends continuously on f and if $f = 1$, then $\bar{f} = 1$.*

Recall that we first defined collars in Section **1.7**. If M is a manifold, then a collar for ∂M may be regarded as an embedding $\sigma: \partial M \times [0, 1] \to M$ such that $\sigma(x, 0) = x$ for all $x \in \partial M$. The existence of such collars was proved in Section **1.7**. The following lemma is proved by an elementary application of one of the techniques of that section. Throughout this section we will adopt the custom of identifying a collar with its image.

Lemma 5.6.3. *Let M be a manifold with a collar $\partial M \times [0, 1]$ and let C_0 and V_0 be compact subsets of ∂M such that $C_0 \subset \operatorname{Int}_{\partial M} V_0$. Let U be a subset of M such that $V_0 \times [0, 1] \subset \operatorname{Int} U$ (see Fig. **5.6.3**). Then, there is a neighborhood P of the inclusion $\eta: U \subset M$ in $E(U, V_0 ; M)$ and a deformation*

$$\Phi: P \times I \to E(U, V_0 ; M) \times I$$

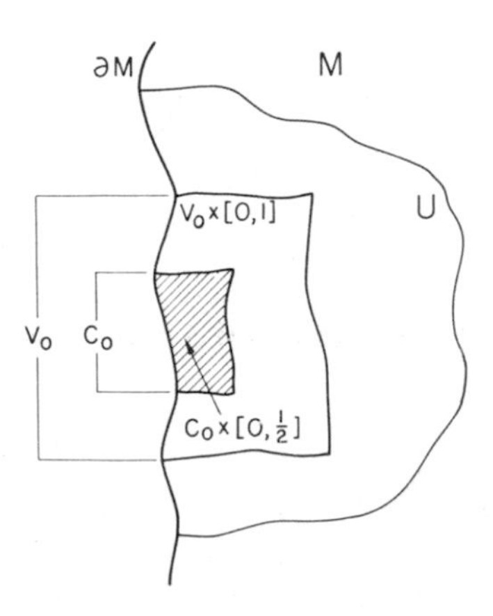

Figure 5.6.3

of P into $E(U, V_0 \cup (C_0 \times [0, \frac{1}{2}]); M)$ such that Φ is modulo the complement of an arbitrarily small neighborhood of $V_0 \times [0, 1]$.

PROOF. Let W be an arbitrary neighborhood of $V_0 \times [0, 1]$ in U and choose P so that $h \in P$ implies that $V_0 \times [0, 1] \subset h(W)$. Let $\lambda: V_0 \to [0,1]$ be a map such that $\lambda(C_0) = 1$ and $\lambda(\mathrm{Fr}_{\partial M} V_0) = 0$. For each $t \in [0, 1]$ let

$$W_t = \{(x, s) \in V_0 \times [0, 1] \mid 0 \leqslant s \leqslant t\lambda(x)\}$$

and define a homeomorphism $\gamma_t: (W_t - \mathrm{Int}\, W_{t/2}) \to W_t$ by linearly stretching the fibers over the boundary points, that is $\gamma_t(x, s) = (x, 2(s - t\lambda(x)/2))$ for each $(x, s) \in W_t - \mathrm{Int}\, W_{t/2}$. Extend γ_t via the identity to a homeomorphism $\pi_t: \mathrm{Cl}(M - W_{t/2}) \to M$. For each $h \in P$, define an isotopy $h_t: U \to M$, $t \in [0, 1)$, by

$$h_t = \begin{cases} \pi_t^{-1} h \pi_t & \text{on } U - W_{t/2}, \\ 1 & \text{on } W_{t/2}. \end{cases}$$

Then, $h_0 = h$, $h_1 \mid V_0 \cup (C_0 \times [0, \frac{1}{2}]) = 1$ and $h_t \mid U - W = h \mid U - W$ for each t. Thus, $\Phi: P \times I \to E(U, V_0; M) \times I$ defined by $\Phi(h, t) = (h_t, t)$ is the desired deformation.

The next lemma was first proved as Local Theorem 2.16 of [Černavskiĭ, 12]. As mentioned earlier, it was nicely reproved as Lemma 4.1 of [Edwards and Kirby, 1], and the proof which we shall present is due to Edwards and Kirby.

Handle Straightening Lemma 5.6.4. *There is a neighborhood Q of the inclusion $\eta: I_1^k \times I_4^n \subset I_1^k \times E^n$ in $E(I_1^k \times I_4^n, \partial I_1^k \times I_4^n; I_1^k \times E^n)$ and a deformation ψ of Q into*

$$E(I_1^k \times I_4^n, (\partial I_1^k \times I_4^n) \cup (I_1^k \times I_1^n); I_1^k \times E^n)$$

modulo $\partial(I_1^k \times I_4^n)$ such that $\psi(\eta, t) = (\eta, t)$ for all t.

PROOF. Let $C = (I_1^k - \mathrm{Int}\, I_{1/2}^k) \times I_3^n$ (see Fig. **5.6.4**). It is convenient to work with embeddings which are the identity on C. This can be arranged by applying Lemma **5.6.3** which says that there exists a deformation

$$\psi_0: Q_0 \times I \to E(I_1^k \times I_4^n, \partial I_1^k \times I_4^n; I_1^k \times E^n) \times I$$

of Q_0 into $E(I_1^k \times I_4^n, (\partial I_1^k \times I_4^n) \cup C; I_1^k \times E^n)$ such that ψ_0 is modulo $\partial(I_1^k \times I_4^n)$. Note that $\psi_0(\eta, t) = (\eta, t)$ for all t.

The main construction of the lemma is as follows. Given an embedding $h \in E(I_1^k \times I_4^n, (\partial I_1^k \times I_4^n) \cup C; I_1^k \times E^n)$ which is sufficiently close

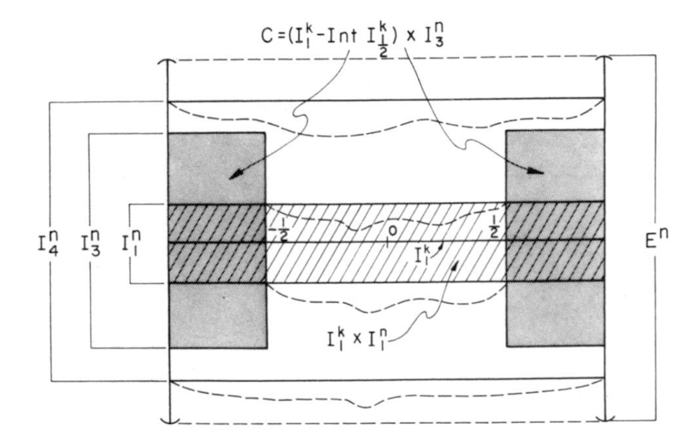

Figure 5.6.4

to the inclusion η, we construct a homeomorphism $g: I_1^k \times E^n \to I_1^k \times E^n$, continuously dependent upon h, such that

$$g \,|(\partial I_1^k \times E^n) \cup (I_1^k \times (E^n - \operatorname{Int} I_3^n) = 1$$

and $g \mid I_1^k \times I_1^n = h \mid I_1^k \times I_1^n$. The deformation of the lemma is then defined by composing an "Alexander isotopy" of g with h. The homeomorphism g is produced by successively lifting maps as indicated in Diagram **5.6.1** and Fig. **5.6.5**.

$$\begin{array}{ccc}
I_1^k \times E^n & \xrightarrow{g=h_5} & I_1^k \times E^n \\
\gamma^{-1}\uparrow & & \uparrow\gamma^{-1} \\
I_1^k \times E^n & \xrightarrow{h_4} & I_1^k \times E^n \\
P\downarrow & & \downarrow P \\
I_1^k \times T^n & \xrightarrow{h_3} & I_1^k \times T^n \\
\cup & & \cup \\
(I_1^k \times T^n) - (B_3^k \times B_3^n) & \xrightarrow{h_2|} & (I_1^k \times T^n) - (B^k \times B^n) \\
\cap & & \cap \\
(I_1^k \times T^n) - (B_2^k \times B_2^n) & \xrightarrow{h_2} & (I_1^k \times T^n) - (B^k \times B^n) \\
\cup & & \cup \\
I_1^k \times (T^n - B_2^n) & \xrightarrow{h_1} & I_1^k \times (T^n - B^n) \\
\alpha|\downarrow & & \downarrow\alpha \\
I_1^k \times I_4^n & \xrightarrow{h} & I_1^k \times E^n
\end{array}$$

Diagram 5.6.1

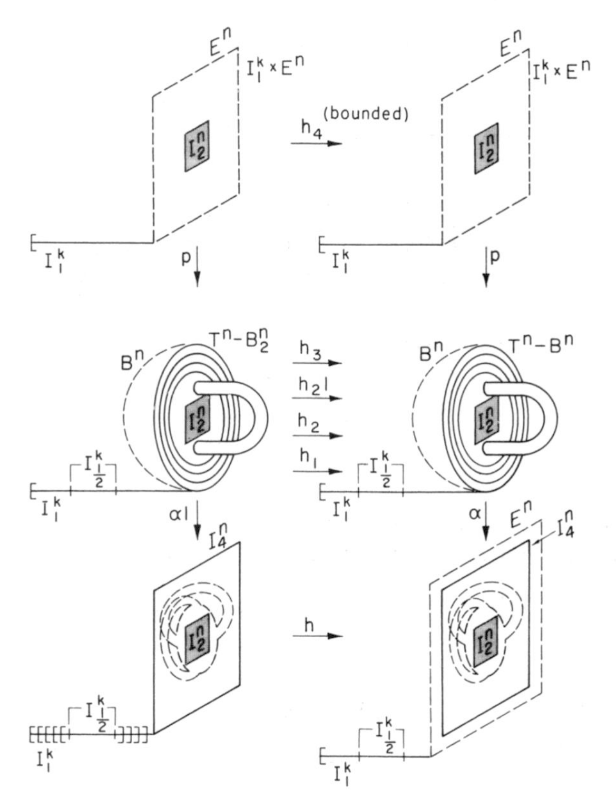

Figure 5.6.5

The neighborhood Q_1 of η which appears in the proof will always be understood to be a neighborhood of the inclusion map η in the space $E(I_1^k \times I_4^n, (\partial I_1^k \times I_4^n) \cup C; I_1^k \times E^n)$.

Let B^n, B_2^n, B_3^n, and B_4^n be four concentric n-cells in $T^n - I_2^n$ such that $B_j^n \subset \operatorname{Int} B_{j+1}^n$ for each j. Likewise let B^k, B_2^k, B_3^k, and B_4^k be four concentric k-cells in $\operatorname{Int} I_1^k$ such that $I_1^k \subset B^k$ and $B_j^k \subset \operatorname{Int} B_{j+1}^k$ for each j. By Immersion Lemma **5.6.1**, there exists an immersion $\alpha_0\colon T^n - B^n \to \operatorname{Int} I_3^n$. By Generalized Schoenflies Theorem **1.8.2**, we can assume that $\alpha_0 \mid I_2^n$ is the identity. Let α denote the product immersion

$$1 \times \alpha_0\colon I_1^k \times (T^n - B^n) \to I_1^k \times \operatorname{Int} I_3^n.$$

If $h \in E(I_1^k \times I_4^n, (\partial I_1^k \times I_4^n) \cup C; I_1^k \times E^n)$ is close enough to η, then h can be covered in a natural way by an embedding

$$h_1\colon I_1^k \times (T^n - B_2^n) \to I_1^k \times (T^n - B^n)$$

(see Diagram **5.6.1** and Fig. **5.6.5**). This is done by defining h_1 to agree locally with $\alpha^{-1}h\alpha$. Thus, let $\{U_i\}_{i=1}^r$ be a finite cover of $I_1^k \times (T^n - \operatorname{Int} B_2^n)$ by open subsets of $I_1^k \times (T^n - B^n)$ such that for any two members U_i, $U_{i'}$ which have a nonempty intersection, $\alpha \mid (U_i \cup U_{i'})$ is an embedding. Let $\{W_i\}_{i=1}^r$ be a cover of $I_1^k \times (T^n - B_2^n)$ by compact subsets of $I_1^k \times (T^n - B^n)$ such that $W_i \subset U_i$ for each i. If ϵ is chosen small enough and if $Q_1 = N_{\alpha(\bigcup_{i=1}^r W_i),\epsilon}(\eta)$, then $h \in Q_1$ implies that $h\alpha(W_i) \subset \alpha(U_i)$ for each i. For such h we can define the lifted map $h_1: I_1^k \times (T^n - B_2^n) \to I_1^k \times (T^n - B_2^n)$ by letting $h_1 \mid W_i = (\alpha \mid U_i)^{-1} h\alpha \mid W_i$ for each i. Then h_1 is an embedding which lifts h and depends continuously on h, and is such that if h is the inclusion then so is h_1. Furthermore, $h_1 \mid (I_1^k - B^k) \times (T^n - B_2^n) = 1$.

From this latter property, it follows that h_1 can be extended via the the identity to an embedding

$$h_2: (I_1^k \times T^n) - (B_2^k \times B_2^n) \to (I_1^k \times T^n) - (B^k \times B^n).$$

Let $B^m = B^k \times B^n$. We can now apply Weak Canonical Schoenflies Lemma **5.6.2** to extend $h_2 \mid (I_1^k \times T^n) - B_3^m$ to a homeomorphism of $I_1^k \times T^n$. For if Q_1 is sufficiently small then $h \in Q_1$ implies that $h_2 \mid (B_{7/2}^m - B_{5/2}^m)$ is close to the identity and therefore $h_2 \mid \partial B_3^m: \partial B_3^m \to \operatorname{Int} B_4^m$ extends to an embedding $\bar{h}_2: B_3^m \to \operatorname{Int} B_4^m$. Define a homeomorphism $h_3: I_1^k \times T^n \to I_1^k \times T^n$ by letting

$$h_3 \mid (I_1^k \times T^n) - B_3^m = h_2 \mid (I_1^k \times T^n) - B_3^m \qquad \text{and} \qquad h_3 \mid B_3^m = \bar{h}_2 .$$

By the construction, h_3 depends continuously on h and if h is the inclusion, then h_3 is the identity.

Now if h_3 is sufficiently close to the identity, then h_3 lifts in a natural way to a bounded homeomorphism $h_4: I_1^k \times E^n \to I_1^k \times E^n$ (where **bounded** means the set $\{\| h_4(x) - x \| \mid x \in I_1^k \times E^n\}$ is bounded). We can define h_4 so that it agrees locally with $p^{-1}h_3 p$, similar to the way that h_1 was defined. For if U is any subset of $I_1^k \times E^n$ of diameter < 4, then $p \mid U$ is an embedding. For each $x \in I_1^k \times E^n$, let $h_4 \mid U_1(x) = (p \mid U_2(x))^{-1} h_3 p \mid U_1(x)$, where $u_\delta(x)$ denotes the open δ-neighborhood of x. Then h_4 depends continuously on h_3, $h_4 \mid \partial I_1^k \times E^n = 1$, and $h_4 = 1$ if $h_3 = 1$.

Let γ: Int $I_3^m \to E^m$ be a homeomorphism which is a radial expansion and which is the identity on $I_2^m = I_2^k \times I_2^n$. Extend h_4 via the identity to a homeomorphism $h_4': E^k \times E^n \to E^k \times E^n$ and define a homeomorphism $h_5: I_1^k \times E^n \to I_1^k \times E^n$ by

$$h_5 = \begin{cases} \gamma^{-1} h_4' \gamma & \text{on} \quad I_1^k \times \operatorname{Int} I_3^n, \\ 1 & \text{on} \quad I_1^k \times (E^n - \operatorname{Int} I_3^n). \end{cases}$$

The continuity of h_5 follows from the fact that h_4' is bounded. Now h_5 has the following properties:

(1) $h_5 \mid (\partial I_1^k \times E^n) \cup (I_1^k \times (E^n - \operatorname{Int} I_3^n)) = 1$,

(2) $\alpha p \gamma h_5(x) = h \alpha p \gamma(x)$ for $x \in (I_1^k \times I_2^n) \cap h_5^{-1}(I_1^k \times I_2^n)$,

(3) h_5 depends continuously on h, and if $h = \eta$, then $h_5 = 1$.

Property 3 implies that if Q_1 is small enough, then $h_5(I_1^k \times I_1^n) \subset I_1^k \times I_2^n$ whenever $h \in Q_1$. Thus, since $\alpha p \gamma \mid I_1^k \times I_2^n = 1$, Property 2 implies that $h_5 \mid I_1^k \times I_1^n = h \mid I_1^k \times I_1^n$. Therefore, h_5 is the desired map g mentioned at the beginning of the proof.

To complete the proof. We show how to use $g = h_5$ to deform h to be the identity on $I_1^k \times I_1^n$. Extend g via the identity to a homeomorphism $g: E^k \times E^n \to E^k \times E^n$ and define an isotopy $g_t: I_1^k \times E^n \to I_1^k \times E^n$, $t \in [0, 1]$, by using the Alexander trick on g. That is,

$$g_t(x) = \begin{cases} tg\left(\dfrac{1}{t}x\right) & \text{if } t > 0, \\ x & \text{if } t = 0. \end{cases}$$

Define a deformation

$$\psi_1: Q_1 \times I \to E(I_1^k \times I_4^n; \partial I_1^k \times I_4^n; I_1^k \times E^n) \times I$$

by $(\psi_1)_t(h) = g_t^{-1}h: I_1^k \times I_4^n \to I_1^k \times E^n$. Then, ψ_1 deforms Q_1 into

$$E(I_1^k \times I_4^n, (\partial I_1^k \times I_4^n) \cup (I_1^k \times I_1^n); I_1^k \times E^n).$$

If Q_1 is small enough so that $h \in Q_1$ implies that

$$h(I_1^k \times \partial I_4^n) \cap (I_1^k \times I_3^n) = \emptyset,$$

then ψ_1 is modulo $\partial(I_1^k \times I_4^n)$. Note that $\psi_1(\eta, t) = (\eta, t)$ for all t. Finally, let Q be a neighborhood of η in $E(I_1^k \times I_4^n, \partial I_1^n \times I_4^n; I_1^k \times E^n)$, $Q \subset Q_0$, such that $\psi_0(Q \times 1) \subset Q_1 \times 1$, and let $\psi = \psi_1 * \psi_0 \mid Q \times I$. Then, ψ is the desired deformation of the lemma.

Although we will prove a stronger result later (Theorem **5.6.3**), let us use Handle Straightening Lemma **5.6.4** for the special case of zero handles to prove the following result. (A similar proof of this result appears in [Kirby, 4].)

Theorem 5.6.1. *$\mathscr{H}_c(E^n)$ is locally contractible.*

Proof. Notice that for zero handles, Handle Straightening Lemma **5.6.4** says the following: *There is a neighborhood Q of the inclusion*

$\eta: I_4{}^n \subset E^n$ *in* $E(I_4{}^n; E^n)$ *and a deformation* ψ *of* Q *into* $E(I_4{}^n, I_1{}^n; E^n)$ *modulo* $\partial I_4{}^n$ *such that* $\psi(\eta, t) = (\eta, t)$ *for all* t. Let Q_* be a neighborhood of $e \in \mathscr{H}_c(E^n)$ such that $h \in Q_* \Rightarrow h \mid I_4{}^n \subset Q$. If $h \in Q_*$, define an isotopy $h_t: E^n \to E^n$, $t \in [0, 1]$, by

$$h_t = \begin{cases} h & \text{on} \quad E^n - I_4{}^n, \\ \psi_t(h \mid I_4{}^n) & \text{on} \quad I_4{}^n. \end{cases}$$

Then, $h_0 = h$ and $h_1 \in E(E^n, I_1{}^n; E^n)$. Let $\psi': Q_* \times I \to \mathscr{H}_c(E^n) \times I$ be defined by $\psi'(h, t) = (h_t, t)$. Then, ψ' is a deformation of Q_* into $E(E^n, I_1{}^n; E^n)$. Now let us define a deformation ψ'' of $E(E^n, I_1{}^n; E^n)$ to e (the identity on E^n). If $h \in E(E^n, I_1{}^n; E^n)$ define $h_t: E^n \to E^n$, $t \in [0, 1]$ by

$$h_t(x) = \begin{cases} \dfrac{1}{t} h(tx) & \text{if} \quad t > 0, \\ x & \text{if} \quad t = 0. \end{cases}$$

Define $\psi'': E(E^n, I_1{}^n; E^n) \times I \to E(E^n, I_1{}^n; E^n) \times I$ by $\psi''(h, t) = (h_t, t)$. Finally, we see that $\psi'' * \psi': Q_* \times I \to \mathscr{H}_c(E^n) \times I$ is the desired deformation of Q_* to e.

Exercise 5.6.7. If M is a compact, PL manifold, then use Handlebody Decomposition Theorem **1.6.12** and Handle Straightening Lemma **5.6.4** to show that $\mathscr{H}_c(M)$ is locally contractible. (In the pages that follow, we will see that this result is more involved for arbitrary topological manifolds M. The technique of proof of this exercise is developed as a step in the proof of the more general result, and so one may want to sneak a glimpse at the following pages before doing this exercise.)

All of the main results of this section will follow rather quickly from the next lemma. The proof of this lemma uses Handle Straightening Lemma **5.6.4** in a fundamental way. This lemma is essentially Proposition (B) of [Černavskiĭ, 12] and is a statement on p. 71 of [Edwards and Kirby, 1].

Lemma 5.6.5. *Given subsets* C, D, U *and* V *of* M *such that* C *is compact,* D *is closed,* U *is a neighborhood of* C *and* V *is a neighborhood of* D, *then there is a neighborhood* P_η *of* η *in* $E(U, U \cap V; M)$ *and a deformation*

$$\varphi: P_\eta \times I \to E(U, U \cap D; M) \times I$$

of P_η *into* $E(U, U \cap (C \cup D); M)$ *such that* φ *is modulo the complement of a compact neighborhood of* C *in* U *and* $\varphi(\eta, t) = (\eta, t)$ *for all* t.

Henceforth, it will be understood that all deformations of subsets of $E(U, M)$ fix the inclusion and are modulo the complement of a compact neighborhood of C in U.

The proof of the lemma is divided into two cases.

Case 1 ($\mathrm{Cl}(C - D) \cap \partial M = \emptyset$). Let $\{(W_i, h_i)\}_{i=1}^{r}$ be a finite cover of $\mathrm{Cl}(C - D)$ by coordinate neighborhoods which lie in U, where $h_i: W_i \to R^m$ is a homeomorphism. Express $\mathrm{Cl}(C - D)$ as the union of r compact subsets $C_1, \ldots, C_r$ such that $C_i \subset W_i$, and let $D_i = D \cup_{j<i} C_j$ for $0 \leqslant i \leqslant r$. The proof of Case 1 is by an inductive argument on i. At the ith step we assume that there exists a neighborhood P_i of $\eta: U \subset M$ in $E(U, U \cap V; M)$ and a deformation $\varphi_i: P_i \times I \to E(U, U \cap D; M) \times I$ of P_i into $E(U, U \cap V_i; M)$, where V_i is some neighborhood of D_i. The induction starts trivially at $i = 0$ by taking $V_0 = V$, $P_0 = E(U, U \cap V; M)$ and φ_0 to be the identity deformation. We show how in general the inductive assumption can be extended to hold true for $i + 1$.

Identify W_{i+1} with E^m in order to simplify the notation. Then, C_{i+1} is a compact subset of E^m and $V_i \cap E^m$ is a neighborhood in E^m of the closed subset $D_i \cap E^m$. By the technique of proof of Handlebody Decomposition Theorem **1.6.12**, there exists a pair of polyhedra (L, K), with $L \subset K$, such that (K, L) has a handlebody decomposition with the following properties (see Fig. **5.6.6**):

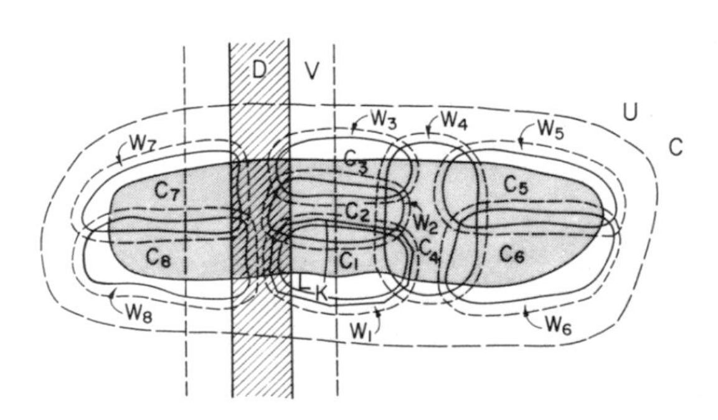

Figure 5.6.6

(1) $D_i \cap C_{i+1} \subset L \subset \mathrm{Int}(V_i \cap E^m)$,

(2) $C_{i+1} \subset K$,

(3) $\mathrm{Cl}(K - L) \cap D_i = \emptyset$, and

(4) if H is a handle of $K - L$ and if k is the index of H, then there is an embedding $\mu: I_1{}^k \times E^n \to E^m$, $m = k + n$, such that $\mu(I_1{}^k \times I_1{}^n) = H$ and $\mu(I_1{}^k \times E^n) \cap (D_i \cup L \cup \mathrm{Cl}(K^k - A)) = \mu(\partial I_1{}^k \times I^n)$, where K^k

denotes the union of all handles of K of index $\leqslant k$. (The embedding μ can be obtained by using a collar of ∂H in $E^m - \text{Int } H$.)

Assume that $H_1, ..., H_j, ..., H_s$ are the handles of $K - L$ subscripted in order of nondecreasing index. We proceed by induction on j to alter the embeddings in $E(U, U \cap V_i; M)$ a step at a time in neighborhoods of the H_j. For each j, $0 \leqslant j \leqslant s$, let $D_j' = D_j \cup L \cup_{1<j} H_1$ and assume inductively that for some neighborhood P_j' of $\eta: U \to M$ in $E(U, U \cap V; M)$ there exists a deformation $\varphi_j': P_j' \times I \to E(U, U \cap D; M) \times I$ of P_j' into $E(U, U \cap V_j'; M)$ where V_j' is some neighborhood of D_j' in M. (If $j = 0$, the main inductive assumption gives precisely the information that is needed.) Consider H_{j+1} and the embedding $\mu: I_1^k \times E^n \to E^m$ given by Property 4 above. By reparametrizing the E^n coordinate if necessary, keeping I_1^n fixed, we can further assume that $\mu(\partial I_1^k \times I_4^n) \subset \text{Int } V_j'$.

According to Handle Straightening Lemma **5.6.4** (replacing $I_1^k \times I_1^n$ by $I_1^k \times I_2^n$) there is a neighborhood Q of the inclusion η_0 in $E(I_1^k \times I_4^n, \partial I_1^k \times I_4^n; I_1^k \times E^n)$ and a deformation ψ of Q into $E(I_1^k \times I_4^n, (\partial I_1^k \times I_4^n) \cup (I_1^k \times I_2^n); I_1^k \times E^n)$ modulo $\partial(I_1^k \times I_4^n)$ such that $\psi_0(\eta_0, t) = (\eta_0, t)$ for all t. Let Q' be a neighborhood of η in $E(U, U \cap V_j'; M)$ such that $h \in Q'$ implies that $h\mu(I_1^k \times I_4^n) \subset \mu(I_1^k \times E^n)$ and $\mu^{-1}h\mu \mid I_1^k \times I_4^n \in Q$. Then, ψ can be used to define a deformation $\psi': Q' \times I \to E(U, U \cap D_j'; M) \times I$ of Q' into $E(U, U \cap V_{j+1}; M)$ as follows, where V_{j+1}' is a neighborhood of D_{j+1}' to be defined. If $h \in Q'$, define an isotopy $h_t: U \to M$, $t \in [0, 1]$, by

$$h_t = \begin{cases} h & \text{on} \quad U - \mu(I_1^k \times I_4^n), \\ \mu(\psi_t(\mu^{-1}h\mu))\, \mu^{-1} & \text{on} \quad \mu(I_1^k \times I_4^n). \end{cases}$$

Then $h_0 = h$ and $h_1 \in E(U, U \cap V_{j+1}'; M)$, where

$$V_{j+1}' = (V_j' \cup \mu(I_1^k \times I_2^n)) - \mu(I_1^k \times (I_4^n - \text{Int } I_2^n)).$$

Let $\psi'(h, t) = (h_t, t)$. By the continuity of φ_j' there is a neighborhood P_{j+1}' of η in $E(U, U \cap V; M)$, $P_{j+1}' \subset P_j'$, such that $\varphi_j'(P_{j+1}' \times 1) \subset Q' \times 1$. Let

$$\varphi_{j+1}' = \psi' * (\varphi_j' \mid P_{j+1}' \times I): P_{j+1}' \times I \to E(U, U \cap D; M) \times I.$$

Then φ_{j+1}' is the desired deformation, that is, φ_{j+1}' deforms P_{j+1}' into $E(U, U \cap V_{j+1}'; M)$.

At the completion of the subinductive argument on j, the main inductive argument can be continued by taking $P_{i+1} = P_s'$, $V_{i+1} = V_s'$, and $Q_{i+1} = Q_s'$. This completes the proof of Case 1.

Case 2 ($\text{Cl}(C - D) \cap \partial M \neq \emptyset$). The idea of the proof of Case 2 is to use a boundary collar for M and Case 1 of the proof to initially deform the embeddings to the identity on a neighborhood of $\text{Cl}(C - D) \cap \partial M$. The deformation can then be completed by applying Case 1.

Let $\partial M \times [0, 1]$ be a boundary collar for M. Without loss of generality we can assume that $C = \text{Cl}(C - D)$ and that D is compact (since we can assume that U is compact and that $D \subset U$). Let C_0, D_0, U_0, and V_0 be subsets of ∂M and let $\epsilon > 0$ be such that C_0 is compact, D_0 is closed, U_0 is a compact neighborhood of C_0, and V_0 is a neighborhood of D_0 and

(a) $C \cap (\partial M \times [0, 5\epsilon]) \subset \text{Int}_{\partial M}\, C_0 \times [0, 5\epsilon]$,
(b) $U_0 \times [0, 5\epsilon] \subset \text{Int}\, U$,
(c) $D \cap (\partial M \times [0, 5\epsilon]) \subset \text{Int}_{\partial M}\, D_0 \times [0, 5\epsilon]$, and
(d) $V_0 \times [0, 5\epsilon] \subset V$.

Let C_1 be a compact neighborhood of C_0 in ∂M such that $C_1 \subset \text{Int}_{\partial M}\, U_0$.

The deformation φ produced for this case is the composition of three deformations φ_1, φ_2, and φ_3. The first deformation $\varphi_1: P_1 \times I \to E(U, U \cap D; M) \times I$ deforms a neighborhood P_1 of $\eta: U \subset M$ in $E(U, U \cap V; M)$ into $E(U, U \cap (C_1 \cup V_1); M)$ modulo $U - U_0 \times [0, 5\epsilon]$ (see Fig. **5.6.7**), where

$$V_1 = (V - (\partial M \times [0, 5\epsilon])) \cup (D_0 \times [0, 5\epsilon]),$$

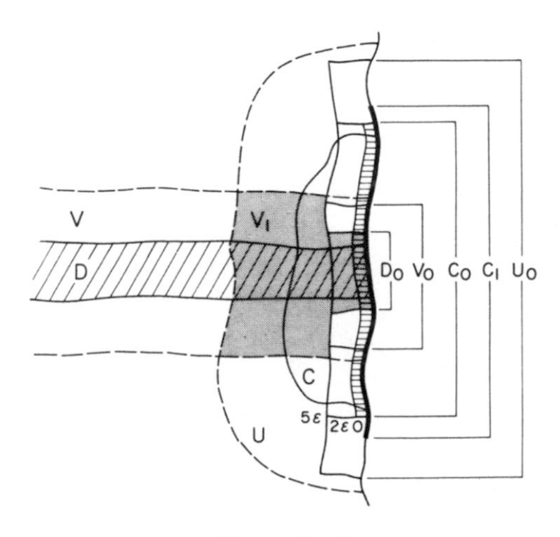

Figure 5.6.7

which is a neighborhood of D. The second deformation φ_2, which is defined using Lemma **5.6.3**, deforms a neighborhood P_2 of η in $E(U, U \cap (C_1 \cup V_1); M)$ into

$$E(U, U \cap ((C_0 \times [0, 2\epsilon]) \cup V_1); M) \quad \text{modulo } U - U_0 \times [0, 5\epsilon].$$

At this stage the problem can be redefined so that Case 1 of the proof applies, which leads to the definition of φ_3.

DEFINITION OF φ_1. It follows from Case 1 that there is a neighborhood P_0 of the inclusion $\eta_0: U_0 \subset \partial M$ in $E(U_0, U_0 \cap V_0; \partial M)$ and a deformation $\varphi_0: P_0 \times I \to E(U_0, U_0 \cap D_0; \partial M) \times I$ of P_0 into $E(U_0, U_0 \cap (\mathrm{Cl} \cup D_0); \partial M)$ modulo $\mathrm{Fr}_{\partial M} U_0$. Let P_1 be a neighborhood of $\eta: U \subset M$ in $E(U, U \cap V; M)$ such that $h \in P_1$ implies that $h \mid U_0 \in P_0$. Given $h \in P_1$, use the deformation φ_0 to define a level preserving homeomorphism $\sigma: U_0 \times [0, 5\epsilon] \to U_0 \times [0, 5\epsilon]$ be letting

$$\sigma \mid U_0 \times t = (h \mid U_0)^{-1}((\varphi_0)_{(5\epsilon - t)/5\epsilon}(h \mid U_0)).$$

Then σ is the identity on $(\mathrm{Fr}_{\partial M} U_0 \times [0, 5\epsilon]) \cup (U_0 \times 5\epsilon)$ and $\sigma \mid C_1 \times 0 = (h \mid U_0)^{-1} \mid C_1$. Extend σ to all of M via the identity. Then $\sigma: M \to M$ is isotopic to 1_M modulo $M - (U_0 \times [0, 5\epsilon])$ by the isotopy $\sigma_t: M \to M$, $t \in [0, 1]$, where σ_t is defined by

$$\sigma_t = \begin{cases} 1 & \text{on } M - (\partial M \times [0, t]), \\ \delta_t^{-1} \sigma \delta_t & \text{on } \partial M \times [0, t], \end{cases}$$

where $\delta_t: \partial M \times [0, t] \to \partial M \times [1 - t, 1]$ is the homeomorphism which send (x, s) to $(x, s + 1 - t)$. Let $\varphi_1(h, t) = (h\sigma_t, t)$.

DEFINITION OF φ_2. It follows from Lemma **5.6.3** that there exists a neighborhood P_2 of η in $E(U, U \cap (C_1 \cup V_1); M)$ and a deformation $\varphi_2: P_2 \times I \to E(U, U \cap V_1; M) \times I$ of P_2 into

$$E(U, U \cap ((C_0 \times [0, 2\epsilon]) \cup V_1); M) \quad \text{modulo } U - (U_0 \times [0, 5\epsilon]).$$

One may take the $V_0 \times [0, 1]$ of the lemma to be $C_1 \times [0, 4\epsilon]$ and may choose $U_0 \times [0, 5\epsilon]$ to be the arbitrarily small neighborhood of $V_0 \times [0, 1]$. It follows from the proof of the lemma that an embedding which is the identity on $U \cap V_1$ remains so during the deformation.

DEFINITION OF φ_3. Let $C_2 = C - (\partial M \times [0, \epsilon])$ and let $D_2 = D \cup (C \cap (\partial M \times [0, \epsilon]))$. Then $C_2 \cap D_2 = C \cup D$ and $C_2 \cap \partial M = \emptyset$. Let $V_2 = (C_0 \times [0, \epsilon]) \cup V_1$, which is a neighborhood of D_2. By Case 1, there is a neighborhood P_3 of $\eta: U \subset M$ in $E(U, U \cap V_2; M)$

and a deformation $\varphi_3: P_3 \times I \to E(U, U \cap D_2; M) \times I$ of P_3 into $E(U, U \cap (C_2 \cup D_2); M)$. This defines φ_3.

To conclude Case 2, let $P_\eta \subset P_1$ be a neighborhood of $\eta: U \subset M$ in $E(U, U \cap V; M)$ such that $\varphi_1(P_\eta \times 1) \subset P_2 \times 1$ and $\varphi_2(\varphi_1(P_\eta \times 1) \times 1) \subset P_3 \times 1$ and define

$$\varphi = \varphi_3 * \varphi_2 * (\varphi_1 \mid P_\eta \times 1): P_\eta \times I \to E(U, U \cap D; M) \times I.$$

Then φ is the desired deformation.

The next theorem is Theorem 1 of [Černavskiĭ, 12] and Corollary 1.1 of [Edwards and Kirby, 1].

Theorem 5.6.2. *If the manifold M is compact, then $\mathscr{H}_c(M)$ is locally contractible.*

Proof. Apply Lemma **5.6.5** where $D = V = \emptyset$ and where $U = M$. In this case Lemma **5.6.5** reduces to Theorem **5.6.2**, since $\mathscr{H}(M) = E(M; M)$ and $\{1_M\} = E(M, M; M)$.

The following theorem is Theorem 2 of [Černavskiĭ, 12] and is Corollary 6.1 of [Edwards and Kirby, 1].

Theorem 5.6.3. *If the manifold M is the interior of a compact manifold N, then $\mathscr{H}_c(M)$ is locally contractible.*

Proof. Let $\partial Q \times (0, 1]$ be an open collar for M induced by a collar for ∂Q in Q. Let $K = M - (\partial Q \times (0, 1])$, which is compact. By Lemma **5.6.5**, there is a neighborhood P of the identity in $\mathscr{H}_c(M)$ and a deformation of P into $\Delta(K)$, where $\Delta(K)$ denotes the subgroup of homeomorphisms that are fixed on $K \subset M$. There is a natural deformation of $\Delta(K)$ into $\{1_M\}$ by making use of the open collar. By composing these deformations, it follows that P deforms into $\{1_M\}$ and therefore $\mathscr{H}_c(M)$ is locally contractible.

Exercise 5.6.8. Write down the naturel deformations of P into $\Delta(K)$ mentioned above. (This easy deformation is given in 1.20 of Černavskiĭ [12].)

Our final theorem is the Fundamental Theorem of [Černavskiĭ, 12] and was proved as Corollary 6.2 of [Edwards and Kirby, 1].

Theorem 5.6.4. *For any manifold M, $\mathscr{H}_m(M)$ is locally contractible.*

Proof. Assume without loss of generality that M is connected. Let $\{U_i, C_i\}_{i=1}^{\infty}$ be a collection of pairs of compact subsets of M such that for each i, U_i is a neighborhood of C_i (see Fig. **5.6.8**), $M = \bigcup_{i=1}^{\infty} \operatorname{Int} C_i$

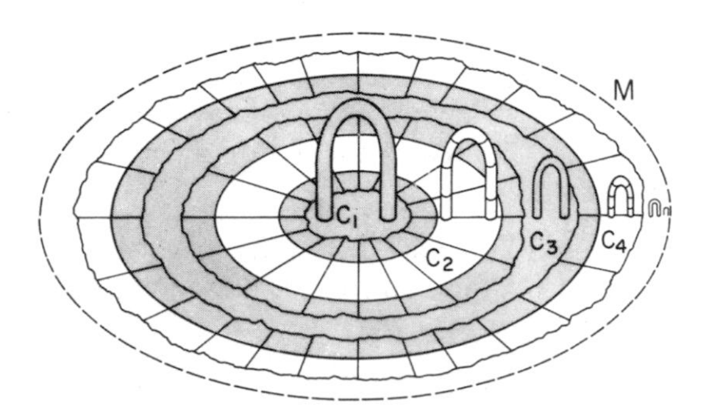

Figure 5.6.8

and $U_i \cap U_j = \emptyset$ if $|i - j| \geqslant 1$. It follows from Lemma **5.6.5** (letting $U = U_{2i}$, $C = C_{2i}$, $V = C_{2i-1} \cup C_{2i+1}$, and $D = \mathrm{Cl}(U_{2i} - C_{2i})$) that there is a sequence $\{\delta_{2i}\}$ of positive numbers such that if P_{2i} is defined to be the neighborhood $N_{U_{2i},\delta_{2i}}(\eta) \cap E(U_{2i}, U_{2i} \cap (C_{2i-1} \cup C_{2i+1}); M)$ of η: $U_{2i} \subset M$, then there is a deformation

$$\varphi_{2i}: P_{2i} \times I \to E(U_{2i}, U_{2i} \cap (C_{2i-1} \cup C_{2i+1}); M) \times I$$

of P_{2i} into $E(U_{2i}, C_{2i}; M)$, hence into $\{\eta\}$, such that φ_{2i} is modulo $\mathrm{Fr}_M U_{2i}$. Likewise, there is a sequence $\{\delta_{2i-1}\}$ of positive numbers such that if P_{2i-1} is defined to be the neighborhood $N_{U_{2i-1},\delta_{2i-1}}(\eta)$ of η: $U_{2i-1} \subset M$ in $E(U_{2i-1}; M)$, then there is a deformation $\varphi_{2i-1}: P_{2i-1} \times I \to E(U_{2i-1}; M) \times I$ of P_{2i-1} into $E(U_{2i-1}, C_{2i-1}; M)$ such that φ_{2i-1} takes place in

$$N_{U_{2i-1,\min\{\delta_{2i-2},\delta_{2i}\}}}(\eta)$$

and φ_{2i-1} is modulo $\mathrm{Fr}_M U_{2i-1}$. Let $\delta: M \to (0, \infty)$ be such that $\sup \delta \mid U_i \leqslant \delta_i$ for each i and let P be the $\delta(x)$-neighborhood of 1_M in $\mathscr{H}_{\mathrm{m}}(M)$. Define a deformation $\varphi: P \to \mathscr{I}_{\mathrm{m}}(M) \subset \mathscr{H}_{\mathrm{m}}(M \times I)$ by

$$\varphi(h)_t = \left\{ \begin{array}{ll} (\varphi_{2i-1})_{2t}\,(h \mid U_{2i-1}) & \text{on} \quad U_{2i-1} \\ h & \text{on} \quad M - \bigcup_{i=1}^{\infty} U_{2i-1} \end{array} \right\} \quad \text{for} \quad t \in [0, \tfrac{1}{2}],$$

$$\varphi(h)_t = \left\{ \begin{array}{ll} (\varphi_{2i})_{2t-1}(\varphi(h)_{1/2} \mid U_{2i}) & \text{on} \quad U_{2i} \\ \varphi(h)_{1/2} & \text{on} \quad M - \bigcup_{i=1}^{\infty} U_{2i} \end{array} \right\} \quad \text{for} \quad t \in [\tfrac{1}{2}, 1].$$

Then, φ is the desired deformation.

We are now ready to give a proof of Immersion Lemma **5.6.1**. The proof presented here was communicated to this author by R. D. Edwards. It was originated by Barden [2] and was formulated in the following picturesque form by Siebenmann. (Immersion Lemma **5.6.1** also follows from [Hirsch, 1].)

Proof of Immersion Lemma 5.6.1. We will work with the following inductive statement which is stronger than Lemma **5.6.1**.

n-DIMENSIONAL INDUCTIVE STATEMENT: *There exists an immersion f of $T^n \times I$ into $E^n \times I$ such that $f \mid T_0^n \times I$ is a product map, $f = \alpha \times 1$, where T_0^n is T^n minus an n-cell.*

We adopt the following notation for this proof: Let $I = [-1, 1] = J$, $J^n = (J)^n$, $S^1 = I \cup_\partial J$, $T^n = (S^1)^n$ and $T_0^n = T^n - \operatorname{Int} J^n$. It is easy to see that $E^n \times S^1$ can be regarded as a subset of E^{n+1} where the I-fibers of $E^n \times I$ are straight and vertical in E^{n+1} (see Fig. **5.6.9**).

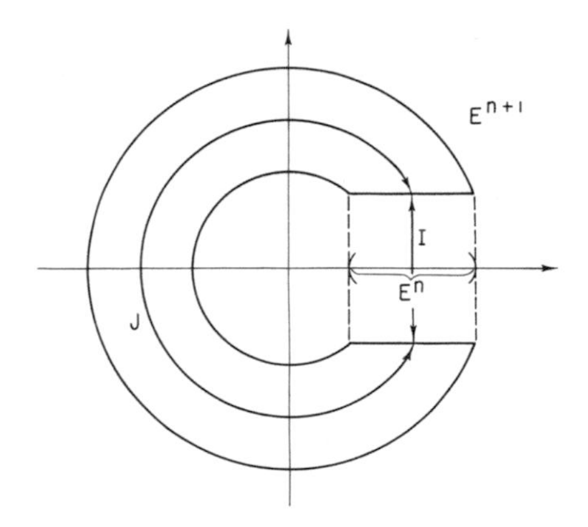

Figure 5.6.9

Assume that f and α are given by the inductive statement in dimension n. It is a simple matter to extend f to an immersion of $T^{n+1} \times I$ into $E^{n+1} \times I$, that is, just let

$$f \times 1_{S^1}\colon T^n \times S^1 \times I \to E^n \times S^1 \times I \subset E^{n+1} \times I$$

be the extension (see Fig. **5.6.10**). However, $f \times 1_{S^1}$ is not a product on $T_0^{n+1} \times I$, but merely on $T_0^n \times S^1 \times I$. The way to correct this is to conjugate $f \times 1_{S^1}$ with a 90° rotation (on the $I \times I$ factor) of the missing plug $(T_0^{n+1} \times I) - (T_0^n \times S^1 \times I) = \operatorname{Int} J^n \times I \times I$. The fact that $f \times 1_{S^1} \mid T_0^n \times I^2$ is a product in the I^2 factor allows one to do this.

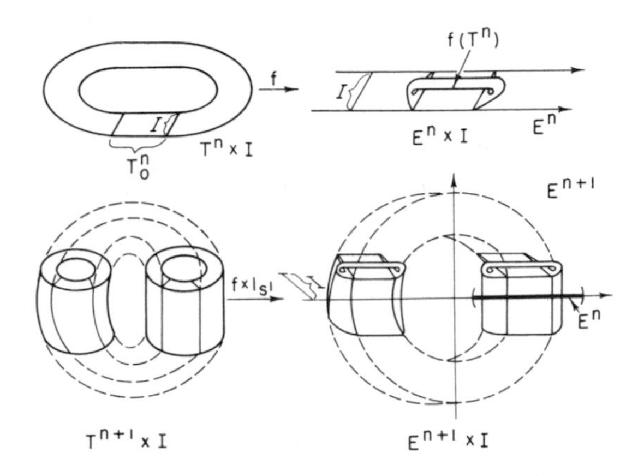

Figure 5.6.10

Assume without loss of generality that $f(T^n \times [-\frac{1}{2}, \frac{1}{2}]) \subset E^n \times [-\frac{2}{3}, \frac{2}{3}]$. Let λ be a homeomorphism of I^2 that is the identity on $\operatorname{Bd} I^2$ and is a $\pi/2$-rotation on $[-\frac{2}{3}, \frac{2}{3}] \times [-\frac{2}{3}, \frac{2}{3}]$ (see Fig. **5.6.11**). Extend λ via the identity to a homeomorphism $\bar{\lambda}$: $S^1 \times I \rightarrow S^1 \times I$ (see Fig. **5.6.12**).

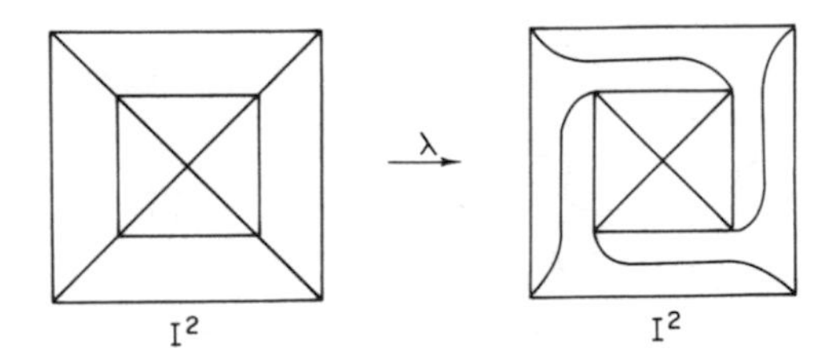

Figure 5.6.11

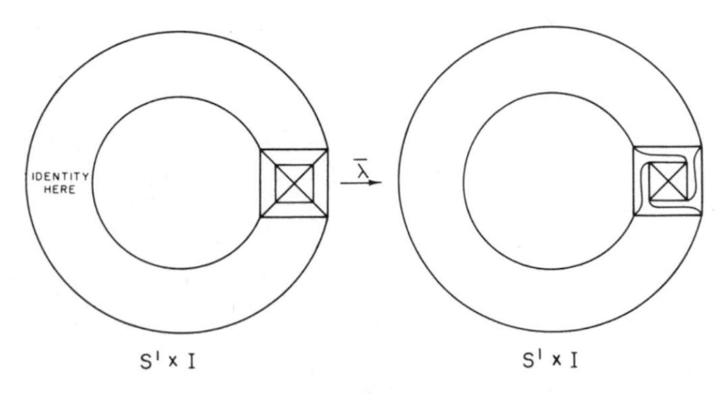

Figure 5.6.12

Consider now the following immersion h of $T^{n+1} \times I$ into $E^{n+1} \times I$,

$$h = (1_{E^n} \times \bar{\lambda}^{-1})(f \times 1_{S^1})(1_{T^n} \times \bar{\lambda}).$$

If we let $g = h \mid T^{n+1} \times [-\frac{1}{2}, \frac{1}{2}]$, then it can easily be checked that g is a product on $(T_0{}^n \times S^1) \cup (J^n \times [-\frac{1}{2}, \frac{1}{2}])$ which is a deformation retract of $T_0^{n+1} \times [-\frac{1}{2}, \frac{1}{2}]$. Thus, without loss of generality we can assume that g is a product on $T_0^{n+1} \times [\frac{1}{2}, \frac{1}{2}]$ (see Figs. **5.6.13** and **5.6.14**).

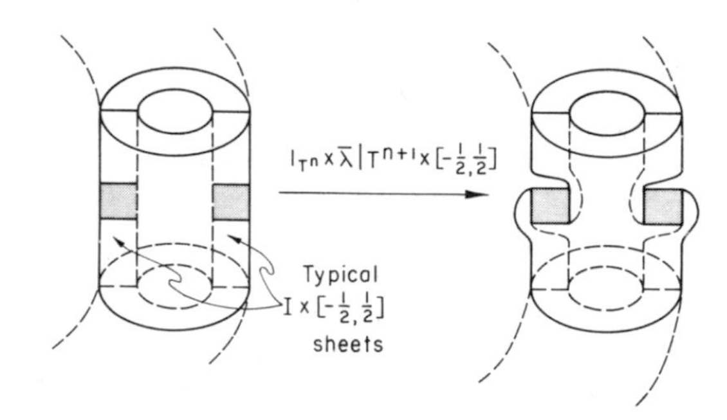

Figure 5.6.13

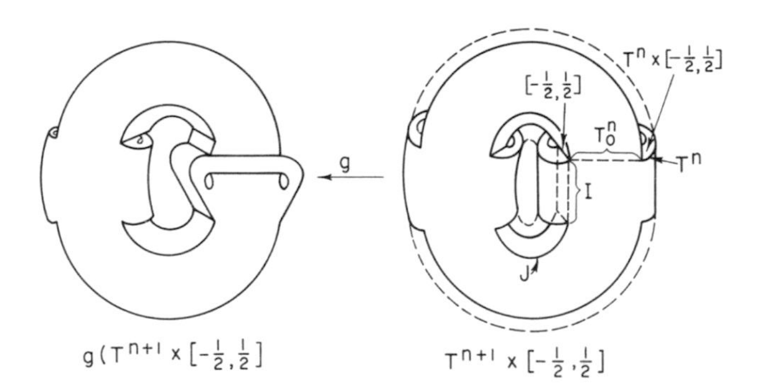

Figure 5.6.14

It is now easy to see that such a g gives rise to an immersion as desired in the theorem. This would be a trivial matter of reparametrizing the I coordinate if we knew that $g(T^{n+1} \times [-\frac{1}{2}, \frac{1}{2}]) \subset E^{n+1} \times [-\frac{1}{2}, \frac{1}{2}]$. To get such an inclusion, one can shrink T_0^{n+1} a little, with the help of an interior collar, to T_1^{n+1}, and using the fact that $g \mid T_0^{n+1} \times [-\frac{1}{2}, \frac{1}{2}]$ is a product, isotop $g(T^{n+1} \times [-\frac{1}{2}, \frac{1}{2}])$ into $E^{n+1} \times [-\frac{1}{2}, \frac{1}{2}])$ keeping $g \mid T_1^{n+1} \times [-\frac{1}{2}, \frac{1}{2}]$ fixed.

Let us conclude this section with a couple of remarks concerning how the preceding results on local contractibility relate to codimension zero taming. First note that if M^n and $\tilde{M}^n$ are PL manifolds and if $h: M \to \tilde{M}$ is a topological homeomorphism which can be approximated arbitrarily closely by PL homeomorphisms, then the results on local contractibility imply that h is ϵ-tame. For instance, it follows from Theorem **4.11.1** that stable homeomorphisms of E^n are ϵ-tame. (For an example of another use of this observation, see Theorem 1 of [Cantrell and Rushing, 1].) A strong form of the *hauptvermutung* for PL manifolds (Question **1.6.5**) is just the following codimension zero taming question: Can every topological homeomorphism of a PL manifold M^n onto a PL manifold $\tilde{M}^n$ be ϵ-tamed? By using some of the techniques presented in this section as well as some work of Wall, it has recently been established by Kirby and Siebenmann that this codimension-zero taming theorem holds for many manifolds and fails for others.

APPENDIX

Some Topics for Further Study

The area of topological embeddings is now an active research area. Thus, although the subject matter of this book should provide a sound basis for development for some time to come, the serious student is advised to make a study of the current literature after covering this book. The purpose of this appendix is to mention a few important recent developments which are particularly appropriate for study as follow-up material at this time. Most of the papers cited below have already been discussed in appropriate places in the text.

In Section **4.9**, we presented McMillan's result that locally nice codimension one spheres in S^n, $n \not\geq 5$, are weakly flat. Daverman [3] has recently shown that such spheres are, in fact, flat. Seebeck [3] made the first significant progress on the problem by showing that such spheres are flat if they can be approximated by locally flat ones. It follows from [Price and Seebeck, 1, 2] that such spheres can be approximated by locally flat ones if they have a locally flat spot. Daverman used very strongly in his proof the fact that a locally nice codimension one sphere with a flat spot is flat. The key to Daverman's proof is a covering space argument similar to that of [Kirby, 4]. (A generalization of that covering space argument was presented in Section **5.6**.) Daverman's proof also uses an infinite engulfing argument as suggested in [Rushing, 10] and a result on homotopy tori given in [Hsiang and Wall, 1].

Homma's metastable range PL approximation theorem for embeddings of PL manifolds was presented in Section **5.4** and related results were discussed. Significant work which brings Homma's approximation theorem up to codimension three is given in [Černavskiĭ, 7, 8, 9] and in [Miller, 1] (see Section **5.4**). Another major approximation theorem has been proved recently by Štan'ko [1]. In particular, Štan'ko shows that any embedding of a codimension three compactum into E^n can be approximated by a locally nice embedding.

Remark **5.2.4** indicated how a technique of Černavskiĭ suffices to prove the γ-statements in the metastable range. The γ-statements are established through codimension three in an important paper by Bryant and Seebeck [1]. The engulfing techniques (see Theorem 2.1) of [Bryant and Seebeck, 1] are fundamental to the paper and they prove to be useful for other purposes. (See, for example [Bryant and Seebeck, 3].) Černavskiĭ has seen how to extend the work of [Bryant and Seebeck, 1] to all codimensions. (See Section 2.2 of [Černavskiĭ, 5].)

Piecewise linear unknotting theorems for close embeddings in codimensions greater than two were proved for PL manifolds and for polyhedra in [Miller, 2] and [Connelly, 1], respectively. Each of these unknotting theorems can be used to prove a taming theorem. In particular, Connelly (Theorem 1_n) shows that close PL embeddings of polyhedra are ϵ-isotopic in codimensions greater than two. He then uses that result to show that locally tame embeddings of finite polyhedra are ϵ-tame in codimensions greater than two (Theorem 2).

Finally, let us mention once again the amazing recent work of Kirby, Siebenmann,and Wall on triangulation, on the hauptvermutung, and on taming. For announcements of the results see [Kirby, Siebenmann, and Wall, 1] and [Kirby and Siebenmann, 1, 3]. Kirby [4] led to the work indicated in those announcements. (A generalization of the basic technique of [Kirby, 4] was presented in Section **5.6**.) For some of the details of proof refer to [Kirby and Siebenmann, 2] and [Kirby, 5].

Bibliography†

Alexander, J. W.

[1] A proof of the invariance of certain constants of analysis situs, *Trans. Amer. Math. Soc.* **16**, 148–154 (1915).

[2] An example of a simply connected surface bounding a region which is not simply connected, *Proc. Nat. Acad. Sci. U.S.A.* **10**, 8–10 (1924).

[3] Remarks on a point set constructed by Antoine, *Proc. Nat. Acad. Sci. U.S.A.* **10**, 10–12 (1924).

[4] On the deformation of an n-cell, *Proc. Nat. Acad. Sci. U.S.A.* **9**, 406–407 (1923).

[5] The combinatorial theory of complexes, *Ann. of Math.* **31**, 292–320 (1930).

Alexandroff, P. S.

*[1] "Combinatorial Topology" (English translation, 3 Vols.). Graylock Press, Rochester, New York, 1956.

Alford, W. R.

[1] Some "nice" wild 2-spheres in E^3. "Topology of 3-Manifolds." Prentice-Hall, Englewood Cliffs, New Jersey, 1962.

[2] Some wild embeddings of the one and two dimensional spheres in the three sphere, Ph.D. Dissertation, Tulane Univ., New Orleans, Louisiana, 1963.

Alford, W. R., and Ball, B. J.

[1] Some almost polyhedral wild arcs, *Duke Math. J.* **30**, 33–38 (1963).

Andrews, J. J., and Curtis, M. L.

[1] n-Space modulo an arc, *Ann. of Math.* **75**, 1–7 (1962).

Antoine, M. L.

[1] Sur la possibilité d'étendre l'homeomorphic de deux figures à leurs voisinages, *C. R. Acad. Sci. Paris* **171**, 661–663 (1920).

[2] Sur l'homeomorphic de deux figures à leurs voisinages, *J. Math. Pures Appl.* **86**, 221–225 (1921).

Arens, R.

[1] Topologies for homeomorphism groups, *Amer. J. Math.* **68**, 593–610 (1946).

Armentrout, S.

[1] Cellular decompositions of 3-manifolds that yield 3-manifolds, *Mem. Amer. Math. Soc.* **107**, 1–72 (1971).

† Books are denoted by *.

Barden, D.
[1] Structure of manifolds, Dissertation, Cambridge Univ., 1963; published in part as: Simply connected five-manifolds, *Ann. of Math.* **82**, 365–385 (1965).
[2] A Quick Immersion, to be published.

Berkowitz, H. W.
[1] Piecewise linear approximation of a homeomorphism from one combinatorial manifold into another, Ph.D. Dissertation, Rutgers Univ., New Brunswick, New Jersey, 1968.
[2] P.L. approximations of embeddings of polyhedra in the metastable range, Mimeographed paper. Univ. of Georgia, Athens, Georgia, 1969.
[3] A counterexample to a proof of Homma, *Yokohama Math. J.* **19**(2), 1–3 (1971).

Berkowitz, H. W., and Dancis, J.
[1] PL approximations and embeddings of manifolds in the 3/4-range, *in* "Topology of Manifolds" (J. C. Cantrell and C. H. Edwards, eds.), pp. 335–340. Markham Publ., Chicago, Illinois, 1970.
[2] PL approximations of embeddings and isotopies of polyhedra in the metastable range, *in* "Topology of Manifolds" (J. C. Cantrell and C. H. Edwards, eds.), pp. 341–352. Markham Publ., Chicago, Illinois, 1970.

Bing, R. H.
[1] A simple closed curve that pierces no disk, *J. Math. Pures Appl.* **35**, 337–343 (1956).
[2] A set is a 3-cell if its cartesian product with an arc is a 4-cell, *Proc. Amer. Math. Soc.* **12**, 13–19 (1961).
[3] Some aspects of the topology of 3-manifolds related to the Poincaré conjecture, "Lectures on Modern Mathematics," Vol. 2, pp. 93–128. Wiley, New York, 1964.
[4] Locally tame sets are tame, *Ann. of Math.* **65**, 456–483 (1967).
[5] An alternative proof that 3-manifolds can be triangulated, *Ann. of Math.* **69**, 37–65 (1959).
[6] A surface is tame if its complement is 1-ULC, *Trans. Amer. Math. Soc.* **101**, 294–305 (1961).
[7] Retractions onto spheres, *Amer. Math. Monthly* **71**, 481–484 (1964).
[8] A wild surface each of whose arcs is tame, *Duke Math. J.* **28**, 1–16 (1961).
[9] Spheres in E^3, *Amer. Math. Monthly* **71**, 353–363 (1964).
[10] Stable homeomorphisms on E^5 can be approximated by piecewise linear ones, *Notices Amer. Math. Soc.* **10**, 666, Abstract 607-16 (1963).
[11] Radial engulfing, *Conf. Topology Manifolds*, 1967, (J. G. Hocking, ed.), pp. 1–18. Prindle, Weber, and Schmidt, 1968.
[12] The Cartesian product of a certain nonmanifold and a line is E^4, *Ann. of Math.* **70**, 399–412 (1959).

Bing, R. H., and Kister, J. M.
[1] Taming complexes in hyperplanes, *Duke Math. J.* **31**, 491–511 (1964).

Blankenship, W. A.
[1] Generalization of a construction of Antoine, *Ann. of Math.* **53** (2), 276–297 (1951).

Blankenship, W. A., and Fox, R. H.
[1] Remarks on certain pathological open subsets of 3-space and their fundamental groups, *Proc. Amer. Math. Soc.* **1**, 618–624 (1950).

Brown, M.
[1] Wild cells and spheres in high dimensions, *Michigan Math. J.* **14**, 219–224 (1967).

[2] The monotone union of open n-cells is an open n-cell, *Proc. Amer. Math. Soc.* **12**, 812–814 (1961).
[3] Locally flat embeddings of topological manifolds, *Ann. of Math.* **75**, 331–341 (1962).
[4] Locally flat embeddings of topological manifolds, "Topology of 3-Manifolds and Related Topics," pp. 83–91. Prentice-Hall, Englewood Cliffs, New Jersey, 1962.
[5] A proof of the generalized Schoenflies theorem, *Bull. Amer. Math. Soc.* **66**, 74–76 (1960).

Brown, M., and Gluck, H.
[1] Stable structures on manifolds: I Homeomorphisms of S^n, *Ann. of Math.* **79** (1), 1–17 (1964).

Bryant, J. L.
[1] Euclidean space modulo a cell, *Fund. Math.* **63**, 43–51 (1968).
[2] Taming polyhedra in the trivial range, *Michigan Math. J.* **13**, 377–384 (1966).
[3] Approximating embeddings of polyhedra in codimension three, to be published.

Bryant, J. L., and Seebeck, C. L. III.
[1] Locally nice embeddings of polyhedra, *Quart. J. Math. Oxford Ser.* **19**, 257–274 (1968).
[2] Locally nice embeddings in codimension three, *Bull. Amer. Math. Soc.* **2**, 378–380 (1968).
[3] Locally nice embeddings in codimension three (complete), *Quart. J. Math. Oxford Ser.* **21**, 265–272 (1970).

Burgess, C. E., and Cannon, J. W.
[1] Embeddings of surfaces in E^3, *Rocky Mountain J. Math.*, **1**(2), 259–344 (1971).

Cannon, J. W.
[1] ULC properties in neighborhoods of embedded surfaces and curves in E^3, to be published.

Cantrell, J. C.
[1] Almost locally flat embeddings of S^{n-1} in S^n, *Bull. Amer. Math. Soc.* **69**, 716–718 (1963).
[2] Separation of the n-sphere by an $(n-1)$-sphere, *Trans. Amer. Math. Soc.* **108**, 185–194 (1963).
[3] Non-flat embeddings of S^{n-1} in S^n, *Michigan Math. J.* **10**, 359–362 (1963).
[4] Almost locally flat embeddings of S^{n-1} in S^n, Mimeographed paper.
[5] Some embedding problems for topological manifolds, Lecture notes. Univ. of Georgia, Athens, Georgia, 1967–1968.
[6] n-Frames in Euclidean k-space, *Proc. Amer. Math. Soc.* **15**, 574–578 (1964).
[7] Some results concerning the union of flat cells, *Duke Math. J.* **32**, 673–677 (1965).

Cantrell, J. C., and Edwards, C. H.
[1] Almost locally flat embeddings of manifolds, *Michigan Math. J.* **12**, 217–223 (1965).
[2] Almost locally polyhedral curves in Euclidean n-space, *Trans. Amer. Math. Soc.* **107**, 451–457 (1963).

Cantrell, J. C., and Lacher, R. C.
[1] Some conditions for manifolds to be locally flat, Mimeographed paper.
[2] Local flattening of a submanifold, *Quart. J. Math. Oxford Ser.* **20**, 1–10 (1969).

Cantrell, J. C., and Rushing, T. B.
[1] On low co-dimensional taming, *in* "Topology of Manifolds," (J. C. Cantrell and C. H. Edwards, eds.), pp. 353–357. Markham Publ., Chicago, Illinois, 1970.

Cantrell, J. C., Price, T. M., and Rushing, T. B.

[1] A class of embeddings of S^{n-1} and B^n in R^n, *Proc. Amer. Math. Soc.*, **29** (1), 208–210 (1971).

Černavskiĭ, A. V.

[1] Singular points of topological imbeddings of manifolds, *Soviet Math. Dokl.* **7**, 433–436 (1966).

[2] Homeomorphisms of euclidean space and topological imbeddings of polyhedra in Euclidean spaces II, *Mat. Sb.* **72** (114), 4, (1967); English transl. *Amer. Math. Soc. Math. USSR-Sb.* **1**, 519–541 (1967).

[3] Locally homotopic unknotted imbeddings of manifolds, *Soviet Math. Dokl.* **9**, 835–839 (1968).

[4] The k-stability of homeomorphisms and the union of cells, *Soviet Math. Dokl.* **9**, 729–731 (1968).

[5] Homeomorphisms of Euclidean space and topological imbeddings of polyhedra in Euclidean spaces, III, *Mat. Sb.* **75** (117) (1968); English transl. *Amer. Math. Soc. Math. USSR-Sb.* **4**, 241–266 (1968).

[6] Homeomorphisms of Euclidean space and topological imbeddings of polyhedra in Euclidean spaces, *Mat. Sb.* **68** (110), 581–613 (1965); English transl. *Amer. Math. Soc. Transl.*, **78** (2), 1–38 (1968).

[7] Topological imbeddings of manifolds, *Soviet Math. Dokl.* **10**, 1037–1041 (1969).

[8] Piecewise linear approximations of embeddings of cells and spheres in codimensions higher than two, *Math. Sb.* **9**, 321–343 (1969).

[9] Supplement to the article "Piecewise linear approximations of cells and spheres in codimensions higher than two," to be published.

[10] Topological imbeddings of polyhedra in Euclidean spaces, *Dokl.* **165**, 1606–1610 (1965).

[11] Local contractibility of the homeomorphism group of a manifold, *Soviet Math. Dokl.* **9**, 1171–1174 (1968).

[12] Local contractibility of the group of homeomorphisms of a manifold, *Math. Sb.* **8**, 287–333 (1969).

Coelbo, R. P.

[1] On the groups of certain linkages, *Portugal. Math.* **6**, 57–65 (1947).

Cohen, M. M.

[1] A general theory of relative regular neighborhoods, *Trans. Amer. Math. Soc.* **136**, 189–229 (1969).

[2] Simplicial structures and transverse cellularity, *Ann. of Math.* **85**, 218–245 (1967).

[3] A proof of Newman's theorem, *Proc. Cambridge Philos. Soc.* **64**, 961–963 (1968).

[4] Simplicial and piecewise linear collapsibility, *Proc. Amer. Math. Soc.* **24**, 649–650 (1970).

Connell, E. H.

[1] Approximating stable homeomorphisms by piecewise linear ones, *Ann. of Math.* **78**, 326–338 (1963).

[2] A topological H-cobordism theorem for $n > 5$, *Illinois J. Math.* **11**, 300–309 (1967).

Connelly, R.

[1] Unknotting close embeddings of polyhedra in codimensions other than two, to be published.

[2] A new proof of Brown's collaring theorem, *Proc. Amer. Math. Soc.* **27**, 180–182 (1971).

Crowell, R. H.

[1] Invertible isotopies, *Proc. Amer. Math. Soc.* **14**, 658–664 (1963).

Crowell, R. H., and Fox, R. H.
*[1] "Introduction to Knot Theory." Ginn, Boston, Massachusetts, 1962.

Curtis, M. L.
[1] Cartesian products with intervals, *Proc. Amer. Math. Soc.* **12**, 819–820 (1961).

Curtis, M. L., and McMillan, D. R.
[1] Cellularity of sets in products, *Michigan Math. J.* **9**, 299–302 (1962).

Dancis, J.
[1] Some nice embeddings in the trivial range, *Topology Seminar, Wisconsin, 1965.* Princeton Univ. Press, Princeton, New Jersey, 1966.
[2] Isotopic submanifolds are ambient isotopic when the codim $\geqslant 3$, to be published.

Daverman, R. J.
[1] On weakly flat 1-spheres, *Proc. Amer. Math. Soc.*, to be published.
[2] On the scarcity of tame disks in certain wild cells, *Fund. Math.*, to be published.
[3] Locally nice codimension one manifolds are locally flat, *Bull. Amer. Math. Soc.* (to appear).

Debrunner, H., and Fox, R.
[1] A mildly wild imbedding of an n-frame, *Duke Math. J.* **27**, 425–429 (1960).

Doyle, P. H.
[1] A wild triod in three space, *Duke Math. J.* **26**, 263–267 (1959).
[2] Unions of cell pairs in E^3, *Pacific J. Math.* **10**, 521–524 (1960).
[3] On the embeddings of complexes in 3-space, *Illinois J. Math.* **8**, 615–621 (1964).

Doyle, P. H., and Hocking, J. G.
[1] Proving that wild cells exist, *Pacific J. Math.* **27**, 265–266 (1968).
[2] Some results on tame disks and spheres in E^3, *Proc. Amer. Math. Soc.* **11**, 832–836 (1960).

Duvall, P. F., Jr.
[1] Weakly flat spheres, *Michigan Math. J.* **16**, 117–124 (1969).

Edwards, C. H.
[1] Piecewise linear topology notes. Mimeographed paper, Univ. of Georgia, Athens, Georgia, 1966.
[2] Taming 2-complexes in high-dimensional manifolds, *Duke Math. J.* **32**, 479–494 (1965).

Edwards, R. D., and Glaser, L. C.
[1] A method for shrinking decompositions of certain manifolds, to be published.

Edwards, R. D., and Kirby, R. C.
[1] Deformations of spaces of imbeddings, *Ann. of Math.* **93**, 63–88 (1971).

Eilenberg, S., and Steenrod, N.
*[1] "Foundations of Algebraic Topology." Princeton Univ. Press, Princeton, New Jersey, 1952.

Eilenberg, S., and Wilder, R. L.
[1] Uniform local connectedness and contractibility, *Amer. J. Math.* **64**, 613–622 (1942).

Fisher, G. M.
[1] On the group of all homeomorphisms of a manifold, *Trans. Amer. Math. Soc.* **97**, 193–212 (1960).

Flores, G.
[1] Über n-dimensionale Komplexe die im E^{2n+1} absolute Selbstverschlungen sind, *Ergeb. Math. Colloq.* **6**, 4–7 (1934).

Floyd, E. E., and Fort, M. K., Jr.
[1] A characterization theorem for monotone mappings, *Proc. Amer. Math. Soc.* **4**, 828–830 (1953).

Fort, M. K., Jr.
[1] A wild sphere which can be pierced at each point by a straight line segment, *Proc. Amer. Math. Soc.* **14**, 994–995 (1963).
[2] A proof that the group of all homeomorphisms of the plane into itself is locally-arcwise connected, *Proc. Amer. Math. Soc.* **1**, 59–62 (1950).

Fox, R. H.
[1] A quick trip through knot theory, "Topology of 3-Manifolds," pp. 120–167. Prentice-Hall, Englewood Cliffs, New Jersey, 1962.
[2] On topologies for function spaces, *Bull. Amer. Math. Soc.* **51**, 429–432 (1945).

Fox, R. H., and Artin, E.
[1] Some wild cells and spheres in three-dimensional space, *Ann. of Math.* **49**, 979–990 (1948).

Fox, R. H., and Harrold, O. G.
[1] The Wilder arcs, "Topology of 3-Manifolds," pp. 184–187. Prentice-Hall, Englewood Cliffs, New Jersey, 1962.

Gauld, D.
[1] Mersions of topological manifolds, Ph.D. Dissertation, Univ. of California, Los Angeles, 1969.

Gillman, D. S.
[1] Sequentially 1-ULC tori, *Trans. Amer. Math. Soc.* **111**, 449–456 (1964).
[2] Note concerning a wild sphere of Bing, *Duke Math. J.* **31**, 247–254 (1964).

Glaser, L. C.
[1] Contractible complexes in S^n, *Proc. Amer. Math. Soc.* **16**, 1357–1364 (1965).
[2] Bing's house with two rooms from a 1–1 continuous map onto E^3, *Amer. Math. Monthly* **74**, 156–160 (1967).
*[3] "Geometrical Combinatorial Topology" (Van Nostrand-Reinhold Math. Studies). Van Nostrand-Reinhold, Princeton, New Jersey, 1970.
[4] Uncountably many almost polyhedral wild $(k-2)$-cells in E^k for $k > 4$, *Pacific J. Math.* **27**, 267–273 (1968).
[5] Intersections of combinatorial balls and of Euclidean spaces, *Bull. Amer. Math. Soc.* **72**, 68–71 (1966).

Glaser, L. C., and Price, T. M.
[1] Unknotting locally flat cell pairs, *Illinois J. Math.* **10**(3), 425–430 (1966).

Gluck, H.
[1] Embeddings in the trivial range, *Bull. Amer. Math. Soc.* **69**, 824–831 (1963).
[2] Embeddings in the trivial range (complete). *Ann. of Math.* **81**, 195–210 (1965).

Goodrick, R. E.
[1] Non-simplicially collapsible triangulations of I^n, *Proc. Cambridge Philos. Soc.* **64**, 31–36 (1968).

Greathouse, C. A.
[1] Locally flat, locally tame, and tame embeddings, *Bull. Amer. Math. Soc.* **69**, 820–823 (1963).

Greenberg, M.
*[1] "Lectures on Algebraic Topology." Benjamin, New York, 1967.

Gugenheim, V. K. A. M.
[1] Piecewise linear isotopy and embeddings of elements and spheres (I), *Proc. London Math. Soc.* **3**(3), 29–53 (1953).

Hamstrom, M. E.
[1] Regular mappings and the space of homeomorphisms on a 3-manifold, *Mem. Amer. Math. Soc.* **40** (1961).

Hamstrom, M. E., and Dyer, E.
[1] Regular mappings and the space of homeomorphisms on a 2-manifold, *Duke Math. J.* **25**, 521–531 (1958).

Harley, P. W.
[1] The product of an n-cell modulo an arc in its boundary and a 1-cell is an $(n + 1)$-cell, *Duke Math. J.* **35**, 463–474 (1968).

Harrold, Jr., O. G.
[1] Euclidean domains with uniformly abelian local fundamental groups, *Trans. Amer. Math. Soc.* **67**, 120–129 (1949).
[2] Euclidean domains with uniformly abelian local fundamental groups, II, *Duke Math. J.* **17**, 269–272 (1950).

Hempel, J. P., and McMillan, D. R., Jr.
[1] Locally nice embeddings of manifolds, *Amer. J. Math.* **88**, 1–19 (1966).

Henderson, D. W.
[1] Relative general position, *Pacific J. Math.* **18**, 513–523 (1966).

Hilton, P. J., and Wylie, S.
*[1] "Homology Theory." Cambridge Univ. Press, London and New York, 1960.

Hirsch, M. W.
[1] On embedding differentiable manifolds in Euclidean space, *Ann. of Math.* **73**, 566–571 (1961).

Hirsch, M. W., and Zeeman, E. C.
[1] Engulfing, *Bull. Amer. Math. Soc.* **72**, 113–115 (1966).

Hocking, J. G., and Young, G. S.
*[1] "Topology." Addison-Wesley, Reading, Massachusetts, 1961.

Homma, T.
[1] On the embedding of polyhedra in manifolds, *Yokohama Math. J.* **10**, 5–10 (1962).
[2] Piecewise linear approximations of embeddings of manifolds. Mimeographed notes, Florida State Univ., Tallahassee, Florida, November 1965.
[3] A theorem of piecewise linear approximations, *Yokohama Math. J.* **14**, 47–54 (1966).
[4] A theorem of piecewise linear approximations, *Yokohama Math. J.* **16**, 107–124 (1969).

Hsiang, W. C., and Wall, C. T. C.
[1] On homotopy tori II, *Bull. London Math. Soc.* **1**, 341–342 (1969).

Hu, S.-T.
*[1] "Homotopy Theory" (Smith and Eilenberg Ser.). Academic Press, New York, 1959.

Hudson, J. F. P.
*[1] "Piecewise Linear Topology" (Math. Lecture Note Ser.). Benjamin, New York, 1969.
[2] Piecewise linear embeddings and isotopies, *Bull. Amer. Math. Soc.* **72**(3), 536–537 (1966).
[3] Piecewise linear embeddings, *Ann. of Math.* **85**(1), 1–31 (1967).
[4] Concordance and isotopy of PL embeddings, *Bull. Amer. Math. Soc.* **72**(3), 534–535 (1966).
[5] Concordance, isotopy and diffeotopy, *Ann. of Math.* **91**(3), 425–488 (1970).

Hudson, J. F. P., and Zeeman, E. C.
[1] On regular neighborhoods, *Proc. London Math. Soc.* **14**(3), 719–745 (1964).

Huebsch, W., and Morse, M.
[1] The dependence of the Schoenflies extension on an accessory parameter, *J. Analyse Math.* **8**, 209–222 (1960–1961).

Hurewicz, W., and Wallman, H.
*[1] "Dimension Theory." Princeton Univ. Press, Princeton, New Jersey, 1941.

Husch, L. S.
[1] On relative regular neighborhoods, *Proc. London Math. Soc.* [3], **19**, 577–585 (1969).

Husch, L. S., and Rushing, T. B.
[1] Restrictions of isotopies and concordances, *Michigan Math. J.* **16**, 303–307 (1969).

Hutchinson, T.
[1] Two-point spheres are flat, *Notices Amer. Math. Soc.* **14**, 364, Abstract 644-18 (1967).

Irwin, M. C.
[1] Embeddings of polyhedral manifolds, *Ann. of Math.* **82**, 1–14 (1965).

Kirby, R. C.
[1] On the set of non-locally flat points of a submanifold of codimension one, *Ann. of Math.* **88**, 281–290 (1968).
[2] The union of flat $(n - 1)$-balls is flat in R^n, *Bull. Amer. Math. Soc.* **74**, 614–617 (1968).
[3] Stable homeomorphisms, *Notices Amer. Math. Soc.* **15**, 7, 1046 (1968).
[4] Stable homeomorphisms and the annulus conjecture, *Ann. of Math.* **89**, 575–582 (1969).
[5] Lectures on triangulations of manifolds. Notes, Univ. of California, Los Angeles, 1969.

Kirby, R. C., and Siebermann, L. C.
[1] A straightening theorem and a hauptvermutung for pairs, *Notices Amer. Math. Soc.* 582, Abstract 69T-G40 (1969).
[2] On the triangulation of manifolds and the hauptvermutung, *Bull. Amer. Math. Soc.* **75**, 242–749 (1969).
[3] A Triangulation Theorem, *Notices Amer. Math. Soc.* 433, Abstract 69T-630 (1969).

Kirby, R. C., Siebenmann, L. C., and Wall, C. T. C.
[1] The annulus conjecture and triangulation, *Notices Amer. Math. Soc.* 432, Abstract 697-G27 (1969).

Kister, J. M.
[1] Small isotopies in Euclidean spaces and 3-manifolds, *Bull. Amer. Math. Soc.* **65**, 371–373 (1969).
[2] Isotopies in 3-manifolds, *Trans. Amer. Math. Soc.* **97**, 213–224 (1960).
[3] Microbundles are fibre bundles, *Ann. of Math.* **80**, 190–199 (1964).

Klee, V. L.
[1] Some topological properties of convex sets, *Trans. Amer. Math. Soc.* **78**, 30–45 (1955).

Kneser, H.
[1] Die Deformationsätze der einfach zusammenhängenden Flächen, *Math. Z.* **25**, 362–372 (1926).

Kwun, K. W., and Raymond, F.
[1] Factors of cubes, *Amer. J. Math.* **84**, 433–440 (1962).

Lacher, R. C.
[1] Locally flat strings and half-strings, *Proc. Amer. Math. Soc.* **18**, 299–304 (1967).
[2] Almost combinatorial manifolds and the annulus conjecture, *Michigan Math. J.* **14**, 357–363 (1967).
[3] Topologically unknotting tubes in Euclidean space, *Illinois J. Math.* **12**, 483–493 (1968).
[4] Cellularity criteria for maps, *Michigan Math. J.*, **17**, 385–396 (1970).
[5] *Math. Reviews* **43**(1), no. 1201, p. 228 (1972).
[6] A disk in n-space which lies on no 2-sphere, *Duke Math. J.* **35**, 735–738 (1968).

Lees, J. A.
[1] h-Cobordism and locally flat imbeddings of topological manifolds. Dissertation, mimeographed, Univ. of Chicago, Chicago, Illinois, 1967.
[2] An engulfing theorem for topological manifolds, to be published.
[3] A classification of locally flat imbeddings of topological manifolds, to be published.
[4] Immersions and surgeries of topological manifolds, *Bull. Amer. Math. Soc.* **75**, 529–534 (1969).

Lickorish, W. B. R.
[1] The piecewise linear unknotting of cones, *Topology* **4**, 67–91 (1965).

Lickorish, W. B. R., and Martin, J. M.
[1] Triangulations of the 3-ball with knotted spanning 1-simplexes and collapsible ith derived subdivisions, to be published.

Machusko, A. J.
[1] See remark in proof of Theorem **3.6.2**.

McMillan, D. R., Jr.
[1] A criterion for cellularity in a manifold, *Ann. of Math.* **79**, 327–337 (1964).

McMillan, D. R., Jr., and Zeeman, E. C.
[1] On contractible open manifolds, *Proc. Cambridge Philos. Soc.* **58**, 221–224 (1962).

Mazur, B.
[1] A note on some contractible 4-manifolds, *Ann. of Math.* **73**, 221-228 (1961).
[2] On embeddings of spheres, *Bull. Amer. Math. Soc.* **69**, 91–94 (1963).
[3] Differential topology from the point of view of simple homotopy theory, *Inst. Hautes Études Sci. Publ. Math.* **No. 15** (1963).
[4] Corrections to: Differential topology from the point of view of simple homotopy theory and further remarks, *Inst. Hautes Études Sci. Publ. Math.* **No. 22** (1964).

Michael, E. A.
[1] Local properties of topological spaces, *Duke Math. J.* **21**, 163–171 (1954).

Miller, R. T.
[1] Approximating codimension 3 embeddings, to be published.
[2] Close isotopies on piecewise-linear manifolds, *Trans. Amer. Math. Soc.* **151**, 597–628 (1970).

Milnor, J.
[1] Two complexes which are homeomorphic but combinatorially distinct, *Ann. of Math.* **74**, 575–590 (1961).
*[2] "Lectures on the H-Cobordism Theorem" (notes by L. Siebenmann and J. Sondow) (Princeton Math. Notes). Princeton Univ. Press, Princeton, New Jersey, 1965.

Moise, E. E.
[1] Affine structures in 3-manifolds, V. The triangulation theorem and Hauptvermutung, *Ann. of Math.* **56**(2), 96–114 (1952).

[2] VIII. Invariance of knot-type; Local tame imbedding, *Ann. of Math.* **59**, 159–170 (1954).
[3] Affine structures in 3-manifolds, IV. Piecewise linear approximations of homeomorphisms, *Ann. of Math.* **55**, 215–222 (1952).

Morse, M.
[1] A reduction of the Schoenflies extension problem, *Bull. Amer. Math. Soc.* **66**, 113–117 (1960).

Newman, M. H. A.
[1] On the superposition of n-dimensional manifolds, *J. London Math. Soc.* **2**, 56–64 (1926).
[2] The engulfing theorem for topological manifolds, *Ann. of Math.* **84**, 555–571 (1966).
[3] On the foundations of combinatory analysis situs, *Proc. Roy. Acad. Amsterdam* **29**, 610–641 (1926).

Penrose, R., Whitehead, J. H. C., and Zeeman, E. C.
[1] Imbeddings of manifolds in Euclidean space, *Ann. of Math.* **73**, 613–623 (1961).

Poenaru, V.
[1] Les décompositions de l'hypercube en produit topologique, *Bull. Soc. Math. France* **88**, 113–119 (1960).

Price, T. M., and Seebeck, C. L. III.
[1] Approximations of somewhere nice codimension one manifolds having 1-ULC complements, I, to be published.
[2] Approximations of somewhere nice codimension one manifolds having 1-ULC complements, II, to be published.

Rado, T.
[1] Über den Begriff der Riemanschen Fläche, *Acta Univ. Szeged.* [2], 101–121 (1924–1926).

Roberts, J. H.
[1] Concerning homeomorphisms of the plane into itself, *Bull. Amer. Math. Soc.* Abstract 44-9-402 (1938).
[2] Local arcwise connectivity in the space H^n of homeomorphisms of S^n onto itself, Summary of Lectures, *Summer Inst. on Set Theor. Topology, Madison, Wisconsin*, p. 100, 1955.

Rosen, R. H.
[1] The five dimensional polyhedral Schoenflies theorem, *Bull. Amer. Math. Soc.* **70**, 511–515 (1964).

Rourke, C. P., and Sanderson, B. J.
[1] On topological neighborhoods, *Compositio Math.* **22**, 387–424 (1970).

Rushing, T. B.
[1] Everywhere wild cells and spheres, *Rocky Mountain J. Math.* **2**(2), 249–258 (1972).
[2] Taming embeddings of certain polyhedra in codimension three, *Trans. Amer. Math. Soc.* **145**, 87–103 (1969).
[3] Taming codimension three embeddings, *Bull. Amer. Math. Soc.* **75**, 815–820 (1969).
[4] Locally flat embeddings of PL manifolds are ϵ-tame in codimension three, *in* "Topology of Manifolds" (J. C. Cantrell and C. H. Edwards, eds.), pp. 439–452. Markham Publ., Chicago, Illinois, 1970.
[5] Unknotting unions of cells, *Pacific J. Math.* **32**, 521–525 (1970).
[6] Two-sided submanifolds, flat submanifolds and pinched bicollars, *Fund. Math.*, **74**, 73–84 (1972).

[7] On countably compact non-locally compact spaces, *Amer. Math. Monthly* [3], **74**, 280–283 (1967).
[8] Adjustment of topological concordances and extensions of homeomorphisms over pinched collars, *Proc. Amer. Math. Soc.* [1], **26**, 174–177 (1970).
[9] Realizing homeomorphisms by ambient isotopies, *Proc. Amer. Math. Soc.* [3], **23**, 723–724 (1969).
[10] Infinite engulfing, to be published.

Sanderson, D. E.
[1] Isotopy in 3-manifolds. III. Connectivity of spaces of homeomorphisms, *Proc. Amer. Math. Soc.* **11**, 171–176 (1960).

Seebeck, C. L. III.
[1] ϵ-taming in codimension 3, *Michigan Math. J.* **14**, 89–93 (1967).
[2] Tame arcs on wild cells, *Proc. Amer. Math. Soc.* **29**, 197–201 (1971).
[3] Collaring an $(n-1)$-manifold in an n-manifold, *Trans. Amer. Math. Soc.* **148**, 63–68 (1970).

Sher, R. B.
[1] A result on unions of flat cells, *Duke Math. J.* **37**, 85–88 (1970).
[2] Tame polyhedra in wild cells and spheres, *Proc. Amer. Math. Soc.* **30**, 169–174 (1971).

Siebenmann, L. C.
[1] Pseudo-annuli and invertible cobordisms, *Arch. Math. (Basel)* **19**, 528–535 (1968).
[2] Approximating cellular maps by homeomorphisms, to be published.

Siebenmann, L. C., and Sondow, J.
[1] Some homeomorphic spheres that are combinatorially distinct, *Comment. Math. Helv.* **41**, 261–272 (1967).

Singer, I. M., and Thorpe, J. A.
*[1] "Lecture Notes on Elementary Topology and Geometry." Scott, Foresman, Glenview, Illinois, 1967.

Smale, S.
[1] Generalised Poincaré conjecture in dimensions greater than four, *Ann. of Math.* **74**, 391–406 (1961).
[2] On the structure of manifolds, *Amer. J. Math.* **84**, 387–399 (1962).

Smith, A.
[1] Extending embeddings of R^{n-1} in R^n, *Proc. Camb. Phil. Soc.* **71**, 5–18 (1972).

Spanier, E. H.
*[1] "Algebraic Topology." McGraw-Hill, New York, 1966.

Stallings, J. R.
[1] On infinite processes leading to differentiability in the complement of a point, "Differential and Combinatorial Topology," pp. 245–254. Princeton Univ. Press, Princeton, New Jersey, 1965.
[2] On topologically unknotted spheres, *Ann. of Math.* **78**, 501–526 (1963).
[3] The piecewise linear structure of Euclidean space, *Proc. Cambridge Philos. Soc.* **58**, 481–488 (1962).
[4] "Lectures on Polyhedral Topology." Tata Inst. of Fundamental Res., Bombay, 1968.

Štan'ko, M.
[1] Approximation of imbeddings of compacta in codimensions greater than two, *Soviet Math. Dokl.* **12**, 906–909 (1971).

Tindell, R.
[1] Some wild embeddings in codimension two, *Proc. Amer. Math. Soc.* **17**(3), 711–716 (1966).
[2] A counterexample on relative regular neighborhoods, *Bull. Amer. Math. Soc.* **72**, 892–893 (1966).
[3] The knotting set of a piecewise linear imbedding, Mimeographed paper, 1967.

Weber, C.
[1] Plongements de polyèdres dans le domaine métastable, Dissertation, Univ. of Geneva, 1967.
[2] Embeddings of polyhedra in the metastable range, *Comment. Math. Helv.* **42**, 1–27 (1967).

Whitehead, J. H. C.
[1] Simplicial spaces, nuclei, and n-groups, *Proc. London Math. Soc.* **45**, 243–327 (1939).
[2] Combinatorial homotopy, I, *Bull. Amer. Math. Soc.* **55**, 213–245 (1949).

Wilder, R. L.
[1] A converse of the Jordan-Brouwer separation theorem in three dimensions, *Trans. Amer. Math. Soc.* **32**, 632–657 (1930).
*[2] Topology of manifolds, *A.M.S. Colloq. Publ.* **32** (1949).

Wright, P.
[1] Radial engulfing in codimension three, *Duke Math. J.* **38**, 295–298 (1971).
[2] A uniform generalized Schoenflies theorem, *Ann. of Math.* **89**, 292–304 (1969).

Zeeman, E. C.
[1] Seminar on combinatorial topology, Mimeographed notes. Inst. des Hautes Études Sci., Paris, 1963.
[2] Relative simplicial approximation, *Proc. Cambridge Philos. Soc.* **60**, 39–43 (1964).
[3] The Poincaré conjecture for $n \geqslant 5$, "Topology of 3-Manifolds and Related Topics," pp. 198–204. Prentice-Hall, Englewood Cliffs, New Jersey, 1962.
[4] Polyhedral n-manifolds, I. Foundations, "Topology of 3-Manifolds and Related Topics," pp. 57–64. Prentice-Hall, Englewood Cliffs, New Jersey, 1962.
[5] Polyhedral n-manifolds, II. Embeddings, "Topology of 3-Manifolds and Related Topics," pp. 64–70. Prentice-Hall, Englewood Cliffs, New Jersey, 1962.
[6] Unknotting combinatorial balls, *Ann. of Math.* **78**, 501–526 (1963).

Author Index

Numbers in italics refer to the pages on which the complete references are listed.

T

W

Y

Z

Subject Index

Pure and Applied Mathematics

A Series of Monographs and Textbooks

Editors **Paul A. Smith and Samuel Eilenberg**

Columbia University, New York

1: Arnold Sommerfeld. Partial Differential Equations in Physics. 1949 (Lectures on Theoretical Physics, Volume VI)
2: Reinhold Baer. Linear Algebra and Projective Geometry. 1952
3: Herbert Busemann and Paul Kelly. Projective Geometry and Projective Metrics. 1953
4: Stefan Bergman and M. Schiffer. Kernel Functions and Elliptic Differential Equations in Mathematical Physics. 1953
5: Ralph Philip Boas, Jr. Entire Functions. 1954
6: Herbert Busemann. The Geometry of Geodesics. 1955
7: Claude Chevalley. Fundamental Concepts of Algebra. 1956
8: Sze-Tsen Hu. Homotopy Theory. 1959
9: A. M. Ostrowski. Solution of Equations and Systems of Equations. Third Edition, in preparation
10: J. Dieudonné. Treatise on Analysis: Volume I, Foundations of Modern Analysis, enlarged and corrected printing, 1969. Volume II, 1970. Volume III, 1972
11: S. I. Goldberg. Curvature and Homology. 1962.
12: Sigurdur Helgason. Differential Geometry and Symmetric Spaces. 1962
13: T. H. Hildebrandt. Introduction to the Theory of Integration. 1963.
14: Shreeram Abhyankar. Local Analytic Geometry. 1964
15: Richard L. Bishop and Richard J. Crittenden. Geometry of Manifolds. 1964
16: Steven A. Gaal. Point Set Topology. 1964
17: Barry Mitchell. Theory of Categories. 1965
18: Anthony P. Morse. A Theory of Sets. 1965
19: Gustave Choquet. Topology. 1966
20: Z. I. Borevich and I. R. Shafarevich. Number Theory. 1966
21: José Luis Massera and Juan Jorge Schaffer. Linear Differential Equations and Function Spaces. 1966
22: Richard D. Schafer. An Introduction to Nonassociative Alegbras. 1966
23: Martin Eichler. Introduction to the Theory of Algebraic Numbers and Functions. 1966
24: Shreeram Abhyankar. Resolution of Singularities of Embedded Algebraic Surfaces. 1966

25: FRANÇOIS TREVES. Topological Vector Spaces, Distributions, and Kernels. 1967

26: PETER D. LAX AND RALPH S. PHILLIPS. Scattering Theory. 1967.

27: OYSTEIN ORE. The Four Color Problem. 1967

28: MAURICE HEINS. Complex Function Theory. 1968

29: R. M. BLUMENTHAL AND R. K. GETOOR. Markov Processes and Potential Theory. 1968

30: L. J. MORDELL. Diophantine Equations. 1969

31: J. BARKLEY ROSSER. Simplified Independence Proofs: Boolean Valued Models of Set Theory. 1969

32: WILLIAM F. DONOGHUE, JR. Distributions and Fourier Transforms. 1969

33: MARSTON MORSE AND STEWART S. CAIRNS. Critical Point Theory in Global Analysis and Differential Topology. 1969

34: EDWIN WEISS. Cohomology of Groups. 1969

35: HANS FREUDENTHAL AND H. DE VRIES. Linear Lie Groups. 1969

36: LASZLO FUCHS. Infinite Abelian Groups: Volume I, 1970. Volume II, 1973

37: KEIO NAGAMI. Dimension Theory. 1970

38: PETER L. DUREN. Theory of H^p Spaces. 1970

39: BODO PAREIGIS. Categories and Functors. 1970

40: PAUL L. BUTZER AND ROLF J. NESSEL. Fourier Analysis and Approximation: Volume 1, One-Dimensional Theory. 1971

41: EDUARD PRUGOVEČKI. Quantum Mechanics in Hilbert Space. 1971

42: D. V. WIDDER: An Introduction to Transform Theory. 1971

43: MAX D. LARSEN AND PAUL J. MCCARTHY. Multiplicative Theory of Ideals. 1971

44: ERNST-AUGUST BEHRENS. Ring Theory. 1972

45: MORRIS NEWMAN. Integral Matrices. 1972

46: GLEN E. BREDON. Introduction to Compact Transformation Groups. 1972

47: WERNER GREUB, STEPHEN HALPERIN, AND RAY VANSTONE. Connections, Curvature, and Cohomology: Volume I, De Rham Cohomology of Manifolds and Vector Bundles, 1972. Volume II, Lie Groups, Principal Bundles, and Characteristic Classes, in preparation

48: XIA DAO-XING. Measure and Integration Theory of Infinite-Dimensional Spaces: Abstract Harmonic Analysis. 1972

49: RONALD G. DOUGLAS. Banach Algebra Techniques in Operator Theory. 1972

50: WILLARD MILLER, JR. Symmetry Groups and Their Applications. 1972

51: ARTHUR A. SAGLE AND RALPH E. WALDE. Introduction to Lie Groups and Lie Algebras. 1973

52: T. BENNY RUSHING. Topological Embeddings. 1973

In preparation

GERALD J. JANUSZ. Algebraic Number Fields

SAMUEL EILENBERG. Automata, Languages, and Machines: Volume A

JAMES W. VICK. Homology Theory: An Introduction to Algebraic Topology

E. R. KOLCHIN. Differential Algebra and Algebraic Groups

H. M. EDWARDS. Riemann's Zeta Function

WAYNE ROBERTS AND DALE VARBERG. Convex Functions